D1175801

The Penultimate Curiosity

The Penultimate Curiosity

How Science Swims in the Slipstream
of Ultimate Questions

ROGER WAGNER
Artist and Writer

AND

ANDREW BRIGGS
Professor of Nanomaterials, University of Oxford, UK

OXFORD
UNIVERSITY PRESS

OXFORD
UNIVERSITY PRESS

Great Clarendon Street, Oxford, ox2 6DP,
United Kingdom

Oxford University Press is a department of the University of Oxford.
It furthers the University's objective of excellence in research, scholarship,
and education by publishing worldwide. Oxford is a registered trade mark of
Oxford University Press in the UK and in certain other countries

First Edition published in 2016

Reprinted 2016

Impression: 2

Published in the United States of America by Oxford University Press
198 Madison Avenue, New York, NY 10016, United States of America

British Library Cataloguing in Publication Data

Data available

Library of Congress Control Number: 2015942568

ISBN 978–0–19–874795–6

Printed and bound by
CPI Group (UK) Ltd, Croydon, CR0 4YY

PRAISE FOR *The Penultimate Curiosity*

'Our species should be called *Homo spiritualis* rather than *sapiens*. Asking "Why?" about the world gave rise to Religion, Philosophy, and Science. The interactions and entanglements are outlined in this book of amazing scope and interest.'

Jean Clottes (Leader of the Chauvet Cave research team)

'This book offers a fascinating perspective on the perennial human quest for understanding and meaning. Its two distinguished authors—with contrasting backgrounds—have meshed their expertise together to create a thought-provoking and original synthesis.'

Lord Rees (Astronomer Royal, President of the Royal Society 2005–2010)

'Evidence-based scientific rationality is very good at finding answers to the how questions. How did the Universe evolve from the Big Bang? How does matter arrange itself into objects ranging from atomic nuclei to human beings, planets and stars? But when it comes to the why questions, science does not necessarily have the answers. Instead of putting science and religion in opposition to each other, we should therefore be asking if dialogue can exist between the two, whether they can respect each other and accept each other's points of view. In *The Penultimate Curiosity*, Andrew Briggs and Roger Wagner demonstrate that it is not only possible, but also enriching to follow such a course.'

Rolf Heuer (Director General, CERN)

'The achievements of science are breathtaking. At times so breathtaking that they cause us to lose perspective on the wonderful created world of which we, the most "curious" of animals, are a part. This book is a remarkable achievement in that whilst reaching from prehistory, through ancient Greece to the present day, it draws upon the distinctive intellectual resources of a distinguished artist and art historian and a researcher at the cutting-edge of contemporary science. The resulting, beautifully illustrated volume, is a feast of interdisciplinary thinking at its best. It raises profound questions, 'The Penultimate Curiosity', posed for millennia by philosophers, religious people and more recently scientists, and points to constructive answers.'

Malcolm Jeeves (President of the Royal Society of Edinburgh 1996–1999)

'This book is an excellent account of how human curiosity has struggled to understand the universe from different viewpoints. It shows in considerable detail how tensions between science and religion have been debated in depth by great minds (Leibniz, Newton, Pascal, Herschel, etc.) for centuries, and charts the development of the idea that science could progressively extend our understanding of the universe. It has many fascinating cameos and a magisterial sweep, and is made lively by details of personal involvement and histories in this development.'

George Ellis, FRS (author with Stephen Hawking of *The Large Scale Structure of Space–Time*)

'This is an erudite and fascinating sweep through the development of ideas. Uniquely, it addresses science and religion through both text and illustrations—from the cave paintings and artefacts of the earliest hominids, through the great thinkers who shaped civilisation, and on to the giants of the scientific revolution and the technology of the present day.'

Bob White, FRS (University of Cambridge)

'A stunningly original and wonderfully engaging book, which opens up some of the deepest questions about human identity and purpose.'

Alister McGrath (Andreas Idreos Professor of Science and Religion, University of Oxford)

'Roger Wagner and Andrew Briggs have written a path-breaking account, vast in scope, thrilling in detail, about how our ultimate curiosity as to what lies beyond the visible universe has danced a minuet through time with our penultimate curiosity as to how the elements of the universe relate to one another. A challenging and persuasive account of the sometimes fraught but often mutually enriching relationship between religion and science.'

Rabbi Lord Jonathan Sacks (Chief Rabbi 1991–2013)

'Here is magnificence. This book will magnify the heart and mind, in the sense of enlarging them to appreciate the scope of science and its underpinnings in the pursuit of theology. It depicts how insatiable—yet how creative and constructive—is the human curiosity for understanding and meaning, from prehistoric time to the present day. It leaves me in awe at the "art" of science: for the way it unveils the magnificence of God our Creator who stretches out the canvas.'

Justin Welby (Archbishop of Canterbury)

To
Anne and Diana

Acknowledgements

'We would like to thank … ': at which words most readers immediately turn the page. A book which traverses so many different fields must though incur a very large number of intellectual debts, and particularly so when it is itself the result of a collaboration.

How do an artist and a scientist write a book together? In Dante's *Divine Comedy*, Dante is led through the realms beyond death by an inhabitant of that region: the Roman poet Virgil. As the two poets set out, with Virgil acting as tour guide and Dante recording the journey, an angel descends into the inferno to clear the path and set them on their way. As their travels continue up the purgatorial mountain such heavenly helpers appear more frequently.

As we set out on our journey through the world of science, with the scientist in the role of tour guide and the artist acting as scribe, we too have received welcome outside help at critical stages.

The first such helper was Professor John Hedley Brooke, the inaugural Andreas Idreos Professor of Science and Religion at Oxford, who read a very early draft of some of these chapters, and while encouraging us forward opened our eyes to the difficulty of the journey we had undertaken. More than a decade later Professor Peter Harrison, the second occupant of the Chair, read the completed draft manuscript and gave us equally helpful advice. We are grateful to the Delegates of the Press for approving publication by OUP, and we were aided by the knowledgeable but anonymous OUP readers whose appraisal, both general and detailed, made this a better book than it would otherwise have been. We are also grateful to several other scholars and scientists who graciously responded to our appeals for help with detailed comments and general wisdom and encouragement.

In particular we must thank Dr Jean Clottes both for his advice on Part I and for his generosity in allowing us to use his photographs of the Chauvet Cave; Professor Sir Richard Sorabji, who shared with us his vast knowledge of the classical world in general and of John Philoponus in particular;

Professor Peter Adamson, who at short notice steered us away from some errors and towards recent scholarship in Part IV; Professor Malcolm Jeeves, who provided us with guidance through the world of CSR research; and Professor Hugh Williamson, who gave us very helpful advice on Part X. The responsibility for any errors that remain must, of course, be shared out between the two authors.

In addition to the above we must also thank Professor Tom McLeish, Dr Allan Chapman, and Dr Julia Golding for reading the manuscript, Mark Wagner for suggesting the title, and a number of people who have helped with the practicalities of putting this book together. Among these are Tim Kirtley at Wadham college library; Juliet Chadwick at the Bodleian; Helen Reilly for her tireless picture research; Nikki Macmichael and Marije Zeldenrijk for their heroic work on the footnotes and other practical aspects of the book; Ania Wronski, Viki Mortimer, and Charles Lauder at OUP for their detailed work on preparing the book for publication, and our editor, Sonke Adlung, for his support and encouragement through the whole process.

At the end of the second part of the *Divine Comedy* Virgil can go no further and Beatrice, Dante's long lost love, takes over to conduct the poet up through the circles of Paradise. Our book does not venture into theological realms and leaves readers to fare forward or not as they choose. Nevertheless there can be no doubt that the role of Beatrice in our book has been played by the two long suffering and endlessly supportive companions in this journey, whose names can be found on the dedication page.

Contents

Prologue

A visitor to Oxford in the height of summer, trying to escape from the tourist groups that surge around the main university buildings, might turn down a tree-lined street at the end of the Broad. Following this street would soon bring them to the green oasis of the university parks where, just before arriving at the park gate, a wide lawn stretches away from the road on the right-hand side. Behind the lawn rises an extraordinary Victorian version of a Rhenish gothic palace. This is the Oxford University Museum [Fig. 0.1].

A midsummer visitor to Cambridge seeking to avoid the crowds of visitors milling around in King's Parade might, in a similar way, duck down a small side road that leads past the Eagle pub. Turning sharp right would find them in the relative tranquillity of Free School Lane. Halfway down the lane on the left-hand side stands another grand Victorian building. A new stone inscription informs the visitor that this was the original home of the Cavendish Laboratory [Fig. 0.2].

0.1 The Oxford University Museum in 1860 photographed by Henry Wagner.

0.2 The Cavendish Laboratory in Free School Lane.

These two edifices share the honour of being among the first purpose-built scientific institutions in their respective universities. They were also the sites of two legendary moments in the history of science.

The Cavendish moment took place on the 28 February 1953, when two young researchers from the laboratory walked into the Eagle at lunchtime to tell everyone within hearing distance that they had 'found the secret of life'.

The moment at the museum had happened almost a hundred years earlier. On the 30 June 1860 the newly completed building was the site of a famous meeting in which a bishop was insulted, a lady fainted, and amid scenes of general excitement Darwin's and Wallace's new theory of evolution by means of natural selection was debated in public for the first time.

Both events have acquired a patina of mythology. Contemporary documents suggest that the museum debate was not quite the straightforward clash between religion and science that later accounts represented it as; while even at the time James Watson felt 'slightly queasy' that the claims being made for their new model of DNA by his partner Francis Crick were a little premature.[1] The myth-making, however, can be seen as a tribute to a sense that these two discoveries both had a significance that reached far beyond the normal parish boundaries of science.

This was particularly true of the first. When a French translation of *The Origin of Species* was published in 1862, two years after the events at the museum, it appeared with a preface in which the translator contrasted the 'rational revelation' of science with what she considered the now obsolete revelation of

[1] J. D. Watson, *The Double Helix* (1968), 197.

the Christian religion. Then twelve years later, in 1874, one of the original protagonists of the museum debate published his own *History of the Conflict between Religion and Science* in which the whole development of science was cast as 'a narrative of the conflict' between these two contending powers.[2]

Dr John Draper was professor of chemistry at New York University, and his paper *On the intellectual development of Europe, considered with reference to the views of Mr Darwin and others, that the progression of organisms is determined by law*, had been the advertised subject of the museum meeting. Most listeners found his paper as long and boring as its title (though many years later one lady could still recall 'the American accents' of Dr Draper's opening address when he asked 'air we a fortuitous concourse of atoms?'). In 1874, however, the professor did succeed in holding his audience. His *History* was a publishing phenomenon (comparable both in content and success to Richard Dawkins' *God Delusion* nearly a century and a half later), going through 21 editions in Britain, 50 printings in America, and being translated around the world.

Draper's book was followed in 1896 by a monumental work from the president of Cornell University, Andrew Dickson White, entitled *A History of the Warfare between Science and Theology in Christendom*. White's work enjoyed less popular success but more intellectual influence and was heavily drawn on by the philosopher Bertrand Russell in a book published in 1935 called *Religion and Science*.

Russell's thesis—that there had been a continuous warfare between science and religion in which 'science had invariably proved victorious'[3]—seemed to have been perfectly illustrated by the 'monkey trial' that had taken place in Tennessee ten years earlier. The prosecution of a biology teacher for teaching evolution had resulted in a notorious legal battle (later made into a play and a Hollywood film) in which in a broiling courtroom, the militant agnostic lawyer Clarence Darrow and the three times presidential nominee (and ardent Christian) William Bryant had literally shaken their fists at one another under the gaze of a fascinated press.

Speaking near the end of his life Francis Crick described his whole scientific career as having been shaped by this narrative of conflict. He had asked himself, 'what were the two things that appear inexplicable and are used to support religious belief: the difference between living and non-living things and

[2] J. W. Draper, *History of the Conflict between Religion and Science* (1875).
[3] J. H. Brooke, *Science and Religion* (1991), 41.

the phenomenon of consciousness' and had set out to try and remove those two supports.[4] In 1961 he had made his feelings public by resigning his fellowship of Churchill College when the fellows voted to build a chapel. An institution 'dedicated to advanced knowledge and free speculation' had no business, he believed, being connected with what he regarded as the 'superstitious nonsense' of religion.[5]

The irony of this was that had Crick or Watson happened to close the outer doors of the Cavendish behind them and look back over their shoulders as they headed out towards the pub that February lunchtime, they would have seen an inscription which connected their own laboratory with the 'superstitious nonsense' of religion every bit as explicitly as did the new chapel at Churchill College. Carved across the two doors is a Latin translation of a line from Psalm 111 [Fig. 0.3].

Probably they didn't notice it. 'The conflict between two competing powers' is such a compelling narrative that it is has become extremely easy to overlook alternative stories.

In 2009 a debate was held in the Oxford University Museum between Richard Dawkins and John Lennox (a mathematician and Christian apologist). In the course of the exchanges Lennox asked Dawkins whether the motivation behind the building of the museum had not been in part religious. Dawkins replied that in this case it had not and the debate moved on.

0.3 The doors of the Cavendish Laboratory in Free School Lane. An English translation reads 'The works of the Lord are great, sought out of all them that have pleasure therein'.

[4] R. Highfield, 'Do Our Genes Reveal the Hand of God?', *The Telegraph* (2003).
[5] F. Spalding, *John Piper, Myfanwy Piper* (2009), 421.

Had either participant happened to have glanced up as they entered the building, they would have seen carved over the door the image of an angel holding an open book in one hand and three living cells in the other [Fig. 0.4]. This according to a contemporary's account was 'to signify the intentions of the founders of the museum whose desire it was to bring future generations of men to study the open book of nature and the mysteries of life under the guidance of a higher power which alone could enable them to read the pages of that book with a right understanding'.[6]

0.4 Angel with book and germ cells above the entrance of the Oxford University Museum.

How did two institutions dedicated to science come to have these religious invocations set over their entrances? That question some 16 years ago was the starting point of this book. One of the authors, a scientist, had come across the Cambridge inscription while working at the Cavendish Laboratory. The other, an artist, had noticed the museum relief while studying gothic revival architecture in Oxford. Both of us were intrigued.

Were these invocations (as contemporary thought might anticipate) merely pious gestures? Had they been intended perhaps to soothe the religious sensibilities of the universities at a moment when advances of science were beginning to suggest that the biblical creation narratives belonged to the same category as every other kind of mythology?

It was the discovery that this was not the case: that these two entrances represented the deeply thought-out convictions of two central figures in the Victorian scientific world (whose stories we tell in Part X), which spurred us on to further investigation. If the impulse to integrate the separate domains of religion and science was something more than a gesture, then where had it come from?

Our first step in trying to answer that question was to consider, over a broader canvas, what was happening at the time that these entrances were constructed.

The last decades of the nineteenth century had seen a sudden and shocking expansion of intellectual horizons in a whole variety of different fields. Dramatic discoveries like finding the source of the Nile, uncovering the ruins of Troy, and deciphering long lost, ancient literatures took place in the same years as a whole series of fundamental scientific breakthroughs. The conflicts that assimilating some of these new perspectives sometimes involved (fainting ladies and shaken fists) have a natural

[6] J. B. Atlay, *Sir Henry Wentworth Acland* (1903).

tendency to magnetize our attention. But this has also, we soon realized, tended to make it easy to overlook the significance of discoveries that potentially resolved conflicts rather than creating them, and to ignore those Victorian scientists and others for whom an openness to new data was expressive of religious commitment rather than destructive of it.

In particular we noticed that the late-nineteenth-century discovery and deciphering of ancient texts that parallel the early chapters of Genesis (the story of which—moving backwards through the book—we tell in Part IX) was beginning to reveal that far from belonging in the same category as other creation myths, the Genesis stories were highly critical and radically distinct from them.

Where other mythologies had presented the gods as inhabiting (or even identified with) aspects of the natural world, and had provided picturesque accounts of how the world was physically made, the later biblical writers regarded what to them was a false identification of God and nature as 'idolatry'. They had, it seemed, consciously rejected these mythologies, and begun to present God as standing outside creation, creating all things by simple fiat, or act of will, without addressing the question of physical process.

The newly discovered texts revealed these ideas in dramatically high relief. It was not, however, the first time they had been noticed. An ancient (though contested) tradition within both Jewish and Christian thought had repeatedly drawn attention to such biblical emphases, and had used them for two purposes: to argue that the study of the natural world was a religious duty, and to suggest that because God was the creator of all, we should expect to find his creation governed by simple universal laws.

Here it seemed was a clue which might ultimately provide an answer to our question. A thread which could perhaps be followed back to its beginning.

At the time we both occupied the same building—the artist's studio was located at the top of the scientist's house—and during our joint occupancy we would occasionally meet for breakfast to discuss our findings. Pulling on this single thread we discovered was unravelling a much longer and more far reaching story than either of us had initially imagined.

The first place our unravelling led us back to was (a little surprisingly) the theory of evolution itself. The story of the clash between biblical literalism and scientific secularism has often been retold; but this we soon realized was only part of what had happened. The story (which we tell in Part VIII) of how the idea had developed that it might be possible to discover God-given laws underlying the biological world in the same kind of way Newton had discovered God-given laws governing the physical universe has been rather less frequently recounted.

Something similar indeed seemed to be true of Newton's own story (which we tell in Part VII). Much recent attention has been given either to showing how his preoccupation with alchemy and a heretical theology had isolated him from orthodox Christianity or to describing how his science led some to a self-defeating natural theology and others to enlightenment rationalism. At the centre of his own thought, however, was a sense that there was a deep congruence between 'the beautiful system' of interlocking laws (which could have been arranged differently) and the free choices of an all-wise creator God. From this Newton drew the inference that such a universal being would create universal laws. Whatever the other idiosyncrasies of his thinking, these were not isolated eccentricities. Newton's approach formed, we discovered, part of a much wider story.

That wider story (which we tell in Part VI) took in many of the pioneers of the so-called 'scientific revolution' for whom the study of nature was conceived as an aspect of religious worship. Protestant scientists, in particular, had seen a significant correlation between the open-minded approach that was needed to read the book of God's works accurately and the liberty of judgement that St Paul's injunction to 'test everything' apparently demanded in reading the book of God's word. Just as the translation of the Bible had democratized the reading of scripture and exposed the seeming idolatry of substituting man's thoughts for God's, so it seemed to them the 'experimental philosophy' had democratized the study of nature and allowed the works of the Lord to speak for themselves.

The story of which these early scientists were a part had not, however, begun in the sixteenth and seventeenth centuries. The founders of the Royal Society may have seen the Protestant Reformation as the fountain and origin of their new philosophy (with its emphasis on experiment, mathematics, and mechanism), but twentieth-century research, as we rapidly discovered, had brought to light other tributaries which had flowed into that same stream.

In 1904, some thirty years after John Draper had published his *History*, a French physicist named Pierre Duhem was writing a book on the origin of statics when he came across a reference to a little-known medieval author called Jordanus de Nemore. Following the clue provided by this reference and using in particular the newly published notebooks of Leonardo da Vinci, he embarked on a programme of research which culminated in a massive nine-volume history of medieval science. Duhem's work opened up a whole hitherto unsuspected field of study, and it was not the only such field to be opened up.

In 1883, some twenty years before Duhem began his work, Ernest Renan, the author of a controversial demythologizing life of Christ, had delivered a lecture at the Sorbonne which was later published as a book called *Islam et la Science*. In it he argued that Islam was inherently incapable of producing science and philosophy. This prompted a response by the Iranian-born scholar Jamal al-Din Afghani, and by the Ottoman intellectual Nemek Kemal who wrote a book called *Renan Mudafanamesi*—A Rebuttal of Renan—which highlighted the scientific achievements of the Muslim Arab world. Together their writings began the long process of drawing attention to a history that has, in the late twentieth and early twenty-first centuries, become an immense and growing field of study.

In New York in 1942 Joseph Needham, a distinguished Cambridge biochemist, wrote in the margin of a letter from the BBC, 'Sci. in general in China—why not develop?'[7] Needham's interest had first been sparked by contact with young Chinese scientists. It became all-consuming when on long war-time trips through China he began to collect and read crates of ancient texts that made clear the extent of the scientific developments that had taken place in China's past.

Francis Bacon had identified gunpowder, printing, and magnetic navigation as the three inventions that had changed the world. Needham found that all three had been discovered in China before they were known in Europe. He went on to write fifteen volumes entitled *Science and Civilisation in China* (a further twelve have subsequently been produced).

The issue with which Needham had begun continued to preoccupy him. Indeed the question why, around the sixteenth century, China lost its scientific and cultural pre-eminence, has

[7] S. Winchester, *Bomb, Book and Compass* (2008).

become known as 'the Needham question', and there is no definitive answer on which scholars agree. In China itself the techological triumphs of the past had been largely forgotten. A Chinese book published around the time Needham began his researches had the title *Why China has no Science*. It was in religious literature like the Taoist canon that Needham had made his discoveries, and in Maoist China all such religious texts were regarded with suspicion.

Discoveries like these of the range and extent of scientific interest in the pre-modern world and their close association with religion has created a radically different picture to that painted by John Draper and Andrew White. In particular it has placed the story of Galileo's quarrels with the church authorities, which had been (and perhaps still is) the central event in any version of a 'narrative of two conflicting powers', in a very different perspective.

Galileo's thinking (which we discuss in Part V) with his insistence on 'the freedom of philosophising' could in one way, we discovered, be linked directly with contemporary reformation thought. It could also though be seen as growing out of a much older story. This was one in which the idea of mathematically defined laws of nature had developed, first in the Islamic world and later in medieval Europe, as the Abrahamic religions had struggled to engage with the legacy of Greek philosophy. Their engagement had involved an argument lasting more than a thousand years (the history of which we describe in Part IV) in which the greatest scholars of three religions had attempted to distinguish the truths of Greek philosophy from what seemed to them the false religious ideas that were embedded within it.

Following this argument back to its beginning led us at last to the city of Alexandria in the years before the Muslim invasions. It was here (as we describe in Part III) that Judaism and Christianity first sought to come to terms with Greek natural philosophy in a systematic way, and it was here that an argument took place between a Christian and a Pagan philosopher (who was later to appear as a central figure in Galileo's dialogues) in which the idea of a universal law established by a God who stood outside nature first began to emerge.

Yet even as we arrived here, it became apparent to us that this was not in fact the beginning of the story. The engagement between Abrahamic religion and Greek philosophy may have started in Alexandria but the entanglement between religion and science had not begun there. Spooling still further back into history it soon became evident to us that the remarkable scientific research undertaken by Aristotle and his disciples had itself been intimately connected with an earlier revolution in religious thinking (and in Part II we describe how that came about).

Was this then finally where the story of the invocations over the laboratory in Cambridge and the museum in Oxford really began? In one sense it seemed to us that it was. Because this was the moment when what we now call 'science' began to emerge from the usual practices of what we now call 'religion'; it was also the moment when a need for integration first became apparent.

Talking in such terms must always involve some distortion. Our contemporary categories cannot be simply read back into the past. Peter Harrison, a former professor of science and religion at Oxford, suggests that if a historian were to announce that they had discovered 'a hitherto unknown war that had broken out in the year 1600 between Israel and Egypt', the claim would be treated with some scepticism. This is because 'the states of Israel and Egypt did not exist in the early modern period'.[8] In a similar way the concept of 'religion' as a set of beliefs hardly existed before the seventeenth century, while the word 'scientist' only appeared in the nineteenth century (it was coined by William Whewell at an 1833 meeting of the British Association 'by analogy with *artist*').

For medieval thinkers, Harrison points out, *religio* referred to interior acts of devotion, while *scientia* was considered a habit of mind: the first was a theological, the second an intellectual virtue. The idea of a conflict between them would have been almost meaningless.

The conflict that occurred in ancient Greece (however we describe it) did, all the same, involve the appearance of something new: an approach to the world that could be perceived as threatening established ideas about the gods and which therefore needed thinking about. Before that moment the study of nature had been so tightly woven into religious practice as to be more or less indistinguishable from it.

Yet was not this earlier integration itself a part of the story? Might not the unique range and scope of human curiosity be fundamentally connected with the capacity of the human mind to integrate disparate perceptions of the world? If so then

[8] P. Harrison, *The Territories of Science and Religion* (2015), 1.

the ultimate answer to the question that we posed about the origin of the impulse to integrate religion and science, and the real beginning of the story that we set out to tell, must go back to the first emergence of a distinctive human consciousness.

The fragmentary nature of the evidence makes this difficult territory to negotiate. Before 1850 these earliest beginnings were entirely inaccessible. The threads of enquiry that could be followed ran out at the dawn of civilization. In the latter half of the nineteenth century, however, a series of entirely unexpected discoveries began to open small tantalizing windows into the prehistory of the human race. It was here, where the nature and scope of human curiosity first begins to be glimpsed, that the first beginnings of an answer to our question had to be found.

The small tantalizing windows that opened consisted primarily of the discovery of prehistoric paintings and sculptures. These extraordinary finds, which began in the nineteenth century, have continued into the twentieth and twenty-first centuries. Indeed during the course of writing this book astonishing images have come to light. Their significance in the story of human curiosity is not only the capacity they reveal in the human mind to integrate different kinds of ability and perceptions, but also what they suggest about how our explorations of the world around us are motivated and sustained.

For an artist and a scientist to think about these questions together was it seemed (in a period when 'religion' and 'science' have been notably polarized) beginning to make surprising sense.

It was in those same decades that the invocations we had both first noticed were being placed over the doors in Oxford and Cambridge, that these other discoveries had begun to be made. Continuing then to pull on our single thread and to follow it down into the caves of prehistory, it is (in Part I) with that story of discovery that our larger story therefore begins.

PART I

In the Beginning I

CHAPTER ONE

The First Men

On 16 December 1832 HMS *Beagle* sailed into the Bay of Good Success in Tierra del Fuego [Fig. 1.1]. It was an experience that left an indelible impression on the young Charles Darwin's mind. Nearly 40 years later he could vividly recall the astonishment he felt 'on first seeing a party of Fuegians on a wild and broken shore'.[1] The men were 'absolutely naked and bedaubed with paint, their mouths frothed with excitement and their expression was wild startled and distrustful'.[2] He 'could not have believed how wide was the distance between savage and civilised man'[3] and the thought rushed into his mind: 'such were our ancestors'.[4]

When first contact was made, the strangeness of this experience only increased. 'The language of these people', he wrote, 'according to our notions scarcely deserves to be called articulate. Captain Cook compared it to a man clearing his throat, but certainly no European ever cleared his throat with so many hoarse, guttural and clicking sounds.'[5] He observed once that when putrid blubber was shared out 'an old man cut off thin slices and muttering over them ... distributed them to the famished party'[6] and wondered whether this might be a kind of religious action.

'Every family or tribe', he noted, 'has a wizard or conjuring doctor whose office we could never clearly ascertain.'[7] But although they buried their dead, 'Captain Fitzroy could not ascertain that the Fuegians have any distinct idea in a future life', or that they 'perform any sort of religious worship'.[8] 'We could never discover', Darwin later wrote, 'that the Fuegians believed in what we should call a God.'[9]

Did 'savage' races like the Fuegians even have the same origins as Europeans? At the beginning of the nineteenth century there were those who argued that they did not—proponents of 'polygenesis' rather than 'monogenesis'. The salience of this proposal was that it could provide a justification for using other races as slaves, and both scientific and religious arguments were

[1] C. R. Darwin, *The Descent of Man* (1871), 404.
[2] C. R. Darwin, *The Descent of Man*.
[3] C. R. Darwin, *The Voyage of the Beagle* (1959), 141.
[4] C. R. Darwin, *The Descent of Man*, 404.
[5] C. R. Darwin, *The Voyage of the Beagle*, 142.
[6] C. R. Darwin, *The Voyage of the Beagle*, 226.
[7] C. R. Darwin, *The Voyage of the Beagle*, 226.
[8] C. R. Darwin, *The Voyage of the Beagle*, 226.
[9] C. R. Darwin, *The Descent of Man*, 17.

1.1 The *Beagle* in Murray Narrow, Beagle Channel by Conrad Martens.

vigorously marshalled on either side: to support the idea or to oppose it. A few months before arriving at Tierra del Fuego, Darwin and Fitzroy had quarrelled so fiercely over the treatment of slaves that Darwin, who had been brought up in abolitionist circles, had come close to leaving the expedition, and it has recently been argued that his lifelong commitment to monogenesis was to some extent fuelled by his abhorrence of slavery.[10]

Darwin had been taught taxidermy by a freed slave, and throughout the voyage was generally able to find common ground with the peoples they came across. He was particularly intrigued by the ease with which Jemmy Button (a Fuegian that Captain Fitzroy was returning to his home) could move first from 'savage' to 'civilised' ways of thinking and then back again. Nevertheless the religious experiences of 'savage' people seemed as opaque to him as they did to travellers like Sir Samuel Barber, who in 1861 reported to the Royal Geographical Society that the inhabitants of Southern Sudan 'Like all other tribes of the White Nile . . . have no idea of a deity or even a vestige of superstition, they are mere brutes.'[11]

If Darwin was right in thinking that 'our ancestors' had similar characteristics to the Fuegians, it was reasonable to expect that alongside a lack of technology and lack of clothes, they would have manifested a similar lack of religion. Nearly forty years after that first encounter in the Bay of Good Success he concluded that 'there is ample evidence . . . that numerous races have existed and still exist, who have no idea of one or

[10] A. Desmond and J. Moore, *Darwin's Sacred Cause* (2009).

[11] S. W. Baker, 'Account of the Discovery of the Second Great Lake of the Nile, Albert Nyanza', *Journal of the Royal Geographical Society of London* 36 (1866), 1–18.

more gods, and have no words in their language to express such an idea.'[12]

The Bison in the Cave

For all their primitive appearance, the Fuegians at Good Success were still Darwin's contemporaries. Twenty years after the *Beagle* had sailed into the bay in Tierra del Fuego a kind of first contact with actual primeval human beings took place in a cave on the edge of the Pyrenees.

Although prehistoric artefacts had been discovered for centuries all over the world, no framework existed which enabled their age to be understood. Then in 1852 Édouard Lartet, a retired French magistrate who had devoted himself to palaeontology, was excavating a cave at Aurignac, when he discovered human bones and stone tools alongside the bones of extinct Ice Age mammals. This, for the first time, established a chronology and in 1861 Lartet published 'New Researches respecting the Co-Existence of Man with the Great Fossil Mammals, regarded as Characteristic of the latest Geological Period'.

In the early 1860s, Lartet, together with his friend Henry Christy, an English banker and amateur ethologist, began excavating a whole series of caves in the Dordogne; in 1864, in a cave known as La Madeleine near the town of Les Eyzies, they discovered a piece of mammoth tusk engraved with the image of a mammoth that seemed decisively to prove his point.

Four years later a gang of workmen were clearing the way for a road to the railway station at Les Eyzies. Under a limestone cliff overhang that was known locally as 'Cro-Magnon', they noticed some stone tools and pieces of bone in the soil. Louis Lartet, Édouard's son, discovered in a cavity at the back of the shelter partial skeletons of three adult males, a woman, and a child, buried among artefacts that were stylistically similar to those that his father had found in the cave at Aurignac. The striking thing about these skeletons was that unlike those that had been found in 1856 in the Neander Valley in Germany, these were essentially identical to modern humans.

When the artefacts discovered by Lartet and Christy were exhibited in the Paris fair of 1867 they caused a sensation. The first International Conference of Prehistory was held later that year, and collectors began to offer large sums for more artefacts, creating a kind of excavating goldrush. As a result an even

[12] C. R. Darwin, *The Descent of Man*, 65.

larger collection of artefacts went on display at the great Paris Universal Exposition in 1878, where they were seen by a visiting Spanish amateur prehistorian, Don Marcelino de Sautuola.

Three years before this, Don Marcelino had started excavating a cave, near his estate in Northern Cantabria, which had been discovered by a local hunter on a hill called Altamira. The artefacts in the Expositon and his conversations with the French prehistorian Édouard Piette inspired Sautuola to resume his excavations the following year.

When he restarted his digging in the floor of the cave Sautuola began to find animal bones, oyster shells, ashes, and bone tools similar to those found in France, all of which had begun to suggest that early human beings were not as lacking in technology as might have been supposed. On one of these excavating trips he took his small daughter Maria, who later recalled that while her father was digging by the light of a lamp in the floor of the cave she 'was running about in the cavern playing here and there'. Going into a side chamber too low for an adult to stand upright 'suddenly I made out forms and figures on the ceiling and cried out "mira Papa bueys!" [look, Papa, oxen!]' [Fig. 1.2].[13]

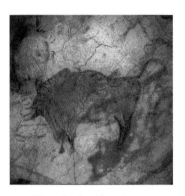

1.2 Bison in the Altamira cave.

Don Marcelino was overcome with amazement: 'what I saw made me so excited I could hardly speak.' Later excavations revealed further chambers filled with drawings paintings and prints of hands. The similarity between the objects and images he had found at Altamira and those he had seen at the Paris exposition convinced Sautuola that they were of a similar date, and having commissioned a French artist to make drawings of the images, he published a booklet: *Brief notes on some prehistoric objects from the province of Santander*.

Despite some brief interest and the visit of King Alfonso XII to see the caves, academic opinion soon turned against the authenticity of the Altamira discovery. Some argued that the cave was too humid and the rock too friable for paint to be preserved. A more fundamental objection rested on an extension of the new evolutionary theory.

In 1866 Ernest Haeckel, a German disciple of Darwin, had stated the so-called 'law of recapitulation'. This held that the juvenile growth stages of an organism—its 'ontogeny'—was 'the short and rapid recaptitulation' of its evolutionary history—its 'phylogeny'. This could most clearly be seen in the development of the embryo as it passed through its different stages, but it was also thought to apply to social development. Bone tools

13 P. G. Bahn, *The Cambridge Illustrated History of Prehistoric Art* (1998), 58.

were one thing, but on this argument if the 'Cro-Magnons', as they had begun to be called, had produced any images at all they would appear like the crude scrawls of children, not like the sophisticated paintings at Altamira.

When Édouard Harlé, a French prehistorian, visited the cave in 1882 he concluded that the images had been made between 1875 and 1879 (with or without the knowledge of Don Marcelino he would not say).

The evidence for the authenticity of the Altamira images took time to accumulate. In the late 1880s Édouard Piette discovered 'painted pebbles' in the cave of Mas d'Azil in the Pyrenees, whose position in the strata indicated that ochre could stick to rock for many thousands of years. Then in 1895, seven years after Don Marcelino's death, a farmer at La Mouthe in the Dordogne cleared out debris from a rock shelter and uncovered the entrance to a tunnel. When four boys explored it they found at the end a painted bison.

The accumulated Palaeolithic deposits which had blocked the entrance meant that it could not have been painted recently. When Émile Rivière excavated it in 1899 he found an artefact that combined both technology and art: a stone lamp with an engraving of an ibex on the underside. The 1901 discovery in caves at Les Combarelles and Font de Gaume of paintings of bison and horses overlain by stalactites which must have taken millennia to form made the case for the authenticity of the images overwhelming. The following year one of the principal denouncers of Don Marcelino, the French prehistorian Émile Cartailhac, published a paper: *Mea Culpa d'un sceptique*.

The Abbé

From then on, examples of Palaeolithic rock paintings and engravings began to be discovered on every continent. The majority, nearly 400 examples, were discovered in Europe, in particular in the north of Spain and the south west of France, and throughout the first half of the twentieth century most of these discoveries were soon accompanied by the appearance of a figure dressed in the black soutane and shovel hat of a French catholic priest. This was the Abbé Breuil [Fig. 1.3].

Almost immediately after the most famous of these discoveries (which took place just outside the village of Montignac in September 1940 when three teenagers rescuing their dog

1.3 Breuil (with candlewax on his soutane) at Altamira.

from a hole in the ground in the Lascaux woods discovered an astonishing sequence of painted chambers) the local schoolmaster sent word to the Abbé, who was staying at the nearby town of Brive. Breuil was able to authenticate the find and begin the excavation. As the original discoverers stood guard over the entrance to ward off the sightseers who had begun to gather, the Abbé would emerge from time to time to give an impromptu lecture on the open hillside about the extraordinary discoveries that were being made in the cave below.

By the time that Lascaux was discovered, Henri Breuil had already spent more than 40 years studying prehistoric cave art (he once estimated that he had spent 700 hours underground). At his ordination in 1900 Breuil had felt that his vocation was to be both priest and scientist. His teacher Abbé Guibert, who introduced him to the theory of evolution, had suggested to the young Breuil that 'there is a lot to be done in prehistory—you ought to tackle it',[14] and Mgr Deramecourt, the Bishop of Soissons, excused him from parochial work for 4 years to pursue his scientific work (this was a dispensation that was to last his whole life).

Breuil's first contribution to prehistory was in analysing the sequence of prehistoric cultures. In addition to the analytic abilities that this revealed, he turned out to be a skilled draughtsman and Emile Rivière employed him to make copies of the engravings he had discovered at the cave of La Moule. While he was in the Vézère Valley in 1901 Breuil was involved in the discovery of the cave paintings at Les Combarelles and Font de Gaume, which he was the first to record.

When Breuil's copies of these paintings were seen by Émile Cartailhac, he invited the young Abbé to accompany him to Altamira, where lying on his back, Breuil laboriously copied the paintings on the low ceiling by the light of a candle. It was this expedition that led to Cartailhac's *Mea Culpa* recantation. For Breuil it was the beginning of decades of exploration. Every year until the outbreak of the Spanish Civil War, he would travel around Spain and Portugal on horse, mule, or foot, identifying and copying prehistoric paintings and engravings at some 250 different sites.

A decade after the Altamira expedition, in October 1912, Breuil and Cartailhac received news from Comte Henri Bégouën that two of his three sons, Max, Louis, and Jacques, had made an extraordinary discovery on his estate at Montesquieu-Avantès in the foothills of the Pyrenees.

[14] A. H. Brodrick, *The Abbé Breuil* (1963), 29.

The Three Brothers

In July of that year Max Bégouën had learnt something about a small stream called the Volp, which flowed through their estate. At one point in its course the Volp goes underground at the entrance to a cavern on the side of a hill and emerges on the other side at an opening known as Tuc d'Audoubert.

François Camel, a local man, had told Max that if you went upstream from Tuc d'Audoubert you came to a series of underground passages blocked eventually by a wall of stalagmites in which there was a 'chatière'—a 'cathole'. The three brothers explored this together and when they came back in October with Camel they hacked their way through the 'chatière', climbed a 40-foot 'chimney', reached a chamber filled with fossilized cave bear bones, and finally at the end of a low gallery in which you had to crouch in some places, discovered two exquisitely modelled clay bison [Fig. 1.4].

1.4 Clay bison at Tuc d'Audoubert Cave.

When the brothers got back to the Chateau de Pujol just after midnight, they roused their father, who with some difficulty followed them back through the tunnels and the next morning sent off a famous telegram to Cartailhac and Breuil: 'The Magdalenians also modelled in clay.'

Cartailhac came from Geneva, Breuil from Paris, and four days later they both made their way along the underground river, up the chimney, and through the long passages to see the new discovery. The news of these clay sculptures, like the news of the paintings at Altamira, was initially greeted with scepticism. It was suggested that the bison had actually been produced by the sculptor Frémiet (who had died two years previously). Four years later, however, on the eve of the Great War, a second discovery by the brothers was to pose an even more fundamental challenge to prevailing thinking.

On 21 July 1914, one week before Gavrilo Princip assassinated Archduke Franz Ferdinand of Austria, the Bégouën brothers with François Camel went down a pothole called L'Aven on the other side of the hill from the Tuc. As far as we can tell this had not been used by prehistoric people. They discovered an 800-yard cavern now named after the brothers: Les Trois Frères.

Halfway through the cave, after a long tunnel through which you have to crawl on hands and knees, is a side chamber in which above a kind of natural 'altar' is a carving of a lion. Below

1.5 'Sorcerer', Les Trois Frères.

the 'altar' bones, teeth, charcoal, and other artefacts were found embedded in the wall like votive objects (similar objects were found elsewhere in the cave near engravings, paintings, and unusual rock formations). Finally, almost at the very end of the cave is another chamber densely covered with images, where 12 feet above the floor is a figure that Breuil was to describe as 'the wizard' or 'god'. Part-human, part-animal, it stands upright on human feet, yet has antlers and a tail [Fig. 1.5].

The assumption that primeval human beings would not have produced art, and the assumption that they would have had no religion were both of a piece. When it became impossible to deny that they did produce art it was easier for atheist prehistorians like Édouard Piette to think of this in terms of the aesthetic philosophy of 1890s as *l'art pour l'art*—art for art's sake. The great archaeologist Gabrielle de Mortillet (who had founded a positivist review) asserted that they could have nothing to do with 'd'idées religieuses'.[15]

The inaccessibility of the images the Bégouën brothers had discovered—located at the far end of long narrow tunnels through which you had to crawl—the 'votive objects' associated with some of these, and the presiding image of a figure half-man, half-beast made this an increasingly difficult position to maintain. It seemed impossible to deny that these images were produced under some kind of religious impulse and Comte Bégouën expounded this in a book: *De la mentalité spiritualiste des premiers hommes.*

Primeval religion, however, was only one facet of the primeval curiosity that the new discoveries were beginning to uncover.

[15] N. Richard, 'De l'art ludique à l'art magique; Interprétations de l'art pariétal au XIXe siècle', *Bulletin de la Société préhistorique française* 90 (1993), 60–68.

CHAPTER TWO

Tentasali

While paintings and sculptures may have suggested a religious impulse, other objects pointed to more down-to-earth explorations and engagements with the natural world. The lamp discovered by Émile Rivière in 1899 demonstrated the use of fire for light. The ashes found by Don Marcelino at Altamira indicated that fires were also used for warmth and for cooking food. Bone needles implied tailoring and (unlike the Fuegians) the wearing of clothes.

A whole variety of stone blades seemed to have specialized functions. Some were used to manufacture antler and bone tools and complex multicomponent tools such as harpoons and spear-throwers with a variety of projectile stone points that seem to have been used for hunting different kinds of animals. In some cases (as recent discoveries have shown) these points were attached to their hafts with glues compounded from several materials that had been ground, mixed, and heated.[1]

The variety of animal bones found at Palaeolithic sites testifies to the detailed knowledge of animal behaviours and the different strategies that would have been required to hunt them. Nor was a knowledge of the natural world confined to animals. At a site in Tell Abu Hureya in Syria that was occupied about 20,000 to 10,000 years ago, some 150 edible plants, together with pestles for grinding and pounding them, have been identified.

The Bone in the Gorge

One of the most extraordinary such discoveries was a small piece of bone dug up by Édouard Lartet and Henry Christy in the Gorge d'Enfer in the Vézère Valley in 1865. The bone was covered with apparently random pitting which Lartet and Christy found 'puzzling' [Fig. 2.1].[2] A quarter-century later a similarly marked bone was discovered in the nearby Blanchard rock shelter. Both objects remained unnoticed in the Musée Antiquités Nationales until 1962 when the archaeologist Jean

2.1 The Lartet bone.

[1] L. Wadley, 'Complex Cognition Required for Compound Adhesive Manufacture in the Middle Stone Age Implies Symbolic Capacity', in *Homo Symbolicus*, ed. C. S. Henshilwood and F. d'Errico (2011).

[2] E. Lartet and H. Christy, *Reliquiæ Aquitanicæ* (1875), 97–99.

2.2 The Ishango bone.

de Heinzelin unearthed a very similar-looking bone in the remains of an 11,000-year-old fishing village he called Ishango on the shores of Lake Edward in central Equitorial Africa [Fig. 2.2].

The pattern of the notches led Heinzelin to suspect that they were more than mere decoration, and he speculated that 'they may represent a mathematical game of some sort'.[3] He was unable to determine what this might be, but the idea that these marks represented some kind of notation was then taken up by a science journalist, Alexander Marshack, who had become interested in the origins of mathematics and astronomy.

By examining the marks on the Lartet and Blanchard bones under a microscope and comparing them with others (some hundreds of similar bones had by then been discovered), Marshack discovered that the markings were in groups made with different tools, at different angles, and with different pressures. It seemed to him 'inconceivable' that 'any man making an ornamental composition 1 3/4 inches in size would have used 24 changes of point and stroke to make 69 close marks'.[4] They must have been made serially, in a clear order and at different times.[5] Marshack went on to argue that these patterns corresponded to a lunar cycle, including a bi-monthly leap day to cover the extra half-day that each cycle requires. This last argument remains controversial, but the idea that these bones constitute some kind of 'artificial memory system' is generally accepted.[6]

Taken together all these different kinds of discoveries might have made Palaeolithic human beings seem impossibly removed from the impression created by the so-called 'savages' encountered by Victorian travellers, but as such encounters continued, these initial impressions slowly began to be modified.

Getting to Know You

The first studies of native peoples living in European colonies tended to describe them in something of the same manner as the local flora and fauna, from an entirely external perspective. Missionaries living with native tribes sometimes achieved a greater understanding, but the first known scientist to adopt a different approach was an American ethologist called Frank Hamilton Cushing. Cushing, having joined an anthropological expedition to New Mexico, decided to 'go native' and spent

[3] J. De Heinzelin, 'Ishango', *Scientific American* 206 (1962), 105.
[4] A. Marshack, *The Roots of Civilization* (1972), 45–48.
[5] A. Marshack, *The Roots of Civilization*, 21–55.
[6] L. Wadley, 'Complex Cognition Required for Compound Adhesive Manufacture in the Middle Stone Age Implies Symbolic Capacity', 50.

2.3 Frank Hamilton Cushing in Zuni costume.

five years, from 1879 to 1884, living with the Zuni tribe and being initiated into their rituals and practices [Fig. 2.3].

Thirty years later a Polish-born anthropologist, Bronisław Malinowski, travelled first to New Guinea where he did field work on Mailu Island, and then onto the Trobriand Islands where, having been trapped by the First World War, he lived among the islanders, getting to know them and studying their way of life [Fig. 2.4]. It was Malinowski who first defined the technique of 'participant-observation', the goal of which was 'to grasp the native's point of view, his relation to life, to realise *his* vision of *his* world'.[7]

This was an approach that soon revealed unsuspected dimensions in apparently primitive cultures. One of the first things it brought to light was the extent of the knowledge of medicinal plants that was common among indigenous peoples.

[7] B. Malinowski, *Argonauts of the Western Pacific* (1961), 25.

2.4 Bronisław Malinowski
with Trobriand islanders.

Cushing himself was given the tribal name 'Tentasali', meaning 'medicine flower'. Matilda Coxe Stevenson, one of the first
women anthropologists, who along with her husband began to
visit and study the Zuni in the same year as Cushing, published
at the end of her life a comprehensive study of *The Ethnobotany
of the Zuñi Indians* (later republished as *The Zuñi Indians and
Their Use of Plants*).

In 1895 an American botanist, John William Harshberger,
coined the term 'ethnobotany' to describe the scientific study
of the relationship between people and plants, and in the twentieth century this discipline was pioneered by the botanist
Richard Evans Schultes, who famously apprenticed himself to
an Amazonian shaman. Studies like those undertaken by Schultes have established that Southeast Asian forest tribes, for instance, use as many as 6500 species of plants altogether, and that
Amazonian forest dwellers employ some 1300 different plants
for medicinal purposes.

In a similar way it soon became apparent that preliterate societies were not, for that reason, innumerate. Although a few
tribal groups like the Piraha Indians have no counting systems,
these are exceptional. Almost every other preliterate society in
the world has been found to employ some kind of counting
system, often based on the digits of the hand and other parts
of the body.[8]

---

[8] B. Butterworth, *The Mathematical
Brain* (2000).

When eventually anthropologists and others began to study the art produced by indigenous peoples, more surprises were in store. It transpired that the 'crudity' and 'primitive violence' admired by artists like Picasso and Braque in the art they saw in the Paris anthropological museum at the beginning of the twentieth century were not quite what they seemed. Tribal peoples who were questioned about their art revealed that (like the Palaeolithic art discovered by the Bégouën brothers) much of it seemed to be motivated by a strong religious impulse.

CHAPTER THREE

Watauinaiwa

In North America the connection between rock art and an idea of 'spiritual power' had first been noticed by settlers in the seventeenth century. Father Jacques Marquette, one of the first Europeans to write about North American rock art, described in 1673 how when descending to the Mississippi 'we saw high rocks with hideous monsters painted on them upon which the bravest Indians dare not look'.[1] In the 1870s J. S. Denison reported that a man from the Klamath people told him that rock paintings 'were made by Indian doctors and inspired fear of the doctor's supernatural power'.[2]

American Dreams

In the twentieth century more detailed studies were made. American ethnographers working with linguistic groups across the continent amassed quantities of material which suggested that many of these rock art sites may have been associated with so-called 'vision quests'. According to native informants, in these quests people seeking supernatural power go to a secluded place to fast, pray, and seek a vision in which a spirit guide in the form of an animal helper appears. The visions often involve a journey, described as entering tunnels that are conceived as 'portals to the supernatural world'.[3]

At the end of the quest, after receiving power, the paintings, which often depict humans and animals and abstract designs, seem to have been made to commemorate the quest by the questers who 'painted their spirits (anit) on rocks ... to let people see what they had done. The spirit must come first in a dream'.[4] At some of these sites, in addition to the paintings pieces of white quartz were wedged into cracks between boulders together with 'offerings of beads, sticks arrows, seeds, berries and special stones'.[5]

Offerings like these seemed to be comparable to the 'votive objects' found in the walls of Les Trois Frères. In Southern

[1] P. A. Armstrong, *The Piasa* (1887), 9.
[2] D. Lewis-Williams, *The Mind in the Cave* (2004), 168.
[3] D. Lewis-Williams, *The Mind in the Cave*, 168.
[4] D. Lewis-Williams, *The Mind in the Cave*, 168.
[5] D. Lewis-Williams, *The Mind in the Cave*, 176.

Africa an even more direct link to Palaeolithic art, however, was discovered in the rock paintings made by the so-called 'Bushmen' or San people.

African Rock

In 1929 the Abbé Breuil went to Africa for the first time, and later spent several years working there in the 1940s and 1950s. Visiting Palaeolithic sites and studying the rock art that he found, he became convinced that some of this was much older than the rock art of the Bushmen and had not, in fact, been produced by them. The so-called 'white lady of Tsiah', which he visited at Brandberg, was, he argued, wearing Cretan costume and must have been of Mediterranean origin.

In the last years of his life Breuil modified these claims, recognizing that there was a lot more to be learnt about this art but it was not until after his death in 1961 that a systematic study was attempted. This located some 50,000 rock art sites in Southern Africa that are now acknowledged to be the work of the /Xam San peoples (today the 'white lady' is thought to be a typical example of this art and is neither 'white' nor indeed a 'lady'). They are difficult to date, but analysis has indicated that the most ancient may be as much as 27,000 years old while the most recent were produced almost within living memory.

Could these images therefore provide a kind of bridge into the ancient past? Might they offer the possibility of gaining some insight into the kind of ideas that had inspired paintings like those at Altamira?

By the time these questions began to be asked it had become difficult to answer them. By the 1960s the /Xam San groups who had produced the rock paintings in places like the Drakensberg Mountains no longer lived there. Driven out by colonial expansion they no longer pursued the old way of life or spoke the old languages. Fortunately, some records of the culture that had created these images still remained.

In the 1870s a German linguist, Wilhelm Bleek, and his sister-in-law, Lucy Lloyd, devised a special orthography to represent the clicks of the /Xam language, and took down some 1200 pages of verbatim accounts in this now-extinct language with Bleek's English translation alongside it. Anthropologists living alongside the /Xam San groups in the Kalahari Desert

meanwhile produced accounts of the culture that obtained in
the 1960s. While these scattered groups no longer practised
rock painting, they still practised rituals like those described in
the Bleek/Lloyd records, which seemed to be connected with
the old rock paintings.

Early investigators of the South African rock paintings had
tended to read them as a kind of 'menu'—a record of hunting and eating. One of the things the survey of all the sites
made clear was that by far the most frequently depicted animal
was the eland (the largest kind of antelope) [Fig. 3.1]. Archaeological evidence, on the other hand, indicated that it was small
antelope and wildebeest that were mainly eaten by the San.
This suggested the possibility that the paintings of eland may
have had a purpose beyond a simple record of hunting and
eating, and the evidence from the Bleek/Lloyd archive and the
Kalahari anthropologists seemed to confirm that.

Both sources described a belief in what was called !gi—an
invisible spiritual energy that pervaded the animal world but
was found especially in the eland and was located particularly in the blood, fat, and sweat. Some of those who remembered the paintings being made said that the pigment was
often mixed with eland blood, and this was confirmed by
chemical tests. The paintings it seemed could be thought of
as a kind of reservoir of !gi—spiritual power which could
be harnessed for healing, for rainmaking, and for fighting off
evil spirits.

An old lady whose father, a !gixa—a person full of !gi, had
made rock paintings was taken back to see the images. When
she saw her father's paintings 'she danced in the rock shelter

3.1 Eland rock paintings,
Drakensberg.

and turned to the painting her father had made so long ago. She lifted her hands and said that power would flow into her.'[6]

The anthropologists meanwhile had discovered that the 'healing' or 'trance' dance was the most important religious ritual among the San groups and the primary means by which the !gi of eland and other animals was harnessed for healing and other purposes. The circular dance, performed around a fire, and often the carcass of a recently killed animal, could go on for hours. In the course of the dance the !gi:ten fell into trances. The !gi from the dead animal 'boiled' in their stomach, transporting them into the world of the spirits where in deep trance they 'go to God's house to plead for the lives of the sick'.[7]

Ethnographic Snap

The association of 'vision quests' with North American rock art and of 'trance dances' with Southern African rock art have led both Thomas Blackburn and David Whitley in America and David Lewis-Williams in South Africa to look for parallels in the laboratory studies of trance states undertaken by neuropsychologists, and to suggest that the dots, zigzags, and grids that are a feature of both recent and prehistoric cave paintings are likely to refer to the flickering 'entoptic' phenomena (from the Greek meaning 'within vision') seen in trances, in the deepest stage of which 'people sometimes feel themselves to be turning into animals and undergoing frightening or exalting transformations'.[8] In the light of this they have argued that the animals depicted on the Palaeolithic rock shelters would have been thought of by the original painters as inhabitants of the spirit world, while 'the walls of the rock shelters were thought of as a "veil" suspended between this world and the spirit realm'.[9]

Not all prehistorians have found this argument entirely persuasive, especially when it seems to become a blanket explanation of prehistoric art. The difficulty of playing what has been described as 'ethnographic snap'—finding apparent similarities between cultures separated by huge geographical or temporal distances—is that such comparisons can easily be misleading. Almost exactly the same abstract marks which in some cultures do indeed refer to 'entoptic phenomena' seen in trances, in others may be numerical tallies, and in others still may be purely decorative.

[6] D. Lewis-Williams, *The Mind in the Cave*, 160.
[7] D. Lewis-Williams, *The Mind in the Cave*, 141.
[8] D. Lewis-Williams, *The Mind in the Cave*, 130.
[9] D. Lewis-Williams, *The Mind in the Cave*, 148–9.

Thus the British archaeologist Paul Bahn argues that since 'the division between the ordinary world and the spirit world scarcely existed in San life, the bald assumption that all rock faces were seen as interfaces between the world of daily life and the spiritual realm is simply a construct of western scholarship'.[10] He points out furthermore that the Bleek/Lloyd records are not always consistent and that one witness who at one time attributed the painting in one rock shelter to !gi:ten later said that the art was a record of the way the /Xam San live and denied that any of the painters with two exceptions were !gi:ten.

The disruption of the San way of life has made the meaning of the older rock paintings difficult to establish with any certainty. At almost the same time that the significance of African rock art was beginning to be recognized, however, a rock art was discovered which was still being practised and had roots of an equal antiquity.

Beagle Three

In 1840 HMS *Beagle* embarked on its third hydrographical survey (this time without Darwin or Fitzroy) under the command of Captain Wickam, Fitzroy's lieutenant on the previous voyage. The task of the expedition was to survey the north-west Australian coast, and in the course of this they landed on Depuch Island off the Pilbara Coast. Wickam was much impressed with the rock engravings that they found on the island [Fig. 3.2]. He was surprised at the 'accuracy of the animals and birds presented . . . the patient perseverance . . . talent and observation of the aborigines',[11] and subsequently published the drawings he made of them.

Some 50 years later a farmer called Joseph Bradshaw reported the discovery of paintings in some caves in the Kimberley region of Western Australia that were among the most remarkable of the many examples of rock art that settlers had begun to encounter. It was not, however, until the middle of the twentieth century that the true extent of this art began to be recognized. A survey of Pilbara, for instance, in the 1960s identified some hundreds of thousands of such engravings at more than 2000 different sites.

At the time of that survey aboriginal artists were still creating rock art. In the late nineteenth century it was sometimes assumed that all such work was of very recent date. An 1895 account of aboriginal cave drawings on the Palmer goldfield

[10] D. Lewis-Williams, *The Mind in the Cave*, 246.
[11] C. Wickham, 'Notes on Depuch Island', *Journal of the Royal Geographical Society of London* 12 (1842), 79–83.

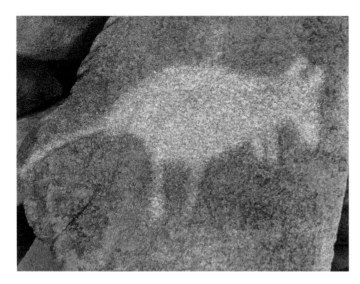

3.2 Burrup Peninsula rock paintings.

describes them as 'examples of a race in the stage of intellectual infancy' which must be the result of European influence and could not be more than 35 years old.[12] The development of scientific dating techniques in the twentieth century however revealed that some of this work was very much older.

When a fossilized wasp's nest on top of some of the paintings in the Bradshaw Caves (known to the Aborigines as Guion Guion) was dated by thermo luminescence to 17,000 years ago it established that the paintings underneath had to be older [Fig. 3.3]. When it became possible to date the Pilbara petroglyphs it became clear that not only did this area possess 'the largest surviving corpus of Pleistocene art' but that it was 'older than any rock art in the Americas or Africa'.[13]

Anthropologists who established a rapport with the tribal elders and the artists who were still producing this work discovered that there was a huge variety and complexity of meaning in the images that continued to be made. Some might have a purely 'secular' meaning, indicating the food and geography of the region, describing the hunt or marking tribal meeting places. Others were connected with sorcery and with death and fertility rituals. 'Religious' ideas were pervasive. Much of this art was connected with sacred sites and referred to 'dreamtime' creation myths and the worship of ancestral figures, and in many cases the ritual of making and touching such images involved the release of sacred power.

3.3 Bradshaw rock paintings.

[12] P. J. Trezise, 'Aboriginal Cave Paintings; Sorcery versus Snider Rifles', *Journal of the Royal Historical Society of Queensland* 8 (1968), 546–51.
[13] R. G. Bednarik, 'First Dating of Pilbara Petroglyphs', *Records of the Western Australian Museum* 20 (2002), 415–29.

This religious dimension came as no surprise. Half a century of field work by dedicated anthropologists and missionaries who had gained the trust of traditional tribal peoples all over the world, spending years living among them, learning their languages, and observing their customs, had transformed the understanding of the place of religion in such societies. While the form that they take might be extraordinarily diverse, the presence of some kind of supernatural beliefs seemed to be all but universal, and often far more sophisticated than had appeared from first impressions.

When for instance Evans-Pritchard, a pupil of Malinowski, published his study of *Nuer Religion* in 1956, he demonstrated that one of the tribes that had been dismissed in 1861 as not having 'even a vestige of superstition' and being 'mere brutes' in fact possessed a kind of religious thought that is 'remarkably sensitive, refined and intelligent. It is also highly complex.'[14]

Likewise when a field worker who had lived among the Fuegians, learnt their language, and studied their customs published his results in 1924, he concluded (flatly contradicting the first impressions of Darwin and Captain Fitzroy) that the Fuegians had a well-developed concept of a god whom they referred to as Watauinaiwa—'the eternal one'.[15]

A part of the difficulty in discerning the place of religion in these cultures had been that it was so pervasive. The modern Western distinction between sacred and secular seemed often to be hardly present. Down-to-earth engagements with the natural world, such as hunting or gathering plants, were as closely entwined with religion as rituals associated with healing or death. Matilda Coxe Stevenson began her account of the ethnobotany of the Zuni Indians by pointing out that 'plants are sacred to the Zuni . . . even those from the heavens are the offspring of the earth mother'.[16]

When, where, and how did this investing of the physical world with a spiritual dimension first begin? Might it have some connection with the peculiar and distinctive development of human curiosity that ultimately gives rise to science?

[14] E. E. Evans-Pritchard, *Nuer Religion* (1956), 322.

[15] J. H. Steward, *Handbook of South American Indians* (1946), 102ff.

[16] M. Coxe Stevenson, *Ethnobotany of the Zuñi Indians*, Thirtieth Annual Report of the Bureau of American Ethnology to the Secretary of the Smithsonian Institution (1908–1909), 36.

Roots and Shoots

Despite the perils of 'ethnographic snap' the parallels between the discoveries of anthropologists and the discovery of Palaeo-lithic paintings and sculptures placed at the end of almost inaccessible tunnels, of 'votive objects' pressed into cave walls, made it hard to escape the conclusion that this kind of awareness went back deep into prehistory.

Henri Breuil, as a Catholic priest, was, in the later part of his career, frequently asked to comment on prehistoric religion. Though not by nature a modest man (he was known in the quarter of Paris where he lived as *le vieux prêtre rogue*—the arrogant old priest) the Abbé was always cautious in responding to this question. He tended to interpret the paintings at Lascaux and elsewhere as being 'related to reproduction-magic (pregnant females, stallions following mares) and to hunting-magic (arrows directed towards or fixed in the bodies of animals)'.[17] He thought that the clearest evidence of religious ideas was to be found in burial practices.

Brueil argued that there was evidence of 'real burial rites' in some sites where skeletons were provided with protective stones over the heads. In early *Homo sapiens* sites in Europe, 'intentional burial' went back some 70,000 years. An excavation of a cave at Es Skhul on Mount Carmel in 1937 uncovered a burial (now dated between 100,000 and 80,000 years ago) where the body, laid on its back, had the jaws of a wild boar placed between its hands. More recent excavations have discovered elaborate grave goods. A burial at Sungir in Russia dated to 28,000 years ago contained the body of a youth covered with many thousands of drilled polished beads made from animal bones [Fig. 3.4].

[17] A. H. Brodrick, *The Abbé Breuil* (1963), 191.

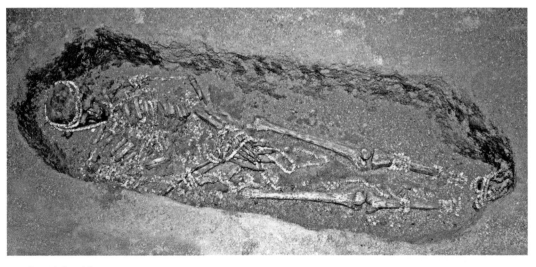

3.4 Sungir burial.

In 1951 Breuil published a paper in which he focussed on a cult of skulls which he suggested might lie at the root of prehistoric religion. When in South Africa in the 1940s he was asked 'to explain his position in regard to the holy scriptures' and produced in response a pamphlet arguing that 'we must not confound religious truths with the symbolical forms by which they are passed on from generation to generation'.[18] He was careful to point out that 'scientific truth founded on facts must not be confounded with working hypotheses',[19] but at the same time insisted that 'working hypotheses, even if incorrect, are *the* essentials of progress'.[20]

In the years since Breuil's death a great many facts have come to light that might make it possible to begin to answer at least the questions of when and where human beings first began to see a religious dimension in the physical world. The question of how such a perspective first arose remains much more mysterious. Given its entanglement with every aspect of human engagement with the world, it seems reasonable to explore, as a working hypothesis, the idea that the curiosity that gives rise to what we describe as "religion" and the curiosity that gives rise to what we think of as 'science' share some kind of common root.

---

[18] A. H. Brodrick, *The Abbé Breuil*, 119.
[19] A. H. Brodrick, *The Abbé Breuil*, 119.
[20] A. H. Brodrick, *The Abbé Breuil*, 115.

CHAPTER FOUR

The Garden of Eden Moment

The Bégouën brothers were not the first young cavers to come face to face with art from the remote past. Many years before a child of about ten years old had encountered a sequence of paintings at least as remarkable as those found on the Bégouën estate, in a cave that dominates the canyon of the Ardèche gorges.

There were no sculptures in this cave, but there were paintings on its walls which were executed in a variety of techniques. Drawings in some chambers were done with red ochre, in others with charcoal, and in others still by scraping off the yellow clay that coated the walls to reveal the white limestone underneath. In one of the chambers decorated with this scraping technique the ground was muddy and the child left a track of footprints on the floor and the print of a clay-covered hand on the wall. Moving on through this labyrinth of caverns the child seems to have made regular charcoal marks with a torch to help find the way back to an entrance to the cave, which at some later date suddenly collapsed. Sealed behind tons of rock, the caves remained undisturbed until their rediscovery in 1994.

The most striking feature of the paintings, which had appeared under the flickering light of the child's torch (and were made famous in Werner Herzog's 2010 film *Cave of Forgotten Dreams*), was the skill with which they were executed. One panel of horses' heads exhibits a grasp of anatomy of which a great eighteenth-century equine painter like George Stubbs would not have been ashamed (and which would be outside the scope of more than a few contemporary artists) [Fig. 4.1]. They demonstrated to the archaeologist who took charge of the site after the 1994 rediscovery that art did not have 'the linear evolution from clumsy and crude beginnings' that had 'been believed since the work of the Abbé Breuil'.[1] Where had this skill come from? What curiosity had driven the close observation of which it was the fruit?

4.1 Chauvet Cave panel of the horses.

[1] J.-M. Chauvet, E. B. Deschamps, and C. Hilaire, *Chauvet Cave* (2001), 126.

Human beings and animals are both often curious about one another. The child who walked through the cave with a torch seems to have been accompanied (or followed) by a dog whose paw prints follow the same route as the footprints. One of the largest paintings in the cave depicts a pride of lions, which would have been at least as adept at hunting human beings as human beings would have been at hunting them. But while both animals and human beings may look, observe, hunt, and sometimes eat one another, there is, as far as we know, only one species that spontaneously makes symbolic representations of others. In the biblical story of the Garden of Eden the idea that Adam names the animals, and not visa versa, conforms to observed reality. Whether it developed very gradually or comparatively suddenly, this symbolic ability must have first appeared at some place and time. When and where and how did this 'Garden of Eden moment' occur?

The Beginning of History

The discovery by an American chemist, Willard Libby, in 1947 of the process of radiocarbon dating provided a means by which it became possible to establish a secure framework for prehistory. Libby's method exploits the fact that the radioactive isotope ^{14}C decays at a regular rate. ^{14}C is produced in the upper atmosphere when thermal neutrons generated by cosmic rays are absorbed by nitrogen atoms. It makes its way into all organic matter via photosynthesis. This made it possible to establish with a specified degree of accuracy the date of organic materials like the charcoal used in cave paintings.

Radiocarbon tests revealed that the Lascaux paintings were executed 18,000 years ago and the Altamira paintings 14,000 years ago. When the torch marks made by the ten-year-old child were tested, it became clear that he or she had visited the Ardèche cave nearly 10,000 years before a single mark had been made at Lascaux: somewhere that is between 26,000 and 27,000 years ago. The paintings the child had looked at had been made by artists as remote from the child as the first Egyptian artists are from us. Although the exact dating remains disputed,[2] it is likely that they had been painted some 5,000 years earlier: somewhere between 30,000 and 32,000 years ago.[3] By the time they were rediscovered in 1994, it had, however, already begun to seem possible that 'the Garden of

[2] J. Combier and G. Jouve, 'Nouvelles recherches sur l'identité culturelle et stylistique de la grotte Chauvet et sur sa datation par la méthode du ^{14}C', *L'Anthropologie* 118 (2014), 115–51.

[3] H. Valladas et al., 'Bilan des datations carbone 14 effectuées sur des charbons de bois de la grotte Chauvet', *Bulletin de la Société préhistorique française* 102 (2005), 109–13.

Eden moment' had occurred long before these images were painted, and in a place that was far removed from the Ardèche gorges.

Out of Africa 1

When the first discoveries of primeval remains were found in France and Spain, it was assumed that the cradle of humanity had been in Europe. As the search widened primeval remains began to be discovered all over the place. In 1891 a Dutch physician named Eugène Dubois unearthed at Trinil in central Java the skull cap and thigh of a creature (nowadays known as *Homo erectus*) that seemed to be halfway between ape and man. Davidson Black made a similar discovery at Zhokoudian near Peking in 1921. In the 1920s, evidence began to appear which suggested that the trail of human origins led back (as Darwin had suspected it would) to Africa.

The first clue that pointed in this direction arrived in a box of fossils dug up from a lime mine at Taung in the Northern Cape, delivered to a young anatomist named Raymond Dart. The box arrived on a Sunday afternoon in 1924 just as Dart was dressing to go to a wedding where he was to be the best man. Unable to restrain his curiosity he opened the box and discovered inside a fossilized skull, not big enough for primitive man, but with a brain three times larger than a baboon and considerably bigger that an adult chimpanzee [Fig. 4.2].

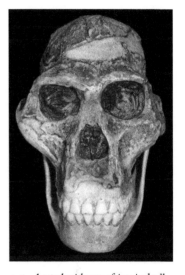

4.2 *Australopithecus africanis* skull.

In the moment before being dragged away by the anxious bridegroom Dart recognized that this 'was one of the most significant finds ever made in the history of anthropology'.[4] He published his discovery in *Nature* in 1925, naming the skull *Australopithecus africanus*—Southern Ape—and arguing that it provided clear evidence that Africa had been the cradle of mankind.[5] This conclusion was met by considerable scepticism from European scientists. So, six years later, were the discoveries of a young Kenyan archaeologist, Louis Leakey.

The son of two missionaries working among the Kikuyu, Leakey had gone up to Cambridge with the intention of following in his parents' footsteps. While there he diverted into palaeontology. In 1925 he was cleaning two ancient skeletons when he noticed similarities with one that had been dug up in Africa twelve years earlier by a German archaeologist named Hans Reck.

[4] R. A. Dart, *Adventures with the Missing Link* (1959).
[5] R. A. Dart, '*Australopithecus africanus*: The Man-Ape of South Africa', *Nature* 115 (1925), 195–9.

Reck's skeleton had been unearthed in the Serengeti from part of the wall of the Olduvai gorge which geologists believed must be 600,000 years old. Reck had argued that the skeleton must be of an equivalent date. Faced with universal incredulity he had been forced to withdraw his claim, but Leakey was convinced he had been right. He bet the German £10 that if he went to Olduvai he would find prehistoric tools there within 24 hours. Six years later in 1931 Leakey led an expedition to the gorge accompanied by Reck. Leakey won his bet.

Neither the skull discovered by Dart nor the tools found by Leakey were conclusive in proving their claim that human beings had originated in Africa. Dart brought his skull to London in 1931. His wife on one occasion accidentally left it in a taxi. His evidence was generally dismissed in part because it was the skull of a child and experts thought that conclusions could only be based on adult specimens. The evidence from Olduvai at first had a more favourable reception, but when Leakey left his wife, an investigation by a Cambridge committee into his morals cut off his funding. When doubts began to be expressed about the provenance of Reck's skeleton, opinion in Cambridge turned against him. His career there was terminated.

Over time nevertheless the claims made by Dart and Leakey began to be vindicated. In South Africa a palaeontologist called Robert Broom initiated a search for an adult specimen of *Australopithecus*. In August 1936 he was handed an adult male skull by the quarry manager of a mine in the Sterkfontein caves near Krugersdorp where it had been unearthed.

In East Africa the Leakeys continued to excavate. In 1959 Mary Leakey, Louis' second wife, unearthed an *Australopithecus*-like skull in a deposit which also contained stone tools. Louis estimated the skull as 600,000 years old. In 1960 two geophysicists from Berkeley, Jack Evernden and Garniss Curtis, stunned experts around the world by dating the deposit with the new radiometric method to 1.75 million years ago.

In the years since 1960 as new fossils have been discovered these new dating methods have consistently strengthened the case for African origins. The earliest 'ape men' found in Europe, classified today as *Homo erectus* and *Homo ergaster* (working man), have been dated to around 1.7 million years ago. In 1974 Donald Johanson and Tom Gray unearthed an almost complete *Australopithecus* skeleton at Hadar, in the Awash valley of Ethiopia. Nicknamed 'Lucy' (after the title of a song by the Beatles) it proved to be 3.5 million years old. In 1976 Mary Leakey discovered the footprints of three *Austalopithecines* (who have so far only been found in Africa) preserved in volcanic ash at Laetoli in Tanzania. The ash layer was dated to 3.7 million years ago, and the footprints confirmed (as Raymond Dart had argued) that *Australopithecus* walked upright.

Meanwhile the earliest anatomically modern human fossils found in Europe have been dated to 42,000 years ago, in Australia to 45,000 years ago, and in Israel to 90,000 years ago (though these fossils have some archaic features). At Herto in Ethiopia *Homo sapiens* fossils have been dated to 160,000 years ago, and in 1967 near Kibish on the Omo River in Ethiopia a team led by Richard Leakey (Louis' son) discovered *Homo sapiens* fossils, which in 2005 were dated to 195,000 years ago.

These findings dovetailed with studies of the genetic history of mitochondrial DNA that were coming out of the University of California in the 1980s. Twenty years earlier Alan Wilson and his then doctoral student Vincent Sarich had shown how what was called a 'molecular clock' could date the divergence between humans and the great apes. In the 1980s Wilson with two new PhD students, Mark Stoneking and Rebecca Cann, turned their attention to mitochondrial DNA.

The DNA in mitochondria (the tiny energy-producing bodies within cells) is passed on from mother to child without any contribution from the father. Because they are non-recombining over the generations, mutations are passed on, making it possible to trace the matrilineal lineage and find connections between geographically remote populations. By sampling 133 living individuals all around the world, the researchers in California were gradually able to construct a giant family tree that showed how the different branches of the human population were related and where they had come from. On the 1st of January 1987 Wilson, Stoneking, and Cann published a paper, again in *Nature*, which concluded that all living human beings were descended from a particular woman (later dubbed 'mitochondrial Eve') who had lived in Africa between 140,000 and 200,000 years ago.[6]

A similar technique applied to the non-recombining portion of the Y chromosome passed on in the male line pointed back to an African male ancestor ('Y chromosomal Adam'), though in this case there has been less consensus about the dating. Assuming a constant mutation rate for mitochondrial DNA the

[6] R. L. Cann, M. Stoneking, and A. C. Wilson, 'Mitochondrial DNA and Human Evolution', *Nature* 325 (1987), 31–6.

migrations from Africa would have occurred around 60,000 years ago, which would accord with the radiometric evidence of human remains appearing in Europe, Asia, and Australia around 40,000 years ago, but not with the fossils found in Israel or India. A more recent recalibration of the mutation rate suggests that this might need to be pushed back to somewhere between 90,000 to 130,000 years ago.[7]

While this evidence seemed to establish that anatomically modern humans originated in Africa, it didn't necessarily follow that 'the Garden of Eden moment' had taken place there. The first clear indications of symbolic thought seemed to be the rock paintings in Europe and Australia, the earliest of which have been dated to around 42,000 years ago. This 'creative explosion' might have been the result of a cognitive development that had taken place after the migration from Africa.

Out of Africa 2

A discovery made in 1999 challenged this perspective. In 1991 Christopher Henshilwood, an archaeologist from Witwatersrand University, was conducting an excavation on some property owned by his grandfather. In the Blombos Cave, high up in a limestone cliff on the southern cape of South Africa, he discovered some stone artefacts known as 'bifacial points' (which look rather like spear tips). This led Henshilwood and his team to conduct a series of further excavations in the cave. In 1999, near an undisturbed deposit of ash and sand close to an ancient hearth, they found a small piece of ochre that had been scraped and ground to form a flat surface and then marked with cross hatches and lines in a distinct geometric pattern [Fig. 4.3]. The following year they found a similar piece together with a set of marine shells which had been drilled to

4.3 Blombos Cave hatched ochre.

[7] A. Scally and R. Durbin, 'Revising the Human Mutation Rate: Implications for Understanding Human Evolution', *Nature Reviews Genetics* 13 (2012), 745–53.

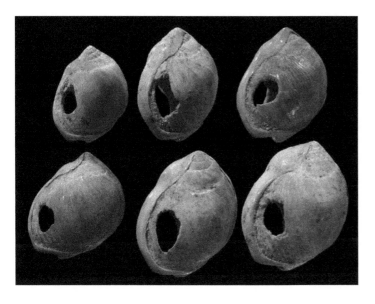

4.4 Blombos Cave necklace.

form a necklace [Fig. 4.4]. Dating of the sand above the ochre
and the burnt stone in the same layer established that the two
pieces had been engraved some 77,000 years ago, making them
the earliest known examples of symbolic activity.

In 2007 a team led by Curtis Marean, a palaeo-anthropologist
from Arizona State University, excavated a cave along the same
coast on a promontory called Pinnacle Point just south of Mos-
sel bay. Alongside examples of so-called 'bladelet technology'—
small blades that were embedded into complex tools—and
evidence of stone tools that had been treated with heat to
make them easier to work, Marean's team found evidence of
scrapped and ground ochre pigment. All of this could be dated
to 164,000 years ago.

The following year Henshilwood and his team made another spectacular discovery at the Blombos Cave. They found
what headlines described as 'the world's oldest artist's studio'.
The 'studio' consisted of two abalone shells that were used to
mix and store pigment, together with grindstones, hammer-
stones, the remains of a small fire pit, and animal bones used to
stir and perhaps apply the pigment. Henshilwood observed that
the makers of this pigment had both 'understood basic chemistry' and apparently 'consciously played with colour' [Fig. 4.5].

The mix in the shells consisted of ground-up ochre chips
and quartzite chips in a liquid that had been gently stirred
together with charcoal and fatty seal bone marrow extracted

4.5 Blombos Cave abalone shell
palate.

with heat, which had served as a crucial binder. This had produced a dramatic red pigment. In one of the shells there was a tiny piece of yellow mineral called goethite, which might have pushed it towards orange. According to the radiocarbon dating, whoever had mixed up this paint had done so around 100,000 years ago.

East of Java

In 2014 two reports from Indonesia strengthened still further the growing impression that symbolic activity goes back deep into the past of *Homo sapiens* (and even beyond it).

In the 1930s, excavations took place in the caves and rock shelters at the base of the Maros Karsts, the remarkable 'stone forests' on the island of Sulawesi. Evidence of prehistoric human occupation was discovered in some ninety of these, and in the 1950s Indonesian researchers recorded the discovery of rock art. These consisted of hand stencils and images of native Sulawesian animals, including the 'warty pig' and the babirusa or 'pig deer'. When these were dated in 2014 they turned out to be contemporary with the oldest rock art found in Europe. A hand stencil was dated to nearly 40,000 years ago and a pig deer to around 35,000 years ago.

The authors of the letter to *Nature* who reported these findings suggested that two conclusions were possible. Either 'rock art emerged independently at around the same time at roughly both ends of the spatial distribution of early modern humans', or else 'cave painting was widely practised by the first *H. sapiens* to leave Africa tens of thousands of years earlier'.[8]

Another letter to *Nature*, published in 2015, raised an even more remarkable possibility. This letter reported a new analysis of some fossil shells excavated in 1891 by Eugène Dubois at Trinil on the Solo river towards the eastern end of Java. The shells had been opened with a tool like a shark's tooth. They were engraved with a zig-zag M shape [Fig. 4.6]. The sediment in the shells suggested they were roughly 500,000 years old. This together with other data led the authors to conclude that 'the engraving was made by *Homo erectus*',[9] an ancestor of *Homo sapiens*.

If symbolic activity can be found in our hominid ancestors, how far back into the human past might it be possible to discover image-making in general and cave painting in particular?

[8] M. Aubert et al., 'Pleistocene Cave Art from Sulawesi, Indonesia', *Nature* 514 (2014), 223–7.

[9] J. C. A. Joordens et al., '*Homo erectus* at Trinil on Java Used Shells for Tool Production and Engraving', *Nature* 518 (2015), 228–31.

4.6 Geometric pattern on shell from Trinil.

The authors of the Sulawesi report conclude that if their second hypothesis is correct then 'we can expect future discoveries of depictions of human hands, figurative art and other forms of image making dating to the earliest period of the global dispersion of our species'.[10]

If *Homo sapiens* evolved biologically as early as 200,000 years ago, what of the forebears and first descendants of 'mitochondrial Eve'? At what point might 'the garden of Eden moment', in the sense of the first appearance of explicit symbolism, be said to have occurred?

The evidence to date remains sparse and hard to come by. Although dating South African rock painting has always been regarded as challenging, progress in this field by David Pearce and his colleagues may in time reveal some remarkable results. At the beginning of 2015 the only securely dated African figurative art from before 9700 BC is seven stone plaques (on one of which is a recognizable image of a zebra) that were found in levels dated to 30,000 years ago in the Huns Mountains of Namibia.[11] While some form of symbolic expression may reach back deep into our human ancestry, the artefacts in the Blombos Cave still remain the earliest known examples of modern human symbolism (though for how long remains to be seen). Symbolism and image-making may not though have been the only customs that human beings took with them in their migration from Africa. The most recent dating of the burial at

[10] M. Aubert et al., 'Pleistocene Cave Art from Sulawesi, Indonesia', 223–7.
[11] R. Rifkind, C. S. Henshilwood, and M. M. Haaland, 'Pleistocene Figurative *Art Mobilier* from Apollo II Cave, Karas Region, Southern Namibia', *South African Archaeological Bulletin* 70(201) (2015), 113–23.

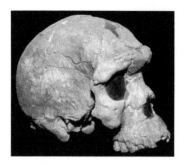

4.7 Herto skull.

Es Skhul on Mount Carmel places it around 90,000 years ago. Did mortuary customs of any kind exist before this in Africa?

Three 160,000-year-old skulls discovered at Herto in Eastern Ethiopia in 1997 consisted of two adults and one child [Fig. 4.7].[12] The child's skull, according to Tim White, who co-lead the team that excavated and analysed the discovery, showed 'cut marks pointing to ancient mortuary practices'.[13] The skull had been defleshed after death 'and the entire cranium worn smooth as if by repeated handling'. One of the adult skulls showed scratches around the perimeter that seemed unrelated to the defleshing. Might this scratching and polishing point to the kind of activity recorded by anthropologists in societies like New Guinea in which the skulls of ancestors were preserved and worshipped?

The evidence for this is tenuous and has been strongly contested.[14] While the heat treatment of rocks at Pinnacle point, and the 'basic chemistry' in the Blombos cave are to date the earliest evidence of what might be called 'proto-science', the Skhul burials are to date perhaps the earliest secure physical evidence of what might be called 'proto-religion'. How had these two extraordinary developments of what we might now call science and religion come about? What was the relationship between them? There is a limit to what can be gleaned from archaeological remains, but in the early 1960s light began to be shed on the question from a different angle.

[12] T. D. White et al., 'Pleistocene *Homo sapiens* from Middle Awash, Ethiopia', *Nature* 423 (2003), 742–7.

[13] R. Sanders, '160,000-Year-Old Fossilized Skulls Uncovered in Ethiopia Are Oldest Anatomically Modern Humans', http://www.berkeley.edu/news/media/releases/2003/06/11_idaltu.shtml. Accessed 7 August 2015.

[14] Recent excavations at Sima de los Huesos (the Pit of Bones) in northern Spain have investigated the burial of middle pleistocence hominins (probably *Homo heidelbergensis*) dating from 430,000 years ago. The researchers interpret this as perhaps 'the earliest funerary behaviour in the human fossil record'. N. Sala et al., 'Lethal Interpersonal Violence in the Middle Pleistocene', *PLoS ONE* 10 (2015), e0126589. doi:10.1371/journal. pone.0126589. Though not decisive, this, if correct, would certainly increase the plausibility of White's interpretation.

CHAPTER FIVE

Primate Parallels

In 1954 just as the Mau Mau rebellion in Kenya was coming to an end, Louis Leakey published a book on the situation. In it he argued that if church leaders 'confuse the fundamental teachings of the New Testament with things that are nothing more than British social custom, or doctrines laid down by men' they would turn Africans away from 'real Christianity' and leave them 'to religions like Mau Mauism and communism'.[1] The question of how to distinguish between 'fundamental teaching' and 'doctrines laid down by men' was not though always easy to decide. In the late 1920s Leakey's father, Canon Harry Leakey, had sparked a major crisis in Kikuyuland with his implacable opposition to the Kikuyu custom of female genital mutilation, which some missionaries had been willing to overlook.

For the younger Leakey a similar kind of issue between 'fundamental teaching' and 'man-made doctrine' turned around the biblical idea of the relationship between human beings and animals. In his 1937 autobiography he described how he 'had become firmly convinced of the truth of the theory of evolution as distinct from creation described in Genesis'.[2] In later life, however, 'Louis would always indignantly say that none of his scientific pursuits ever contradicted the Bible', insisting that 'nothing I've ever found has contradicted the Bible. It's people with their limited minds who misread the Bible.'[3] Christians (including his Catholic mission school workers at Kamba) sometimes felt that any similarity between animals and human beings compromised the idea of human uniqueness. If, however, God was the creator of both and the former were, in the title of his 1934 book, *Adam's Ancestors*, then one would expect to find parallels between them. As the Mau Mau rebellion came to an end it was the nature of such parallels that Leakey wanted to investigate.

In 1931 while on an expedition in the vicinity of Lake Victoria, an associate called Arthur Hopwood discovered the fossil remains of a primate species he named 'Proconsul'. From the mid-1940s in an attempt to understand the environment in

[1] L. S. B. Leakey, *Defeating Mau Mau* (2004), 130.
[2] L. S. B. Leakey, *White African* (1937), 161.
[3] V. Morell, *Ancestral Passions* (1995), 56.

which this species had lived, Leakey set about trying to find observers to study the chimpanzee and gorilla populations that inhabited these lakeside areas.

Leakey's Angels

In 1956 he persuaded his secretary Rosalie Osborn to study gorillas in Uganda, but she returned to Britain after four months. Then in 1957 he was visited by a young Englishwoman called Jane Goodall, who, though at the time without any scientific training, had a deep interest in animals. After visiting the Olduvai gorge and working briefly as Leakey's secretary, Goodall accepted his invitation to study the chimpanzees living in the Gombe Stream Reserve on the shores of Lake Tangynika.

Three years later Dian Fossey, then working as an occupational therapist, came to Leakey's attention when, on a visit to the Olduvai gorge, she sprained her ankle and fell into the excavation. After meeting again in America, Leakey had arranged funding for Fossey to begin in 1967 what was to be a lifelong study of the gorillas in the mountain forests of Rwanda.

The third of this trio was Biruté Galdikas (who coined the name 'Leakey's angels' in preference to being described as 'ape ladies'). Galdikas met Leakey after a lecture at UCLA in 1969, and in 1971 with his support began a long-term field study of orang-utans in the jungles of Borneo.

Within months of arriving at the Gombe Stream Reserve Goodall began to observe chimpanzee behaviours which had never been witnessed before and been assumed to be exclusive to human beings. These involved a variety of different kinds of activity, but the similarity between the social communication of chimpanzees and human beings was particularly striking and offered a clue as to how a central aspect of the human mind might have developed.

Natural Psychologists

When a chimpanzee was suddenly frightened, Goodall noticed, 'he frequently reaches to touch or embrace a chimpanzee nearby, rather as a girl watching a horror film may seize her companion's hand'.[4] When chimpanzees greeted one another they used a whole variety of such gestures which reflected their different position within the social group. The individuals within these

4 J. Goodall, *In the Shadow of Man* (1999), 234.

groups were often closely related to one another but the relationships between them, she discovered, could be highly complex and involved constantly changing associations—associations 'which could best be described as friendship'. Some of these might be 'of relatively short duration' but others, it transpired, 'persisted over the years'.[5]

The complex relationships between chimpanzees described by Goodall was paralleled in those of the silverback gorilla groups observed by Dian Fossey (though less so in the solitary orang-utans studied by Galdikas). When Nicholas Humphrey, a British theoretical psychologist, spent some time with Fossey's gorillas in 1971 he began to wonder whether there was some relationship between the size of the gorilla's brains in relationship to their bodies and the sophistication of these social interactions. Although in the laboratory gorillas could be got to solve all kinds of clever conceptual puzzles, in the wild, with their simple bamboo munching lifestyle, they showed no obvious sign of using or requiring any of this intelligence.

Trying to understand this puzzle, Humphrey began to reflect on his own experience. He had come to Africa to escape from a failed marriage and an on-off affair with another woman: 'my head (when I was not thinking about gorillas) was full of unresolved problems concerning my social relationships'.[6] Might something similar be true for the silverbacks?

The simplicity of the gorillas lifestyle, it seemed, was founded on the cohesion of their family groups, which functioned both as 'a protective mafia and a kind of polytechnic school'.[7] Within these groups young gorillas were guarded and learned the ways of the forest. Maintaining this unity, however, was not so simple. It involved continuous small disputes around grooming, food, and sleeping arrangements, with occasional larger disputes about mating and membership of the group. All of these turned around issues of social dominance which might potentially disrupt the unity of the group. The fact that such disruptions rarely happened reflected the gorillas' skills as natural psychologists in understanding and anticipating one another's behaviour. How did they achieve this?

The Rouge Test

In the 1960s developmental psychologists devised a test to detect self-awareness in children. This was the so-called 'rouge

[5] J. Goodall, *In the Shadow of Man*, 117.
[6] N. Humphrey, *The Inner Eye* (2002), 37.
[7] N. Humphrey, *The Inner Eye*, 40.

test', which involved putting some rouge on the forehead of a young child while they were asleep, and then leaving them in front of a mirror. Up to the age of one children tend to react to the mirror image as though it were another child and try to play with it. Children around 18 months of age begin to react to the mirror as an image of themselves. When they look in the mirror they notice the rouge on their foreheads and try and wipe it away.

In 1970 an experimental psychologist called Gordon Gallup tried this on chimpanzees. He discovered that unlike most other animals, they did the same thing. Other experiments with captive chimpanzees seemed to suggest that they had what psychologists (somewhat misleadingly) called 'a theory of mind'. They could as it were put themselves in someone else's position, imagine their intentions and how they might fulfil them [Fig. 5.1].

Putting the results of these experiments together, Humphrey argued that it was chimpanzees' ability to recognize themselves that made possible 'a theory of mind' by providing a template or model by which to recognize the actions of others. In rather the same way that living in a terraced house in a London square enabled him to 'read' what went on in houses around the square he had never been inside due to his experience of his own house, so creatures who could examine themselves and

5.1 The theory of mind.

what went on in their own minds could use that knowledge as the basis for guessing what went on in someone else's mind.[8]

Some confirmation that there was a link between brain size and social relationships was provided in the 1990s by a British anthropologist, Robin Dunbar, who demonstrated that there was a correlation between the brain size of primates (in proportion to their bodies) and the average size of the groups in which they lived. Dunbar found that he could accurately predict the size of the group in which a given primate would live from the size of their neocortex (though the large-brained orang-utans remain an exception).

He argued from this that primates who live in very large groups need extra processing power to keep track of the larger number of social relationships that arise in these circumstances, and was able to show that the larger the size of the groups in which primates live, the longer the time they spend grooming one another—picking lice and fleas from each other's hair. This was not because there were more lice, but because more time was required for social communication. There is a limit to the amount of time that can be spent in grooming rather than finding food, and Dunbar argued that language may have developed because it made it possible to communicate with more than one individual at a time.[9]

Chimpanzees, it turned out, did have some linguistic ability (though the extent of this has been hotly contested). When two American scientists, Alan and Beatrice Gardner, tried the experiment of teaching a young chimpanzee (they called Washoe) standard American sign language they found that by the age of five, Washoe could recognize and use some 350 different symbols. When Washoe was asked (in sign language) 'Who is that?', as she was looking into a mirror, she signalled back 'Me, Washoe' [Fig. 5.2].[10]

5.2 Washoe.

If the consciousness of self could enable a chimpanzee to become aware of the thoughts and intentions of other chimpanzees, and even to some extent of creatures like human beings, could it go beyond this and effect the chimpanzee's perception of the wider world in which it lives?

The Waterfall Dance

In the late spring of 1961 Jane Goodall was observing a group of seven male chimpanzees in the Gombe Stream Reserve just

[8] N. Humphrey, *The Inner Eye*, 71–2.
[9] R. I. M. Dunbar, 'Coevolution of Neocortical Size, Group Size and Language in Humans', *Behaviour and Brain Sciences* 16 (1993), 681–94.
[10] J. Goodall, *In the Shadow of Man*, 242.

as the long rains were beginning. The group had just reached the crest of a ridge when the storm broke and there was a sudden clap of thunder:

> As if this were a signal one of the big males stood upright and as he swayed and swaggered rhythmically from foot to foot I could hear a rising crescendo of his pant-hoots above the beating rain. Then he charged off, flat-out down the slope towards the trees he had just left. He ran some thirty yards . . . leaped into the low branches and sat motionless.
>
> Almost at once two other males charged after him. One broke off a low branch from a tree as he ran and brandished it in the air before hurling it ahead of him. The other as he reached the end of his run, stood upright and rhythmically swayed the branches of a tree back and forth before seizing a huge branch and dragging it further down the slope. A fourth male as he too charged, leaped into a tree and, almost without breaking his speed, tore off a large branch, leaped with it to the ground, and continued down the slope. As the last two males called and charged down, so the one who had started the whole performance climbed from his tree and began plodding up the slope again. The others . . . followed suit. When they had reached the ridge they started charging down again all over again one after another with equal vigour.[11]

Individual male chimpanzees often reacted to the start of heavy rain by performing this kind of rain dance, though group displays were more rare. In 1970 Goodall saw an even more remarkable group display when she witnessed a troupe of chimpanzees encountering one of the spectacular waterfalls that flow through the Gombe Forest. The display she saw then has been observed many times since. It generally involves a single individual—most often an adult male.

As the chimpanzee approaches one of these falls [Fig. 5.3]:

> his hair bristles slightly, a sign of heightened arousal. As he gets closer, and the roar of falling water gets louder, his pace quickens, his hair becomes fully erect, and upon reaching the stream he may perform a magnificent display close to the foot of the falls. Standing upright, he sways rhythmically from foot to foot, stamping in the shallow,

[11] J. Goodall, *In the Shadow of Man*, 52.

5.3 Gombe Forest waterfall dance.

rushing water, picking up and hurling great rocks. Some-times he climbs up the slender vines that hang down from the trees high above and swings out into the spray of the falling water. This 'waterfall dance' may last for ten or fif-teen minutes.... After a waterfall display the performer may sit on a rock, his eyes following the falling water.[12]

What was going through the chimpanzees' minds? Might it not be possible, Goodall wondered, that these performances were 'stimulated by feelings akin to wonder and awe?'[13] These dances were not, however, the only way in which chimpanzees reacted to the world around them.

Before the long rains began, Goodall had already observed the chimpanzee she called David Greybeard using a long grass stem to fish for termites in a mound [Fig. 5.4]. This was the first of many such observations of chimpanzee tool use. Chim-panzees use a whole variety of different tools for different pur-poses: leaves to gather water or clean themselves, stones as a hammer and anvil to crack nuts, sticks to extract marrow. As new observation projects were begun and compared it became apparent that different groups of chimpanzees appeared to have different cultural traditions of tool use. While the chimpanzees at Gombe used leaves for personal hygiene, those in the Tai Forest did not. The Tai Forest chimps, on the other hand, ex-tract bone marrow with sticks, unlike those at Gombe.

[12] J. Goodall, 'Primate Spirituality', in *The Encyclopedia of Religion and Na-ture*, ed. Bron Taylor (2005), 1034.
[13] J. Goodall, 'Primate Spirituality', 1034.

5.4 Chimpanzee termite fishing at Gombe Stream National Park.

If the waterfall dance was evidence of something like 'primate spirituality', then tool use was evidence of something like 'primate technology'. Could the discovery that chimpanzees could contemplate nature suggest a parallel to 'proto-religion' in early human beings? Could the discovery that they could use tools to manipulate nature suggest a parallel to Palaeolithic 'proto-science'?

Such parallels also posed new questions. When Leakey received an excited telegram from Goodall anouncing her discovery, he telegraphed in reply that 'Now we must redefine "tool", redefine "man" or accept chimpanzees as human.'[14] If tool use was not the defining characteristic of human beings what was? If the intellectual development of chimpanzees and other primates seemed to parallel our own, why had it not followed the trajectory that ultimately produced the paintings in the Ardèche cave? What if anything was the missing ingredient?

[14] J. Goodall, *Reason for Hope* (1999), 67.

CHAPTER SIX

Horizons of Curiosity

Curiosity and the Cat

Anyone who has dangled a ball of string in front of a kitten will have witnessed a very focussed form of curiosity: a fierce tigerish concentration on every slightest twitch of the thread while you, the person dangling the thread, are entirely ignored.

Anyone who has played the same game with a small child will have found something rather different: whatever pleasure there may be in trying to grasp the dangling thread is soon eclipsed by the discovery that there is someone at the other end of it.

The curiosity of the cat and the child might, one say, have different kinds of horizons. What has shaped these?

Curiosity at some level is as old as life itself. The prosperity of everything that exists behind a cell wall must depend on the world that exists beyond that wall. Hence all living things at a minimal level investigate their surroundings by bumping into (or being bumped into by) whatever is near them. For creatures with nervous systems things are more complex. They actively explore their environment by prodding, poking, and mapping their surroundings with whatever means are available to them.

The curiosity of young animals is directly related to acquiring the survival skills of foraging, hunting, or avoiding being hunted. While they are protected and their food is provided by adult parents, they can, however, explore and hone their skills without the direct consequences they will face in later life. The longer this educational period in which animals care for their young is extended, the further this unfocussed curiosity of childhood seems to continue into adult life.

In primates for instance this curiosity is particularly highly developed. Captive capuchin monkeys play with everything they can get their hands on even when they are not hungry: 'they seem to have a passionate desire to discover whatever is inside or behind anything that can be pulled or plucked or dismembered'.[1]

[1] M. Moynihan, *The New World Primates* (1976), 107.

Human infants, before they learn to talk, interrogate the world in very similar ways, by prodding things, by pulling them, and by putting them in their mouths. When children begin to use words, however, their horizon of curiosity is extended far beyond these manual investigations. Once they have learned to use the word 'Why?' they discover that there is no end to what can be interrogated. Curiosity loses its horizon.[2] An obvious answer to the question 'What was the ingredient that distinguished the development of human beings?' would be 'language'.

Is it though entirely a matter of language? The Gardners observed that Washoe, like other chimpanzees that have been taught sign language, tended to use it only to ask for things. Comments on the surrounding world were very rare. Observers of chimpanzees in the wild have found that they also seem to be quite restricted in the nature of their communications with one another.

Two Swiss primatologists, Cristophe and Hedwige Boesche, studied the chimpanzees in the Tai Forest 20 years after Jane Goodall's first discoveries. They discovered that their 'cultural traditions' of tool use were more caught than taught. Young chimpanzees would observe their elders using a tool, like a stone to crack a nut, and attempt to imitate their behaviour [Fig. 6.1]. Occasionally a mother would demonstrate how to position a nut or grip a rock, but the Boesches found that this kind of teaching seemed to be exceptional [Fig. 6.2]. On the whole chimpanzee mothers did not seem to notice the

6.1 Young chimpanzee trying to crack nuts with a stone hammer.

[2] When the sister of one of the authors was applying to Oxford, running out of time in the general paper and coming to the question, 'Should there be limits to curiosity?', she wrote, 'Why ask?' and moved on to the next question.

struggles of their children and allowed them to proceed by trial and error.

Given the time that they devote to social communication, the general failure of chimpanzees to teach their young to use tools was remarkable. In more than 4000 hours of watching chimpanzees using stones to crack nuts, the Boesches only observed mothers helping their young on two occasions. The consequences for cultural development were obvious. Whereas in captivity chimpanzees can become proficient in all kinds of different tool use, in more than 50 years since Jane Goodall first saw chimpanzees using tools in the wild, no technological progress of any kind has been observed. It is rather, as a British archaeologist Stephen Mithen has put it, as though 'each generation of chimpanzees', struggles, 'to attain the technical level attained by the previous generation'.[3] What might be the cause of this?

The Cathedral of Mind

As an archaeological student Mithen had spent his summer vacations excavating the Benedictine Abbey of San Vicenzo in Italy. During the excavation old walls were uncovered and patiently recorded. It gradually became possible to reconstruct the different architectural phases through which the building had passed over the course of a thousand years, as walls were built and demolished, doors were opened and blocked, and new stories were added. When in the 1980s he became interested in the development of the human mind, this experience provided Mithen with a powerful analogy.

In 1983 two influential books had been published on the architecture of the human mind, one by the psychologist Howard Gardner, the other by the philosopher Jerry Fodor. Gardner argued that the mind contained a set of what he described as 'multiple intelligences'; Fodor that the mind contained a combination of generalized 'central processes' like memory and trial and error reasoning, together with a set of specialized automatic 'cognitive modules' like sight, touch, and things like language ability. Drawing on both these ideas Mithen compared the mind to a cathedral with a large central nave—the 'central processes'—and opening off it a whole series of side chapels—the 'cognitive modules'. Using this metaphor he then proposed a way in which the architectural history of this cognitive cathedral might be reconstructed.

6.2 Older chimpanzee demonstrating how to crack nuts with a stone.

[3] S. J. Mithen, *The Prehistory of the Mind* (1996), 84.

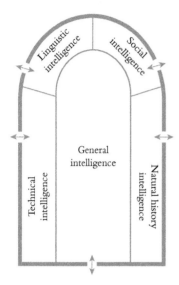

6.3 Steven Mithen's cathedral of intelligence without accessibility between chapels.

The first phase of the development of the cathedral of mind, Mithen argued, was dominated by the central nave of general intelligence. This general intelligence nave was made up of a suite of 'general purpose and decision making rules' which can modify behaviour in the light of experience. Information is delivered here through a series of 'doors'—imputs from 'modules concerned with perception' [Fig. 6.3].

In the second phase a whole series of 'chapels' of specialized intelligences were added to the general intelligence nave in rather the same way that side chapels began to proliferate round the nave in Romanesque cathedrals of the twelfth century. Each of these side chapels was related to a particular domain of behaviour, and three of them were of particular importance. One was a chapel of 'social intelligence' that included the facility of self recognition and the related 'theory of mind'. Another was a chapel of 'natural history intelligence'. The third was a chapel of 'technical intelligence' that housed 'the mental modules for the manufacture of stone and wooden artefacts'.[4]

In this postulated second phase these specialized intelligences are supposed to operate semi-independently. In Mithen's metaphor there is no access between the chapels: 'the walls are thick and almost impenetrable to sound from elsewhere in the cathedral'.[5] Thus the apparent 'brick wall between social and tool behaviour'[6] in chimpanzees, Mithen argued, reflects an underlying neurological architecture. Where there are exceptions—when chimpanzee mothers show their offspring how to use tools—we may think of this 'as sounds emanating from one chapel being heard in a heavily muffled and indistinct form elsewhere in the cathedral'.[7] It was this lack of internal communication in their cognitive architecture which explains, Mithen suggested, why the technology of both chimpanzees and some of our hominid ancestors show such little development.

The critical new development in the third phase of the cathedral is that 'doors and windows' are now inserted between the chapel walls, so that, as in the Gothic architecture that succeeded Romanesque, 'sound and light . . . can flow freely round the building . . . to produce a sense of almost limitless space'.[8] Specialized intelligences no longer operate semi-independently. Instead a 'mapping across knowledge systems' becomes possible [Fig. 6.4]. This new kind of integration or 'cognitive fluidity'

[4] S. J. Mithen, *The Prehistory of the Mind*, 74.
[5] S. J. Mithen, *The Prehistory of the Mind*, 74.
[6] S. J. Mithen, *The Prehistory of the Mind*, 100.
[7] S. J. Mithen, *The Prehistory of the Mind*, 75.
[8] S. J. Mithen, *The Prehistory of the Mind*, 77.

abolishes distinct behavioural domains and brings with it all kind of previously unknown synergies. In particular it creates a 'passion for the analogical'.

The ability to carry over ideas from one domain to another makes it possible to create and use symbols. When the domains for toolmaking and social relationship begin to interact the result is something like the necklace in the Blombos Cave, where a physical object is transformed to give it a social meaning. Mithen found evidence for this new cognitive fluidity in every aspect of what he called 'the human revolution', from the appearance of cave paintings to the mathematical notations on the Lartet bone.

Mithen's book, which was written in 1996, placed this 'revolution' relatively late, 'between 30,000 and 60,000 years ago',[9] and made a sharp distinction between modern human 'reflexive consciousness' and the Neanderthal mind.[10] The discoveries at Blombos in 2008 have made such a late date implausible, while excavations like those at the Gorham Cave in Gibraltar which have revealed that the Neanderthals used black and red pigments and carved abstract designs[11] reveal a more complex situation. These together with the *Homo erectus* engravings at Trinil seem to suggest that at least some of the capacities necessary for symbolic expression were already present in several hominid species, and that the emergence of something like symbolism was, as Francesco d'Errico has put it, 'a multi species multi regional phenomenon'.[12]

How far metaphors of modularity and models of cathedrals help to understand the actual cognitive architecture that produced this has remained a matter of some controversy. A particular criticism turned on the notion of 'general intelligence'. Did this actually exist or is it the case as some have argued that 'there are only specialized intelligences and that general intelligence is a chimera'.[13] Both the study of brain-damaged individuals and techniques of brain imaging have revealed a surprising degree of localization of functions like language or mathematical ability. It has also been shown that localized functions can move around, and that distributed functions (in which different areas of the brain 'light up') can be as important as localized ones. Whatever the limits of such models, and whatever the mechanisms underlying them, they may help to illlustrate the consequences that flowed from the new cognitive situation.

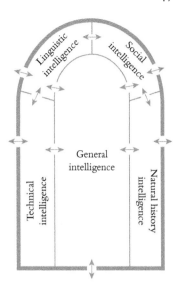

6.4 Steven Mithen's cathedral of intelligence with accessibility between chapels.

[9] S. J. Mithen, *The Prehistory of the Mind*, 222.

[10] S. J. Mithen, *The Prehistory of the Mind*, 167.

[11] J. Rodríguez-Vidal et al., 'A Rock Engraving Made by Neanderthals in Gibraltar', *Proceedings of the National Academy of Sciences of the United States of America* 111 (2014), 13, 301–6.

[12] F. d'Errico, 'The Invisible Frontier: A Multiple Species Model for the Origin of Behavioral Modernity', *Evolutionary Anthropology* 12 (2003), 188–202.

[13] A. Walter, *Evolutionary Psychology and the Propositional-Attitudes* (2012), 60.

Fluidity and Gridlock

Even a relatively simple model of an integrated system, however, like that envisaged by Mithen, would face challenges to
which a less integrated system would be less vulnerable. A more
integrated system faces a greater possibility of system-wide
paralysis if one module jams up (as a result perhaps of damage
or contradictory imputs) than a less integrated system.

If the upside of cognitive fluidity was the emergence of 'previously unknown synergies', the downside might be the possibility of such overall gridlock. Avoiding this outcome would
involve the development of system-wide strategies and some
accounts of the origins of human ethical behaviours have described them in exactly these terms. 'The human revolution'
on these accounts involved a 'garden of Eden moment' in more
than just the narrow sense of the beginning of symbolism. If
they are correct, then the coming together of the technology
evident in the making of pigments and the symbolism apparent
in the cave paintings where they were used were specific features in a much wider landscape.

Darwin's Swallows

In *The Descent of Man*, the book that he wrote ten years after
The Origin of Species, Darwin included a chapter in which he
discusses the origins of what he calls 'the moral sense'. In it he
describes how female swallows and house martins exhibit two
instincts which sometimes compete: an instinct to look after
their young and an instinct to migrate.

When this happens the migratory instinct usually wins and
the birds 'desert their young, leaving them to perish miserably
in their nests'.[14] If swallows had longer memories and more
active imaginations these competing instincts could produce
agonizing conflicts, as for instance they seem to do in migrating female deer who hesitate, going back and forth, when their
young get into trouble. The competition in this and similar
instances seemed to be, Darwin thought, between violent but
temporary impulses, like migration and panic, and what he described as less urgent but more persistent social impulses, like
that of caring for young.

In 1975 a Dutch primatologist, Frans de Waal, began a study
of the world's largest captive colony of chimpanzees at Arnhem

[14] C. R. Darwin, *The Descent of Man*
     (1871), 84.

6.5 A young chimpanzee aiding an elder.

Zoo, which revealed just how similar to human moral behaviour such social impulses could be. De Waal and his team observed in the chimps both 'moral emotions', like gratitude, empathy, sympathy, and indignation at unequal treatment, and 'moral behaviour', like reciprocal food sharing, conflict resolution, and altruistic acts such as young chimps helping an elderly female chimp to climb up on to a bar [Fig. 6.5]. A year before de Waal started his study, however, researchers at the Gombe Reserve had begun to witness an altogether darker side to chimpanzee behaviour.

The Four-Year War

At the beginning of 1974, the reign of one alpha male at Gombe was ended by a younger more aggressive rival. A split then took place in the chimp community. Sixteen males withdrew from the Kasakela region and went to live in the Kahama Valley at the southern end of their range. After the new alpha male was himself displaced, his successor then proceeded to lead the Kasakela chimpanzees into what amounted to a four-year war against the breakaway Kahama community. This involved a strategy of hunting down the Kahama males one by one, brutally attacking them, and leaving them to die of their wounds. Within the Kasakela group meanwhile one of the females developed an abnormal cannibalistic taste for other females' babies. The four-year war only ended when every last one of the Kahama males had been killed [Fig. 6.6].

The capacity of chimpanzees to wage something like sustained warfare against members of their own species (which since the 'four-year war' has been witnessed on several occasions in other places) increased the similarity of their behaviour to the conflicted ethical activities of human beings. There was one element of human morality for which de Waal's team found no evidence among chimpanzees. This was what Adam Smith had called 'the impartial spectator'—the ability to look at a situation and make a judgement about it even though it does not affect you. Although chimpanzees might react with indignation when they saw one chimp mistreating another, there was no sense that they were 'impartial spectators' or 'had a concept about what kind of society they want to live in'.[15]

15 T. Sommers, 'Interview with Frans de Waal', http://www.believermag.com/issues/200709/?read=interview_de-waal. Accessed 7 August 2015.

6.6 Chimpanzee warfare.

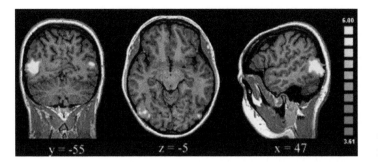

6.7 fMRI image of brain engaged in moral judgement.

Darwin thought it 'exceedingly likely' that any animal which had these kinds of 'well-marked social instincts' would 'inevitably acquire a moral sense or conscience as its intellectual powers become ... anything like as developed as in man.'[16] Such 'intellectual powers' would require both a capacity for abstract thought and the ability to integrate that capacity with 'social instincts' and 'moral emotions'. Like Mithen's description of the 'seamless functioning of multiple intelligences', this ability would allow a 'mapping across knowledge systems'.

A review of recent research undertaken by Thomas Meeks and Dilip Jeste at the University of California School of Medicine in San Diego has shown some physical evidence for this. Neuroimaging of people faced with difficult moral judgements showed brain activity both in the prefrontal cortex—an area linked to conscious abstract thought—and in areas like the limbic system and the cingulate cortex that are linked with emotions [Fig. 6.7].[17]

It may be then that this capacity for neurological integration gives rise to what the British philosopher Mary Midgley describes as the 'integrative struggle': the endeavour 'to act *as a whole* rather than as a peculiar isolated component coming into the control of the rest of the person', which she describes as the 'core of what we mean by human freedom'.[18]

The Integrative Struggle

It was this capacity for integration, Midgley argues, which propelled *Homo sapiens* into a new kind of cognitive world. 'Once you realise you are constantly wrecking your own schemes in the way the migrating swallow does you are forced', she suggests, 'to evolve some kind of priority scheme and try to stick

[16] C. R. Darwin, *The Descent of Man*, 72.
[17] T. W. Meeks and D. V. Jeste, 'Neurobiology of Wisdom: A Literature Overview', *Archives of General Psychiatry* 66 (2009), 355–65.
[18] M. Midgley, *The Ethical Primate* (1994), 168.

to it.'[19] For creatures who are able to remember past events and reflect on their own and other lives, conflicting impulses no longer appear as isolated events. Once the power of thought makes visible 'the fact that motives clash, and clash in the context of a mental life that badly needs to work as a whole',[20] some means must be found to arbitrate these conflicts. 'In order to do this', Midgley concludes, 'they would have to set systematic priorities between different aims, and this means accepting lasting principles or rules.'[21]

Midgley observes that 'the mere fact that a motive occurs persistently in our own or any other species does not give it automatic authority or turn it into a moral rule'.[22] The 'impartial spectator' cannot be simply located *within* the intuitive preferences of an individual or group. To have a lasting purchase either 'horizontally' over varied circumstances and within a group, or 'vertically' over the course of a lifetime and between successive generations, explicit rules would seem to require some kind of anchor in an entrenched understanding of the nature of reality. To give consistent guidance a moral compass, one might say, requires some kind of equivalent to magnetic north.

Could this give a clue to the role of what might be called 'ultimate curiosity' in human societies? The horizonless curiosity of children might seem to be an expensive luxury in adult life. Nevertheless, the evidence of archaeology and anthropology suggests that almost all human societies, from prehistoric times to the present day, have organized themselves around stories and practices which in different ways focus attention on something beyond the horizon of the visible world.

The same evidence suggests that this 'ultimate curiosity' has in its turn shaped and motivated the kind of interest that human societies manifest in the physical world around them. The kind of interest which might for that reason be described as 'penultimate curiosity'.

[19] M. Midgley, *The Ethical Primate*, 178.
[20] M. Midgley, *The Ethical Primate*, 138.
[21] M. Midgley, *The Ethical Primate*, 139.
[22] M. Midgley, *The Ethical Primate*, 138.

CHAPTER SEVEN

Ultimate Curiosity

At the beginning of most summer evenings a flock of geese, which lived near the river a couple of hundred yards from where the first chapters of this book were being written, would take to the air. Settling into their characteristic V pattern they would give an immaculate demonstration of formation flying as they circled above our stretch of the Cherwell. Occasionally a B team of geese from lower down the river would put in an appearance giving a more ragged performance but still maintaining the semblance of a V. These V formations were not accidental. In using them each bird in the flock was able to find a position of maximum uplift from a vortex of the slipstream created by the bird in front.

As our weekly breakfast discussions continued it began to occur to us that this phenomenon of slipstreaming might provide a way of thinking about the peculiar relationship we were trying to understand. The relationship that is between investigating the visible world and reaching out to something beyond it. In particular it suggested a way of relating the connections we were finding between 'ultimate' and 'penultimate' curiosity, to a new way of thinking about the development of religion that has arisen in part from the emerging discoveries of prehistoric art.

Homo religiosus

The 'cognitive science of religion' (CSR) first received a name in the year 2000 (though studies that now come under this banner had been acquiring definition for more than a decade before that). As a multidisciplinary field it draws from anthropology, cognitive science, and evolutionary psychology to study the nature of religious thoughts and actions. A basic starting point for this science has been the question of whether religion (defined for its purposes as something that has to do with supernatural agents) is a fundamental capacity of the mind in the same kind of way as language and speech, or is a human invention like the wheel.

While language and speech are ubiquitous, inventions like the wheel are not. They occur in some cultures but not in others. While we may remember how we learnt to ride a bicycle or use some other invention, language is acquired so early that we do not remember its acquisition.

Religion seems in these respects to be more like language than technology. Despite the first impressions of some nineteenth-century explorers it has been shown to be ubiquitous.[1] Experiments with young children have demonstrated that religious thoughts appear spontaneously from a very early age.[2] These observations have prompted a further question: 'does religion provide some kind of direct adaptive advantage that has been selected for by evolution (suggesting there might even be a "religious gene") or is it a by-product of other things that do provide an adaptive advantage?'

Although there is some evidence that a tendency towards religiosity may be heritable, even proponents of the 'God gene' hypothesis admit that no gene so far identified is able to account for the range of human religion.[3] The fact that something like the ability to cook or make clothes turns out to be advantageous does not necessarily imply that they have been directly selected for by an evolutionary process. It would be rather surprising to find a gene that directly codes for making clothes. If religion is in some sense a 'by-product' of characteristics that do carry an evolutionary advantage, what might these be? This is the central question that CSR sets out to answer.

In seeking to do so CSR researchers have often focussed on so-called 'mental tools', sometimes comparing the mind to a kind of Swiss army knife containing devices that had evolved for one purpose but could be used for another (rather as the pick for getting stones out of horses' hooves turned out to be rather useful for cleaning nails). Chief among these is the so-called 'theory of mind' (discussed in Chapter Five), which may have evolved by allowing advantageous social cooperation.

If the mind has a system for recognizing other minds as well as a system for recognizing inanimate objects, then corpses and bones might bring these two systems into conflict, creating an 'uneasiness'. Hence, it is argued, the preoccupation of some animals like elephants with the bodies and bones of their species,

[1] S. Atran, *In Gods We Trust* (2002).
[2] D. Kelemen, 'Are Children "Intuitive Theists"? Reasoning about Purpose and Design in Nature', *Psychological Science* 15 (2004), 295–301.
[3] D. H. Hamer, *The God Gene* (2004).

the 'cult of skulls' identified by the Abbé Breuil in prehistoric sites, and the ancestor cults described by anthropologists.

If, furthermore, it is suggested, the mind has developed a hair-trigger device for detecting the agency of other minds (whether human or animal) in the surrounding world, that might explain our readiness to attribute sounds or sights that don't have a discernible physical cause to disembodied agencies (spirits, ghosts, and the like). It might also explain the experimental discovery that young children of all cultures have a bias to explain every aspect of the world around them as having been designed.[4]

None of these 'mental tools' has, however, seemed to researchers in this field sufficient to account for the existence of religion as a social phenomenon. For that, a further capacity was required. The nature of this further capacity bears some relation to what theologians have descibed as 'self transcendence'.[5] In scientific terminology it has been variously decribed as a 'supervisory system',[6] 'bifurcation of consciousness',[7] and 'high end metarepresentational theory of mind'.[8] All of these terms refer to the capacity to think about thoughts. The last phrase, coined by Justin Barrett—one of the leading proponents of CSR—is though more specific.

If I have a picture in my mind of a spear, that is a representation. If I turn the spear round in my mind and think about how to make it sharper, that is a 'metarepresentation'. If I deliberately and reflectively think about the contents of my own or someone else's mind, that in Barrett's language is 'high end metarepresentational theory of mind'. This capacity to look at a thought as though it were a kind of object gives rise, Barrett suggests, to the phenomenon that has been described as 'joint attention'. This is the ability of two people to consider a shared thought: 'looking' together at something that neither of them can physically see. Some such ability would seem to be a precondition of any kind of shared religion. It would also seem to be a precondition of the capacity to create and read symbolism.

Homo symbolicus

Birds sing, dogs bark, lions roar. Many kinds of creatures produce signals in the form of sounds, scents, gestures, and dances that may convey information and trigger behavioural responses. Deliberately created symbols seem to involve something beyond

[4] D. Kelemen, 'Are Children "Intuitive Theists"? Reasoning about Purpose and Design in Nature'.

[5] R. Niebuhr, *The Nature and Destiny of Man* (1964).

[6] D. M. MacKay, *Behind the Eye* (1991).

[7] E. S. Savage-Rumbaugh and W. M. Fields, 'The Evolution and the Rise of Human Language: Carry the Baby', in *Homo Symbolicus: The Dawn of Language, Imagination and Spirituality*, ed. C. S. Henshilwood and F. d'Errico (2011), 13–48.

[8] J. L. Barrett, 'Metarepresentation, *Homo religiosus*, and *Homo symbolicus*', in *Homo Symbolicus: The Dawn of Language, Imagination and Spirituality*, ed. C. S. Henshilwood and F. d'Errico (2011), 205–24.

this. They are, Barrett argues, 'metarepresentational' in the sense that they involve the capacity to manipulate a thought by expressing it in physical form or to read and interpret what has been expressed. They involve a 'high end theory of mind' in the sense that they trigger 'mental states (thoughts, ideas, affective states etc) instead of only triggering behavioural routines'. To understand a symbol 'I would need to be able to wonder "what is it intended for?" '. It may be this capacity 'that *changes signalling into linguistic communication and symbolism* more generally'.

These are also very much the same kind of questions that are applied to portents and omens in a religious context. Hence Barrett concludes that 'it may be that when metarepresentation was added to our ancestors tool kits', they quite suddenly became capable of this kind of religious interpretation, of language, and of symbolism '*all at the same time* all because of the addition of *one* incremental change. If so then the tool kit that makes *Homo religiosus* makes *Homo symbolicus* and behaviourally modern humans as well. The three are identical and evolved concurrently.'[9]

Whether or not this turns out to be the case, it is apparent that something like 'metarepresentation' and 'joint attention' might together give rise to the cumulative culture that is the distinctive feature of behaviourally modern humans.

Cultural traditions do exist in other species and some of these may involve something like symbolism (though this seems to be the much more temporally and geographically limited). In 1991, for instance, the Boeschs reported that the chimpanzees drummed on tree roots in a way that seemed to convey to other chimpanzees their location and direction of travel.[10] When, however, the main drummer died, the behaviour disappeared.

By contrast the hand stencils and pictures in Sulawesi seem to suggest an extraordinarily long line of cultural transmission extended over thousands of years and across thousands of miles. It is easy to see what might motivate the passing on of hunting, cooking, and clothes-making skills. What might motivate the transmission of this kind of cultural activity?

Big Picture Thinking

The cognitive anthropologist Pascal Boyer, another leading figure in the CSR field, has argued that gods or spirits who cannot be seen have a particular salience in moral thinking,

[9] J. L. Barrett, 'Metapresentation, *Homo religiosus*, and *Homo symbolicus*', 220.
[10] C. Boesch, 'Symbolic Communication in Wild Chimpanzees?', *Human Evolution* 6 (1991), 81–9.

which is likely to have given them a central position in human communities. Whatever other attributes they may have, invisible beings can potentially see the good or evil we are doing when we are otherwise unobserved. They may even see our hidden thoughts and secret motives. They may know the true causes of problems and the future consequences of actions. In a way that no human being can, such invisible presences could genuinely stand in the position of the 'impartial spectator'.

Indeed they may be more than spectators. They may stand as the ultimate guarantors of moral behaviour, punishing evil and rewarding good when human societies fail to do so. They may, furthermore, control the mysterious force of good or ill fortune, making it possible to relate the way we act to the outcomes of our lives. While not all gods and spirits are conceived as having these kinds of characteristics (so that, for instance, the Baining people of the province East New Britain of Papua New Guinea have two classes of deities, ancestor spirits who read peoples minds and the *sega*, or forest spirits, who have no such special powers, while the deities of Yucatek Maya of southern Mexico include *Chiichi*, 'who do not know much more than a peeping Tom would') generally, as Barrett points out, 'it is the more informed gods that matter most to the religious systems'[11] (as is the case in both these examples).

The significance of invisible gods and spirits, in the endeavour of individuals and groups to 'act as a whole', might help to explain why human communities (apparently from the earliest times) have manifested a continual need to extend the horizon of curiosity beyond the rim of the visible world. The 'integrative struggle' to act as a whole, in other words, would seem to require a larger cognitive struggle. This is a curiosity that employs every sense and capacity in the attempt to create 'a big picture', an integrated image of reality.

The sticks used by chimpanzees in the 'rain dance' and the stones employed in the 'waterfall dance' already show how tool use could be pressed into the service of something 'akin to wonder and awe'. In *Homo sapiens*, however, such synergies had more far-reaching consequences. The evidence of the Ardèche cave suggests that in the attempt to see different aspects of reality as part of a meaningful whole, early human beings discovered radical new ways of perceiving and describing the world.

[11] J. L. Barrett, *Why Would Anyone Believe in God?* (2004), 49.

Making and Matching

In the deepest chambers of the Ardèche cave are two extended compositions, both around 30 feet long, which demonstrate discreet individual perceptions being formed as it were into a big picture of reality [Fig. 7.1]. The exact motivations may remain mysterious, but here as in the Trois Frères Cave, the presence in the end chamber of a bison-headed human figure (if that is what it is) on a rock pendant is strong evidence of some kind of religious purpose [Fig. 7.2]. In pursuit of this purpose, however, the Palaeolithic artists employed techniques of representation that anticipate developments tens of thousands of years later.

In the penultimate chamber there is a panel of four horse heads, which have been drawn one on top of another on a diagonal axis. The first is at the top of the axis, the last at the bottom. The images were made by scraping the clay from the wall to provide a white background, outlining the heads with a charcoal stick, then stumping—that is, mixing the charcoal with white limestone to produce halftones which model the

7.1 The end chamber of the Chauvet Cave.

head—and finally working into the heads with an engraving tool to sharpen the contours and highlight features like the nostril and mouth [Fig. 7.3].

There is a clear progression in accuracy between these four heads. The first head is a little wobbly and generalized. The horse's eye floats unconvincingly in its socket while the muscles are undifferentiated and represented by a single stroke. In the second horse, the eye is still unrelated to the structure of the head but more attention has been given to the muscles of the jaw. In the third the eye is more clearly anchored in the skull, the great sternomastoid muscles of the neck are indicated and the artist has used an engraving tool to outline the nostril and lips. On the fourth head this technique is used to even greater effect to outline the flaring nostril, and the open mouth where the masseter muscles of the jaw are clearly differentiated. Meanwhile the eye is firmly lodged in the horse's head, which displays a grasp of underlying structure that is comparable to the anatomical exactness of the great horse painters of the eighteenth and nineteenth centuries [Fig. 7.4].

Although there are several examples in the cave of what look like second thoughts (where the outline of an animal has been first sketched and then rubbed out), the way that the panel was prepared for a large composition suggests that the first three horses were not conceived in this kind of way as preliminary sketches for the fourth. Nevertheless the differences between the four images does reveal a pattern of what the art historian E. H. Gombrich referred to as the rhythm of 'making and

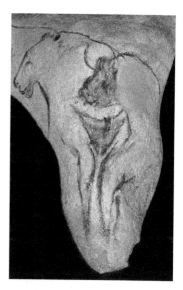

7.2 Composite figure in the Chauvet Cave.

7.3 Detail from the panel of the horses in the Chauvet Cave.

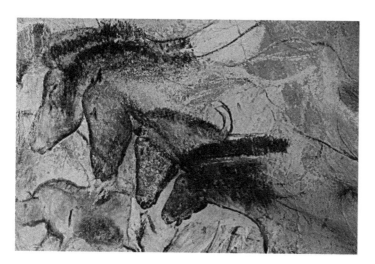

7.4 The panel of the horses in the Chauvet Cave.

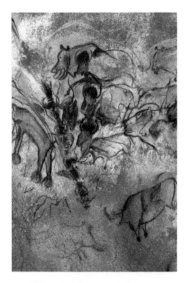

7.5 Bison in front and three-quarter view, in the central niche of the end chamber in the Chauvet Cave.

matching' or 'schema and correction'—the experimental process by which an artist setting out to make a life-like picture makes an image or 'schema' and then progressively corrects and modifies it to achieve a tighter fit with reality.[12]

Ceaseless experimentation of this kind can sometimes lead to the discovery of new techniques of representation. In the end chamber, in a central niche between the lions and the bison, other bison are seen front on with their bodies foreshortened so that they seem to be emerging from the wall. On another ridge in the wall 'a bison was drawn with its head on one plane, seen full face, and its body in profile on another, at 90 degrees'. 'The effect of perspective', according to those lucky enough to have access to the cave 'is astonishing' [Fig. 7.5].[13]

Harnessed to the development of an interest in narrative that changed the religious function of art, 'the illusionistic tricks of foreshortening and modelling' were, Gombrich argues, at the centre of what he calls 'the Greek revolution'. The moment between the sixth and fifth centuries BC when this combination of means and motive produced a 'chain reaction . . . unique in the annals of mankind'[14] which set the course 'for new continents of human experience'.[15]

In one sense the images in the Ardèche cave seem to anticipate this revolution by some 30,000 years, even to the extent in the end cave of depicting a kind of narrative in which 'a whole group of lions appear to be hunting bison and seem to be stretching out towards them' [Fig. 7.6].[16] The difference, however, between individual observations of shadows or foreshortening that appear before the 'Greek revolution' is, as Gombrich points out, 'that they remained without consequences. They do not become part of the tradition to be improved and extended as they do in Greece.'[17] Our knowledge of Palaeolithic art is too fragmentary (based as it is on miraculous survivals like the Ardèche cave) to be sure of its progress, but there is no evidence that the final horse in the Chauvet series intiated a new way of drawing.

'Ultimate curiosity'—the attempt to see beyond the rim of the visible world—may have motivated this kind of experimentation in cave art, but it did not by itself create the kind of 'chain reaction' that occurred in Greece. If it was a necessary condition for such developments it was not a sufficient one. Tens of thousands of years seem to have passed in which although the small groups of hunter-gatherer communities

[12] E. H. Gombrich, *Art and Illusion* (1977), 24, 99.
[13] J. Clottes, *Return to Chauvet Cave* (2003), 140.
[14] E. H. Gombrich, *Art and Illusion*, 120–1.
[15] E. H. Gombrich, *Art and Illusion*, 115.
[16] J. Clottes, *Return to Chauvet Cave*, 137.
[17] E. H. Gombrich, *Art and Illusion*, 123.

7.6 Lions hunting bison in the end chamber of the Chauvet Cave.

scattered around the world evidently possessed the capacity for new kinds of symbolic communication and technological innovation, there is very little evidence of either. Instead there seem to have been prolonged periods of stasis in which entrenched traditions were handed on with little challenge or modification.

When 'the Greek revolution' came, the 'chain reaction' in representation took place alongside a kind of revolution occurring across the whole landscape of thought. This had at its centre a deep entanglement between a radical paradigm shift in religious thought and a new way of looking at the natural world. Nor was it the last time that such a thing was to occur.

The series of revolutions in thinking that ultimately gave birth to what we now call 'science' each seems to have taken place in tandem with some kind of religious paradigm shift. While these religious developments were not necessarily the primary causes of the revolutions they accompanied (though they were sometimes contributory factors) neither were they incidental to them. Rather they were connected in a manner that bore a more than passing resemblance to what can be seen in a physical slipstream.

Moving in a Slipstream

The term 'slipstream' can refer to a variety of phenomena. At the most basic level one object travelling behind another—say, a child walking close behind a parent in the teeth of a howling gale—will experience less force from the wind. Riders in the Tour de France who stick close enough to the lead cyclist (the effect tapers rapidly with distance) can move at the same speed as the leader while exerting less energy. This is because at high speeds the lead bike reduces the drag experienced by the riders behind who are carried along in a wake of forward-moving air.

In a similar way fish swimming in a school derive a significant hydrodynamic advantage from the slipstream of those ahead of them. Depending on their position in the school this is probably obtained by exploiting vortices generated by the leading fish, and enables the followers to achieve the same speed with a slower rate of tail beating, and thus a lower energy consumption.[18] These stronger effects depend rather precisely on the fluid dynamics and the positions of those involved. Hence, the exact V formation of the Cherwell geese.[19]

How might this phenomenon provide us with a way of thinking about human curiosity?

If the expenditure of effort involved in paintings at Lascaux or the Chauvet Cave demonstrates the urgency and importance of their ritual functions in prehistoric societies, it also shows how less urgent interests, like the close observation and representation of animal anatomy, could be borne along in the slipstream of these larger purposes. While this slipstream seems to have remained at a steady level for a prolonged period, when physical, geographical, and social circumstances precipitated radical changes in human society the strength of this kind of motivational current markedly increased.

The first such major developments in human cultures seem to have flowed from the material results of emigration. When small groups of human beings began to migrate from Africa, the different physical circumstances of the continents in which they settled ensured that the cultural construction of different societies proceeded along dramatically different trajectories.

The beginning of agriculture in regions where there were suites of wild plants that were relatively easy to cultivate and species of animals that were relatively easy to domesticate triggered an extraordinary cascade of new developments.[20] The ability to sustain large settled populations which followed the development of agriculture allowed a division of labour in which some could focus on new kinds of tasks, such as the

[18] S. S. Killen et al., 'Aerobic Capacity Influences the Spatial Position of Individuals within Fish Schools', *Proceedings of the Royal Society B: Biological Sciences* 279 (2012), 357–64.
[19] S. J. Portugal et al., 'Upwash Exploitation and Downwash Avoidance by Flap Phasing in Ibis Formation Flight', *Nature* 505 (2014), 399–402.
[20] J. M. Diamond, *Guns, Germs and Steel* (1998).

development of writing systems that seem to have been employed initially to record commercial transactions.

As they developed, these writing systems rapidly began to
be used to record those broader human concerns which, as the
cave paintings suggest, were already there. Creation stories and
other kinds of sacred text begin to appear, allied to attempts
(found in everything from Chinese oracle bones to Mesopotamian instructions on the significance of marks on the liver of
sacrificial animals) to interpret every aspect of reality as part of
a meaningful whole.

Wherever settled civilizations begin to develop, whether in
China, India, South America, or the Middle East, there is evidence of these kind of attempts to make sense of the world as
a whole, creating a slipstream of interest in specific physical
phenomena.

In the British Museum, for instance, there are 70 cuneiform
tablets known as *Enuma Anu Enlil*. First deciphered by a Jesuit priest named Johann Strassmaier, they record astronomical
phenomena, like the appearance of planets, alongside earthly
phenomena like plagues and flood. These have been described
as perhaps the first ancestor of something like scientific writing.

Slipstream effects of this kind, though far from insignificant,
might still be thought of as relatively weak. Certain configurations
of ultimate curiosity seem though to have created much stronger
kinds of slipstream (analogous to that experienced by geese, fish,
and Tour de France riders) in which penultimate curiosities about
the physical world were able to advance more rapidly.

As our breakfast discussions continued it seemed to us that if
we were to answer the question that seemed to be posed by the
doorways in Oxford and Cambridge, it was these strong slipstream effects that we would need to focus on. Along the way
we might hope to gain some insight not only into the factors
that made them strong, but also into the particular hazards that
come with this strength.

The strength of a slipstream, as we noted, increases with the
velocity. In the Tour de France, to gain maximum benefit from
the slipstream of rider in front a following cyclist must come as
close to him as possible—within centimetres of his bike. At these
speeds a clash of wheels can spell disaster—hence the 'chutes' or
pile-ups that are a not uncommon feature of the race.

Particular configurations of 'ultimate curiosity' seem to
have produced a powerful motivational slipstream in which

the consequences of 'penultimate curiosity' became ever more spectacular. The temptation to close the gap between them—to make religion answer scientific questions or vice versa—and the chutes that result have been a part of this story that we would need to consider.

Our objective though, we reminded ourselves over coffee, was not to explain *how* what we now call 'science' had developed nor to decide the importance or otherwise of religion in that development. Still less was it to try and answer the million dollar question of integration: whether and in what way 'science' and 'religion' were or were not compatible. Our more modest, though hardly unimportant, aim was to try rather to understand *why* there had in fact been this long and continuing entanglement.

The first example of the kind of strong slipstream effect in which we were interested occurred when literacy began to spread from the priests or administrators responsible for *Enuma Anu Enlil* to a larger segment of the population. It was only then that a new kind of reflection on religious ideas began to appear. Slipstreaming behind it came a new kind of interest in the physical world.

The original epicentre of this earthquake in thinking was associated with a very small number of individuals, located in just a handful of city-states scattered around the Aegean. These were the men who came in time to describe themselves as 'lovers of wisdom': the philosophers of ancient Greece.

PART II
God-Driven Science

CHAPTER EIGHT

The Lions of Miletus

On a spring day in 423 BC as many as 14,000 people ('women, men, and children, slaves as well as free men', according to the philosopher Plato[1]) crowded into the enormous theatre of Dionysus, just below the Acropolis in Athens, to watch a new comedy by the young playwright Aristophanes. The play (which was called *The Clouds*) was one of three comedies being staged that year as part of the city Dionysia—the annual festival of Dionysus—and came third and last in the festival competition. Its lack of success may have been the result of its novelty. Embedded within some traditionally crude jokes is what has been described as 'the first comedy of ideas': a commentary on an extraordinary revolution in Greek culture just at the moment that it was taking place.

The first scene opened by introducing the audience to Strepsiades ('Twister'), a farmer married to a wife from an upper class family, whose son Pheidippides has developed an aristocratic taste for chariot racing that is rapidly bankrupting his father. Besieged by creditors who are threatening to take him to court, Strepsiades urges his son to go to the house next door, the *phrontisterion*—'the thinking shop'. The residents of this place call themselves 'philosophers' and will, so Strepsiades believes, teach his son to make the wrong argument defeat the right, and so enable him to win his lawsuits and evade his debts. Pheidippides refuses to get involved so Strepsiades himself goes to call.

The second scene took the audience inside the *phrontisterion*. Here a philosopher describes to the farmer some of the bizarre experiments conducted by his colleagues: measuring the distance a flea can leap, studying the orbit of the Moon, using geometry to make a map of the world. Socrates, the leader of the group, swings into view suspended in a basket from which he has been studying the Sun. He tells Strepsiades that 'with us the gods are no longer current'.[2] The philosophers worship 'the clouds': 'Zeus is dead, Awhirl is the new king'.[3]

[1] Plato, *Gorgias*, 502d, trans. Walter Hamilton (1960), 109.
[2] Aristophanes, *The Clouds*, trans. Alan H. Sommerstein (1973), 122.
[3] Aristophanes, *The Clouds*, 129.

The farmer turns out to be too stupid to learn how to make bad arguments defeat good ones, and Pheidippides eventually agrees to let the philosophers teach him. The result is disastrous. When he emerges from his lessons, Pheidippides promptly beats up his father and justifies his actions by arguing that the laws, which forbid such things, are just man-made customs. In the final scene of the play the Athenian audience saw the infuriated farmer, armed with mattocks and torches, take revenge on behalf of the gods by setting fire to the philosophers' roof and burning down their thinking shop.

Twenty-four years later, a few hundred yards from the theatre of Dionysius, the seventy-year-old Socrates stood trial for his life, on a charge of 'not acknowledging the gods the city acknowledges', 'introducing new deities', and 'subverting the young men of the city'.[4] In his defence Socrates, according to Plato, cited Aristophanes' play as the source of these ideas about him. This had not been the only play which mocked him. *Connus* by Ameipsias, the play which came second in the festival competition, had, it seems, been concerned with Socrates' music teacher. *The Clouds*, however, attributed to Socrates the ideas of others, and (though he doesn't mention it) these had been brought to Athens from other cities. Making bad arguments defeat good ones was a skill taught by orators from Sicily and Abdera. The scientific and religious ideas, which gave *The Clouds* its name, had come to Athens across the Aegean from the Greek-speaking Ionian colony of Miletus.

The Three Clues

Five-hundred miles east of Athens, in what is now modern Turkey, the dusty ruins of Miletus offer few obvious indications as to why such a profound revolution in thought should have taken place there. A once-thriving port, the city is now marooned several miles inland. A marble lion, one of a pair that once guarded the harbour entrance, now looks out upon fields of waving cotton [Fig. 8.1]. Overlooking the silted-up harbour, the steps of a second-century *bouleuterion*, or council chamber, and beyond that an impressive fourth-century theatre are among the few easily legible buildings that remain. In the theatre, inscriptions on some of the seats—'for the Jews and God-fearers', 'this was the place of the goldsmiths of the blues'—are haunting traces of the rich cultural life that once existed in the

8.1 One of the Miletus harbour lions.

4 Plato, 'The Apology', in *The Last Days of Socrates*, trans. Hugh Tredennick (1954), 54.

city. Slight though they might appear, each of these fragmentary remains provides a clue as to why it was that such a radical new kind of curiosity should have developed in this place.

The clue from the inscriptions comes from the script in which they are written. Linear B, the syllabic script developed by the Mycenaean Greek civilization around 1400 BC, seems to have been exclusively used by palace bureaucrats (probably because it must have been exceedingly difficult to master). When Linear B was finally deciphered, almost every tablet turned out (rather disappointingly) to consist of accountants' records of sheep, wool, and flax.

8.2 Inscribed wine jug from 740 BC.

The first examples of an alphabetic Greek script occur by contrast on a wine jug (dating from around 740 BC) and on a drinking cup. The text on the jug says 'whoever of all the dancers performs most nimbly will win this jug as prize' [Fig. 8.2]. That on the cup announces 'I am Nestor's delicious drinking cup. Whoever drinks from this cup the desire of the fair-crowned Aphrodite seize him.' Inscriptions like these on domestic objects testify to the appearance of a new kind of literacy. With the introduction of the new alphabet, Greek literacy was no longer restricted to public officials but was available in private homes, and allowed ordinary individuals to fix their thoughts in a form that could be handed around among friends.

The stone benches of the bouleuterion and of the theatre are in their way equally significant [Fig. 8.3]. These were places of assembly and public debate and are reflections of a kind of social and political development within a number of Greek city-states that had begun in the sixth century BC. Driven by various social and economic pressures the power of ordinary citizens within these small city-states had begun to challenge and modify monarchical and aristocratic forms of government. A whole variety of new constitutional arrangements sprang up, ranging from tyrannies to oligarchies to the fully fledged democracy that was established in Athens in 508 BC. The citizens of these different city-states not only participated in government to a greater or lesser extent, they also argued about which kind of constitutional arrangement was best. It is these kind of arguments which may have been crucial.

Miletus itself, according to the historian Herodotus, had an unsettled constitution and was ruled on and off by tyrants. But this, as the historian of Greek science Geoffrey Lloyd has pointed out, 'did not prevent and may even have done a great

8.3 The bouleuterion at Miletus.

deal to encourage the development of political institutions and political awareness.'[5] These kinds of arguments about fundamental principles could move from politics to religion. Lloyd argues that 'the growth of a new critical spirit in the 6th and 5th centuries may be seen as a counterpoint and offshoot of the contemporary habit of free debate and discussion of politics throughout the Greek world'.

Finally, the Lion Harbour of Miletus (along with the city's other three harbours) was the source both of wealth (and the leisure to think that goes with wealth) and of a contact with other civilizations, which must itself have stimulated curiosity [Fig. 8.4]. A trading centre for goods and an exporter of famous woollen garments, Miletus was supposed to have founded 90 colonies (archaeologists have so far found 45) including the commercial settlement Naucratis in Egypt. Located on the eastern edge of the new literacy and public argumentation of the Greek-speaking world, and on the western edge of the ancient civilizations of Persia and Babylon with their long traditions of astronomy, geometry, and mathematics, Miletus provided a natural bridge between the two.

All three of these elements seem to have gone into the making of a book that was published around 547/6 BC by a citizen of Miletus named Anaximandros (described by one ancient historian as 'the first of our Greeks to publish a book on nature'). Only a few lines of Anaximandros' book survive,

[5] G. E. R. Lloyd, *Methods and Problems in Greek Science* (1991), 131.

8.4 A harbour lion at Miletus.

but those who knew it make clear that it was one of the first Greek books to be written in prose, that it both developed and disagreed with ideas first put forward by someone else, and that the author (who was one of the first to draw a map of the known world) had a knowledge of lands beyond those where people spoke Greek. Each one of these factors played into the revolutionary concept contained within it: an idea in which a new kind of religious thinking and a new kind of interest in the natural world were two sides of the same coin.

The Ionian Philosophers

In his classic study of Greek religion, Walter Burkert points out that up to this point in Greek history (as possibly indeed in the history of the world), 'speaking about the gods in public had almost been the exclusive privilege of poets'.[6] Early prose texts tend to be concerned with more mundane things like commercial transactions, laws, and other kinds of practical communication, where it was necessary 'to state in a matter-of-fact manner what is the case'.[7] Expressed in the oral formulas and metaphors of poetry, the tapestry of stories recounted by Homer and Hesiod could sit alongside one another without anyone feeling the need to resolve contradictions or establish any absolute coherence. In the wake of the new literacy, when attempts were made for the first time to write about the gods in 'matter-of-fact prose', a whole series of new questions began to appear over the horizon.

[6] W. Burkert, *Greek Religion* (1985), 305.
[7] W. Burkert, *Greek Religion*, 305.

The most fundamental question concerned origins. Where did everything ultimately come from? The deathless gods were by definition immortal, and Zeus, 'the father of gods and men', could be described in terms that implied he was the source of all things. A line of Orphic poetry declares that 'Zeus is beginning, Zeus is middle, from Zeus all things are finished'. At the shrine of Dodona the priestess sang 'Zeus was, Zeus is, Zeus will be'.[8] On the other hand, the poet Hesiod's *Theogony*, written in the seventh century BC, provides a genealogy for Zeus. This describes how he had deposed his father, Kronos. Kronos had emasculated his own father, Ouranos (heaven). Ouranos had been married to his own mother, Gaia (earth). Gaia had appeared out of the primeval chaos.

Anaximandros' prose radically simplified these stories, trimming off everything that looked like poetic embroidery. For him everything has its origin in a single *arche*—a source or principle—which is 'ungenerated and imperishable'. According to the philosopher Aristotle, who had a copy of Anaximandros' book, this *arche* which 'encompasses all things and directs all things' was referred to by Anaximandros as *apeiron*—the boundless—'and this . . . is the divine' (*theion*).[9]

The new concept of 'the divine', formulated like this in prose, seems to have been arrived at through a process of argument. According to all the ancient historians the first Milesian to start thinking in this way was named Thales; he was the friend and teacher of Anaximandros. Thales wrote no book and almost the only actual statement attributed to him is the claim that 'all things are full of gods'. According to Aristotle, Thales (perhaps drawing on Homer's description of Oceanus as 'first parent of the gods') maintained that 'the *arche* is water'.[10] Anaximandros agreed that there was a single divine principle but seems to have been unhappy with his teacher's idea that this could be identified with anything visible, and so arrived at his own still more abstract notion of 'the boundless'.

This was an argument that continued both in Miletus and in other nearby cities. Thus, Anaxamines, a younger contemporary of Anaximandros, argued that the divine *arche* was in fact 'air' (basing this probably on the association between life and breath). On the nearby island of Samos, Pythagoras and his followers began to argue that 'numbers are the cause of being in everything'.[11] Just up the coast in Ephesus, Heraclitus, who was aware (and somewhat scornful) of these predecessors, argued

[8] W. Burkert, *Greek Religion*, 131.
[9] W. K. C. Guthrie, *A History of Greek Philosophy* (1962), 88.
[10] Aristotle, *The Metaphysics* (1933), 19.
[11] Aristotle, *The Metaphysics*, 45.

that what he describes as 'the divine logos', which 'orders all things and steers all things through all',[12] should rather be described as a 'fire ever living'.[13]

The new culture of argumentation combined with the new literacy, allowed this debate to run over succeeding generations. The spark which ignited it seems to have been the possibility of comparing different religions and beliefs.

Thales, Anaximandros, and Pythagoras are all described by ancient historians as having studied in Egypt, and Pythagoras was often described as having been influenced by Persian Zoroastrian thought. Whatever the truth of these stories it seems likely that Thales' belief that the Earth 'rests on water', was influenced both by the Egyptian myth of the Earth floating on *Nun*—the primordial waters—and by the Babylonian myth where everything begins from the watery chaos of Apsu, Ti'amat and Mummu. In arriving at their abstract notion of a divine *arche* it seems likely, as G. S. Kirk argued, 'that the crucial factor was the comparison of Mesopotamian, Egyptian and Greek versions (of myths) which first became possible in the late 7th and early 6th century in Ionia and especially Miletus'.[14]

Such comparisons are explicitly referred to in the writings of the poet Xenophanes. A contemporary of Pythagoras, Xenophanes came originally from the town of Colophon (north of Miletus), but after Colophon fell to the Medes, he spent much of his long life travelling among the western Greeks, where he seems to have inspired two philosopher/poets, Parmenides and Empedocles.

Xenophanes' cross-cultural experiences are perhaps evident in his observation that 'Ethiopians imagine their gods as black and snub-nosed. Thracians as blue eyed and red haired'.[15] From this he concluded that 'if oxen and horses or lions had hands, and could draw and fashion works as men do, horses would draw the gods shaped like horses, and lions like lions'.[16] In a similar way 'men suppose that gods are brought to birth and have clothes and voice and shape like their own'.[17] And indeed share their vices. 'Homer and Hesiod', he complained, 'have ascribed to the gods all deeds that among men are a reproach and a disgrace'.[18]

The reality, he argued, was 'God is one, greatest among gods and men, in no way like mortals either in body or in mind'.[19] Translating the ideas of the Ionian philosophers back into poetry, he describes God as being 'always . . . in the same place . . .

[12] Heraclitus, *The Art and Thought of Heraclitus* (1979), 45.

[13] W. K. C. Guthrie, *A History of Greek Philosophy*, 454.

[14] G. S. Kirk, *The Nature of Greek Myths* (1974), 200.

[15] Fr. 16, in W. K. C. Guthrie, *A History of Greek Philosophy*, 371.

[16] Fr. 15, in W. K. C. Guthrie, *A History of Greek Philosophy*, 371.

[17] Fr. 14, in W. K. C. Guthrie, *A History of Greek Philosophy*, 371.

[18] Fr. 11, in W. K. C. Guthrie, *A History of Greek Philosophy*, 371.

[19] Fr. 25, in W. K. C. Guthrie, *A History of Greek Philosophy*, 374.

but without toil he makes all things shiver by the impulse of his mind'.[20]

A similar concept of the divine is expressed by Parmenides, who lived in the Greek colony of Elea (south of present-day Naples) and, according to Aristotle, was the pupil of Xenophanes. In the fragmentary remains of Parmenides' poem *On Nature*, 'Being' (which according to a later philosopher called Aetius he identified with God[21]) 'is without beginning and without end, it is the same and rests in the self same place abiding in itself'.[22] Likewise Empedocles (who lived on Sicily and is said to have died by hurling himself into the crater of Mount Etna) emphasized that 'it is not possible to draw near (to god) even with the eyes or to take hold of him with the hands . . . but he is sacred and ineffable darting through the whole world with swift thoughts'.[23]

Empedocles' own poem *On Nature* goes on to propose that nature is composed of four 'roots'—earth, water, fire, and air mixed together in different proportions. An association between these new ideas of the divine and a new kind of interest in the natural world is in fact evident in all these thinkers.

In the stories told by the earlier poets, anything that happened in the natural world could be attributed directly to the actions of one or another god. Earthquakes were caused by Poseidon shaking the ground. Lightning bolts were thrown by Zeus. The Sun was Helios driving his horses and chariot around the sky. Abolishing these poetic images left an uncomfortable gap. If the gods of Olympus did not do these things, why did they happen? How could something like the divine explain the world of experience? If the divine *arche* did indeed order 'all things and steers all things through all', by what means did this take place?

From the standpoint of their novel theological position it was this gap that the Ionian prose writers needed to fill, and these questions that they set out to answer. They did so not by looking at a particular earthquake, a specific lightning bolt, or an isolated astronomical event, but by considering each of these as classes of ordered natural phenomena. Thus, Thales, according to Aristotle, argued that earthquakes were the result of the Earth being rocked by wave tremors in the water on which it floats. Thunder, according to Anaximandros, was caused by wind, and lightning by clouds splitting in two. The Sun, he argued, was not Helios in his chariot, but a circle twenty-eight

[20] Frr. 25, 26, in W. K. C. Guthrie, *A History of Greek Philosophy*, 374.
[21] A. Drozdek, *Greek Philosophers as Theologians* (2007), 50.
[22] L. Táran, *Parmenides* (1965).
[23] Empedocles, *The Poem of Empedocles* (1992).

times the size of the Earth, which rotates around it and is visible through a kind of 'blowhole'.

This idea (which might seem slightly odd) was, as Geoffrey Lloyd points out, 'the first attempt at . . . a mechanical model of the heavenly bodies in Greek astronomy',[24] and in his book Anaximandros went on to propose different kinds of mechanical models for a whole range of natural phenomena from the origin of life to the origin of winds and rain. These may be, as Lloyd remarks, explanations which 'leave the gods out'.[25] What underlay and made plausible the attempt to group apparently separate phenomena together in this way was, however, the belief in a fundamental rationality which linked them together: a divine *arche* that was the origin of all things and whose nature was 'to encompass all things and to direct all things'.[26]

Each of these thinkers proposed a different means by which the underlying order manifested itself in the world of experience. Anaximandros argued (in somewhat 'poetical' terms according to one ancient commentator) that 'things perish into those things out of which they have their being, as is due; for they make just recompense to one another according to their injustice according to the ordinance of time'.[27] Anaxamines agued that everything emerged from the divine air by a process of rarefying or condensation: 'Rarefied it becomes fire; condensed it becomes first wind then cloud and when condensed still further earth and stones.'[28] Pythagoras, starting from the observation 'that numerical properties are inherent in the musical scale in the heavens and many other things',[29] went on, according to Aristotle, to argue that 'the whole heaven' was 'a *harmonia* (a musical scale or a "fitting together") and a number'.[30] Having been forced to leave Samos, Pythagoras made this doctrine the basis of a religious community he founded in Southern Italy.

Pythagoras was the only one of these 'philosophers' (a term he seems to have coined) to have established a new religious cult. Yet as the poetry of Xenophanes makes clear, the new theological thinking was implicitly critical of the traditional temple cults. In the sayings of Heraclitus this became explicit. Heraclitus described praying before statues being 'as if one wanted to hold a conversation with houses'. The traditional ritual of purification from shedding blood by washing in blood was 'as if one who had stepped into mud were to wash it off with mud'.[31]

[24] G. E. R. Lloyd, *Early Greek Science* (1970), 17.
[25] G. E. R. Lloyd, *Early Greek Science*, 9.
[26] Aristotle, *The Metaphysics*, 203b206.
[27] W. K. C. Guthrie, *A History of Greek Philosophy*, 76.
[28] W. K. C. Guthrie, *A History of Greek Philosophy*, 121.
[29] Aristotle, *The Metaphysics*, 987b928.
[30] Aristotle, *The Metaphysics*, 987b928.
[31] W. Burkert, *Greek Religion*, 309.

Heraclitus, however, did not entirely reject traditional religious language. God, he thought, was 'both willing and unwilling to be called by the name of Zeus'.[32] Underlying the traditional cults was a truth that gave a new kind of coherence to the world. 'All human laws', he argued, 'are nourished by a divine one'.[33]

These concepts which initiated a powerful new slipstream of curiosity could also seem to challenge existing beliefs and were likely to divide opinion. Such fractures first became visible when these new ideas began to arrive in Athens, borne across the Aegean on the winds of war.

[32] W. Burkert, *Greek Religion*, 309.
[33] Heraclitus, *The Art and Thought of Heraclitus*, 43.

CHAPTER NINE

The Move to Athens

In 499 BC Aristagoras, the tyrant of Miletus, incited the Greek-speaking cities of Ionia to stage a revolt against their Persian overlords. The rebellion, which lasted for six years, ended in disaster. Miletus, 'the glory of Ionia' as Herodotus describes it, was, along with the other cities, razed to the ground and its inhabitants put to the sword. The Persians then began their pre-parations for the invasion of mainland Greece.

Among the refugees from the Ionian revolt who washed up in Athens was a young philosopher named Anaxagoras. His stay in the city might have been a short one. According to one account he arrived in Athens the year before the entire popu-lation evacuated, leaving their homes to be sacked by Xerxes' two million-strong army. But following the Athenians great na-val victory over the Persians at Salamis, Anaxagoras returned to the city, according to this story, making it his home for the next 30 years and introducing its citizens to the new Ionian way of thinking.

Like Anaximandros before him, Anaxagoras wrote a book. Socrates, at his trial, remarked that it could be bought for a drachma in the orchestra of the theatre (which seems to have doubled as a bookshop). No copy survives, but those who quoted from it made it clear that Anaxagoras carried forward and extended two of the central themes of Ionian thinking: the nature of the divine *arche* and the way in which it or-dered and steered the universe. According to the historian Plutarch, Anaxagoras' nickname in Athens was *Nous*, mean-ing 'mind or intelligence'.[1] This was partly in tribute to his brainpower, but also because he was thought to be 'the first of the philosophers who did not refer the first ordering of the world to fortune or chance but to a pure unadulterated intelligence'.

A few fragments of his book, quoted by a later writer, de-scribe 'mind' both as being that which 'controls all things which possess soul' and as that which 'controlled the whole revolution

[1] Plutarchus, *Plutarch's Lives* (1910), 229.

so that it revolved in the first place'.[2] Although he does not talk directly about the divine, this seems to be implied by his concept of mind. Mind 'arranged everything—what was and what now is and what will be', and in particular it controls 'this revolution in which the sun and the moon and the air and the aether are separating off'.

As with Anaximandros, this conviction of an underlying order provided the rationale for studying whole classes of natural phenomena. Anaxagoras' interests, like those of his predecessor, ranged across the whole natural world, but his central preoccupation was with astronomy. When asked why he thought he had been born he is supposed to have replied, 'for the study of the sun and the moon and the heavens', and this was the field in which he made his most striking contribution to the understanding of nature. Plato refers to Anaxagoras' proposition that 'the moon receives its light from the sun',[3] and Plutarch describes him as the man 'who most clearly and boldly put into writing an account of the illuminations and shadowing of the moon'.[4] This must have involved not only theorizing but also careful observation, and his most controversial theory was similarly bound up with the observation of a particular astronomical event.

In 467 BC 'during the archonship of Demulus', a meteorite fell from the sky in the district of Thrace near Aegospotami. According to Pliny's *Natural History*, 'the stone', which fell in daylight, 'is still shown, the size of a cartload and brown in colour'. This either suggested or confirmed to Anaxagoras the idea that 'the whole firmament to be made of stones; that the rapidity of rotation caused it to cohere; and that if this were relaxed it would fall'.[5] The Sun, he argued, was 'a mass of red hot metal . . . larger than the Pelopennesus'.[6] The shockwaves from this idea were still being felt 60 years later. When Meletus, one of the prosecutors, accused Socrates at his trial of not believing that the Sun is a god and saying that it is a stone, Socrates replied that he must imagine he was prosecuting Anaxagoras. Whether such a prosecution actually took place is uncertain. There are no contemporary records, but Plutarch claimed this did indeed happen.

Written some 500 years after the event, but generally based on good research, Plutarch's *Life of Pericles* describes how a citizen called Diopethes proposed a decree 'that public accusations should be brought against persons who neglected religion or

[2] J. Barnes, *Early Greek Philosophy* (2002), 229.
[3] Plato, *Cratylus; Parmenides* (1926), 91.
[4] Anaxagoras, *Anaxagoras of Clazomenae* (2007), 84.
[5] L. Diogenes, *Life of Eminent Philosophers* (1972), 141–3.
[6] L. Diogenes, *Life of Eminent Philosophers*, 137.

taught new doctrines about things, directing suspicions by this means against Pericles himself'.[7] Pericles was the leading figure in the early Athenian democracy and, according to Plutarch, had a great admiration for Anaxagoras' 'lofty and as they call it "up-in-the-air sort of thought"', which led him to replace 'a wild and timid superstition with the good hope and assurance of an intelligent piety'.[8]

Throughout his career Pericles was a target for attack by rival factions in Athens, so that Plutarch's scenario is not implausible. According to Diogenes Laertius' *Lives of the Philosophers*, Anaxagoras was condemned for impiety 'because he declared the sun to be a mass of red hot metal'.[9] He was forced to leave Athens, returning across the Aegean to the Ionian town of Lampsacus, where he died in 428 BC. The citizens of Lampsacus, it was said, erected an altar to Mind and Truth in his memory, and the school children, as he had requested, had a holiday on the anniversary of his death.

Five years after his death *The Clouds* was produced at the city Dionysia, and it is not difficult to recognize in the play's picture of the philosophers' worship of the clouds and Socrates' study of the Sun a crude parody of Anaxagoras' 'up-in-the-air kind of thought'. Yet despite its title the play's real target was the ambition of 'making the weaker argument defeat the stronger', which was an idea associated not with the Ionian thinkers but with a group of teachers known as 'sophists'. What made this idea so disturbing was its potential effect on the way people actually behaved, and it was in response to this challenge that Athenian philosophers began to develop and extend the ideas of the Ionians.

Spells and Incantations

The teachings of the sophists had been born out of some of the same causes that had given rise to new Ionian ideas. The new kinds of political order that had been appearing in the Greek city-states had created the need for a new kind of education. Wealth, position, and fighting skills were no longer enough. Men who wanted power needed to be able 'to convince by means of a speech ... any gathering of citizens'.[10] This was a skill that needed to be taught, and when Athens established itself as a full democracy it became a magnet for those who could do so. One of Plato's dialogues describes the arrival of

[7] Plutarchus, *Plutarch's Lives*, 256.
[8] Plutarchus, *Plutarch's Lives*, 230.
[9] L. Diogenes, *Life of Eminent Philosophers*, 143.
[10] Plato, *Gorgias* (1960), 28.

one of these teachers as like a 'feast day'; another refers to the excitement among the aristocracy in Athens at the news that Protagoras, the most famous of the sophists, was making a visit to the city.

Protagoras (according to Plato's dialogue of that title) taught a broad syllabus which aimed to equip a man 'to become a real power in the city both as a speaker and a man of action'.[11] The ability to argue a case was central to this and Protagoras is known to have published two books of 'contrary arguments' to demonstrate 'that there are two opposite arguments on every subject'. This realization could easily lead to some scepticism about the possibility of discovering any ultimate truth. Thus, the opening sentence of Protagoras' book *On the Gods* (which is all that survives of it) raises the theoretical possibility of atheism: 'concerning the gods I am not able to discover whether they exist or not or what they are like'.[12]

Previous generations had believed that the laws which governed the city derived ultimately from the gods. Protagoras, by contrast, speaks of 'the inventions of lawgivers of ancient times', and the awareness that both laws and ethical norms varied in different societies seems to have suggested to him that all these things were variable customs. Thus, it has been argued that his famous dictum that 'man is the measure of all things' was a way of saying that all knowledge and ethical values are relative to the individual.

Plato's portrait of Protagoras depicts him as having a deep respect for conventional morality. His principle of relativity had the democratic virtue of suggesting a respect for the opinion of others. The drawback of the principle was the encouragement it might give to those who had rather less respect for the conventions. Thus, in his dialogue *Gorgias* Plato has a character called Callicles describe the rules and conventions of morality as 'spells and incantations and unnatural laws' that have been 'made by the weaklings who form the majority of mankind'[13] to protect themselves against the basic law that might is right. This is the principle that Pheidippides acts on at the end of the *The Clouds* when he beats up his father.

These fictional characters had real-life counterparts. Almost the first thing we learn about Pheidippides in the *The Clouds* is that he races chariots and has grown his hair long.

[11] Plato, *Protagoras* (1956), 50.
[12] W. K. C. Guthrie, *The Sophists* (1971), 234.
[13] Plato, *Gorgias*, 78.

For the Athenian audience this would immediately have placed him among the group of racing-obsessed young aristocrats (led by the glamorous figure of Alcibiades, the ward of Pericles) who affected floppy fashions and unconventional attitudes [Fig. 9.1].

Some of the rowdy dining clubs in which they consorted had titles that explicitly mocked conventional religion. The 'Triballoi' or 'Bongo Bongo' collected offerings left for Hecate at crossroads and used them for their own dinners.[14] The 'Kakodaimonistai', or 'worshippers of bad luck', deliberately met on days that were supposed to be ill-omened. Sometimes these parties spilt out on to the street, and in 415 BC (eight years after the performance of *The Clouds*), a notorious event associated with these clubs involved the mutilation one night of all the statues of Hermes in the city, a subsequent board of enquiry, and the exile of Alcibiades.

9.1 Alcibiades.

It was never finally established who had been responsible for this vandalism, but among those suspected of it were men whose public behaviour revealed attitudes not unlike those of Callicles. The most notorious of these was Alcibiades himself, who the previous summer had sailed an enormous Athenian fleet to the small neutral island of Melos and demanded their unconditional surrender. The Athenians, according to Thucydides, announced that 'justice depends upon the equality of power to compel and that in fact the strong do what they have the power to do',[15] and when the Melians defied them, the Athenians proceeded to kill all the men on the island and enslave the women and children.

Another almost equally notorious suspect in the affair of the herms was Critias, who went on to become one of the leaders of a Spartan puppet regime—the Thirty Tyrants—who established a murderous reign of terror in Athens. A fragment of a satyr play from this time, which one ancient source attributes to Critias, argues (in a way that parallels Callicles' talk of 'spells and incantations') that the gods were no more than a device invented by a clever man to frighten the wicked 'even if they acted or spoke or thought in secret'.[16] Given this open cynicism about the religious foundations of morality, and the possibility that it might have a real effect on behaviour, it is not surprising that some might have felt the need to discover whether it was possible to find a reliable basis for moral action.

[14] S. C. Humphreys, *The Strangeness of Gods* (2004), 63.
[15] Thucydides, *History of the Peloponnesian War* (1972), 402.
[16] W. K. C. Guthrie, *The Sophists*, 243.

9.2 Socrates.

The Socratic Search

Finding a reliable basis for moral action was, according to Plato, the self-appointed mission of Socrates, in which he took as his starting point the writings of Anaxagoras [Fig. 9.2]. In Plato's account of his death, Socrates describes how as a young man he had heard someone reading from Anaxagoras' book the assertion 'that it is Mind that produces order and is the cause of everything'.[17] This idea excited him because it seemed to provide a foundation for establishing the nature of 'the best and highest good'. On buying the books and reading them for himself, he was, however, quickly disappointed.

Although all Anaxagoras' theories were rooted in the idea of an ultimate order, he had only concerned himself with physical causes. These could not by themselves lead to any understanding of goodness or moral obligation, anymore than it was possible to understand Socrates' own moral decisions by examining his 'bones and sinews'.[18] Having become 'worn out with my physical investigations', and feeling that focussing on physical causation could produce a kind of blindness, he then embarked on a very different kind of enquiry.

The starting point for this new sort of investigation was Socrates' conviction that everyone bases their actions on what they think at the time are good reasons. It followed that finding a reliable basis for morality involved trying to discover what a good reason really was. This was the rationale behind his method of buttonholing fellow citizens and trying, through a process of question and answer, to get them to clarify their ideas of virtue. It was also the meaning of his dictum (attested by both Plato and Xenophon, the two disciples who wrote most about him) that 'virtue is knowledge'.

Socrates' discovery that even those with the greatest reputation for wisdom were unable to give satisfactory answers may well, as Plato's *Apology* suggests, have accounted for some of his unpopularity in the city. But whereas a later school of sceptical philosophers who took Socrates as their inspiration saw this as a reason to doubt the possibility of any real kind of knowledge, Socrates himself seems to have thought that the admission of ignorance was the beginning of wisdom: the necessary quay from which all curiosity-driven voyages must embark.

Socrates' optimism that the pursuit of wisdom was a worthwhile quest is portrayed by Plato as being founded in religious

[17] Plato, *The Last Days of Socrates* (2003), 155.
[18] Plato, *The Last Days of Socrates*, 56.

conviction. Both Plato and Xenophon (whose portraits of their teacher vary in many other respects), in their accounts of his trial, describe Socrates referring to a 'daimonion'—'a divine something'[19] which had 'always been my constant companion opposing me even in quite trivial things if I was going to take the wrong course',[20] 'a voice of God . . . made manifest to me indicating my duty'.[21] Plato emphasizes that Socrates saw his role as a moral 'gadfly' as a divine calling, 'the duty I have accepted . . . in obedience to God's commands given in dreams and oracles',[22] and attributes his stubborn refusal to accede to the jurors demand that he give up his investigations to this same sense of calling: 'Gentlemen I am your grateful and devoted servant but I owe a greater obedience to God than to you.'[23]

Anaxagoras and his predecessors had argued that there was an underlying order which gave rise to all the physical processes we perceive. Socrates extended this notion by arguing that this underlying order was also a moral order. His incessant questions about what we would now call moral realism provided his successors with an intellectual slipstream for further physical investigations. In a dialogue in which Xenophon claims 'I heard myself',[24] Socrates argues that 'he who co-ordinates and holds together the universe' has filled it with 'all things fair and good',[25] and persuades his interlocutor to agree that these are 'tokens of loving kindness' which 'shows design at work'.[26]

The conviction that there is a moral order upheld by 'a good and wise God' certainly provided the rationale for Socrates' distinctive ethical position, as described by Plato, that it was better to suffer evil than to do it and that one should not return evil for evil. Justin Martyr, the first-century AD philosophical convert to Christianity, described Socrates as 'a Christian before Christ'. But on the evidence of Xenophon's dialogue, Socrates' conviction also provided a motive for the study of nature which encouraged the beholder to recognize 'the things which are unseen' by 'realising their power in the manifestations of nature' and by doing so to 'honour the godhead' (daimonion).[27]

[19] Plato, *The Last Days of Socrates*, 74; Xenophon, *Memorabilia; Oeconomicus* (2013), 649.
[20] Plato, *The Last Days of Socrates*, 74.
[21] Xenophon, *Memorabilia; Oeconomicus*, 649.
[22] Xenophon, *Memorabilia; Oeconomicus*, 66.
[23] Xenophon, *Memorabilia; Oeconomicus*, 61.
[24] Xenophon, *Memorabilia; Oeconomicus*, 299.
[25] Xenophon, *Memorabilia; Oeconomicus*, 305.
[26] Xenophon, *Memorabilia; Oeconomicus*, 307.
[27] Xenophon, *Memorabilia; Oeconomicus*, 307.

CHAPTER TEN

Through the Academy Door

10.1 A Greek astronomer.

10.2 Plato.

[1] W. K. C. Guthrie, *The Sophists* (1971), 233.
[2] W. K. C. Guthrie, *The Sophists*, 233.
[3] C. A. Huffman, *Archytas of Tarentum* (2005), 342–401.

There is evidence that even within Socrates' own lifetime a kind of religious piety that embraced the scientific study of nature was beginning to gain currency. A fragment from a chorus of Euripides talks of 'the happiness of a man who has learned the ways of scientific enquiry . . . the ageless order and beauty (*kosmos*) of immortal nature and how it was put together . . . such a man . . . will have no part in wicked or injurious deeds'.[1] Another fragment asks (apparently with reference to this beauty), 'Beholding these things, who is not conscious of God?' [Fig. 10.1].[2]

Socrates' willingness to die rather than abandon his intellectual quest powerfully dramatized the idea that the rational pursuit of truth was a religious obligation. In time it became the model for a whole new idea of heroism. In the immediate aftermath of his execution, however, many of his disciples left Athens. Some took refuge in the nearby city of Megara. Xenophon had already left to fight as a mercenary in Persia. Plato, having decided to 'withdraw from the prevailing wickedness', travelled to Italy [Fig. 10.2]. There he formed a lasting relationship with the Pythagorean community at Tarentum led by Archytas, who seems to have had a significant impact on the development of his thinking.

Plato and the Mathematicians

Although he was most celebrated in antiquity for having sent a ship to rescue Plato from the tyrant of Syracuse, Archytas was a substantial figure in his own right. As a mathematician he brought a new kind of mathematical rigour and sophistication to the Pythagoreans' interest in the numerical ratios of musical harmony, and was the first person to solve the most famous mathematical problem in the ancient world: the duplication of the cube.[3] This solution cannot be implemented with the conventional instruments of a compass and a straightedge (it can be proved that there is no such solution), but it is nevertheless very ingenious.

Archytas' talents were not only mathematical. As a general he was never defeated. Breaking all precedents Tarentum elected him as *stratêgos* seven times in a row, and under his leadership became one of the most powerful cities in the Greek world.

Both his leadership of the city and his mathematics were united in a philosophy that emerges in the few fragments of his writing that are generally accepted as authentic, in which he argues that *logisitic*—the science of number—is the basis of all the sciences, and that *logismos*—rational calculation which can be expressed numerically—produces the fairness on which the state depends.[4]

In Plato's *The Republic*, his longest work where he describes the idea of a 'philosopher-king', it is hard to doubt that Archytas was in his mind. In his description of the education that would be necessary for this role Plato lays great emphasis on the importance of mathematics. The Pythagorean idea that divine number was the cause of harmony and order in the universe seems to have provided Plato with a model of the kind of universal moral truths that Socrates had sought to uncover. Numbers were the clearest instance of something like 'the forms'—the ultimate realities which he argued lay behind the appearance of things—and over the entrance to the Academy, the school that he founded in Athens, there was said to be an inscription which read 'let no one un-versed in geometry enter'.

For Plato the principle value of mathematics and geometry was that they provided the easiest way for the mind 'to pass from the world of becoming to that of truth and reality'.[5] Pythagoreans might 'tease and torment their strings' and astronomers might 'gape at the heavens', but to attain true knowledge it was necessary to recognize the primacy of mathematics and geometry. Thus, we should treat astronomy and geometry, he argued, 'as setting problems for solutions and ignore the visible heavens'.[6]

This, according to one ancient source, is precisely what he did. A philosopher in the sixth century AD called Simplicius quoted a writer called Sosigenes from the second century AD as saying that Plato had initiated what was in effect the first programme of scientific research by offering a problem to astronomers and mathematicians: 'by the assumption of what uniform and orderly motions can the apparent motions of the planets be accounted for?'[7]

[4] C. A. Huffman, 'Archytas', in *The Stanford Encyclopedia of Philosophy* (2011).
[5] Plato, *The Republic* (1955), 292.
[6] Plato, *The Republic*, 298.
[7] G. E. R. Lloyd, *Early Greek Science* (1970), 84.

A pupil of both Archytas and Plato named Eudoxos attempted to solve this problem. Born in Cnidus in Asia Minor, Eudoxos studied mathematics in Tarentum before sailing to Athens to 'learn from the Socratics'. Diogenes Laertius' *Lives of the Philosophers* describes how, having arrived at Piraeus 'in straightened circumstances', Eudoxos had walked to Athens each day to hear Plato's lectures, returning at night to sleep in the boat. When years later he returned to Athens bringing a large group of his own pupils to visit the Academy, Plato gave a lavish banquet in his honour.

There was much to celebrate. In the intervening years Eudoxos had both laid the foundations of axiomatic geometry and devised an ingenious model of concentric spheres, one within another, whose combined motions could account for the apparently irregular motions of the Moon and planets in the night sky.

Eudoxos' model was soon seen not to cover all the facts. Subsequent Greek astronomers replaced it with models of so-called 'epicycles' and 'eccentric circles', which 'to save the phenomena' became ever more elaborate until some two thousand years later they were eventually discarded. Nevertheless Eudoxos' application of geometry to the movement of the heavens was the starting point from which all such later astronomy proceeded. In showing that the planets moved in ways that mathematicians could calculate, he provided apparent confirmation of the religious and philosophical conviction that the world was formed on a rational design.

This conviction that the world was formed on a rational design lies at the core of the *Timaeus*, Plato's philosophical rewriting of Hesiod's mythological account of the generation of the gods and the creation of the world. Timaeus, the principal speaker, is an Archytas-like figure who has both held 'the most important political offices' in an 'exceptionally well-governed city in Italy'[8] and been an astronomer 'who has specialised in natural science'.[9] Although glancing at the stories of the generation of the gods he says, 'we must follow custom' and 'trust the accounts of our predecessors . . . even when what they say is implausible and illogical',[10] behind them he sees what he refers to as the *Demiurgos*—the Craftsman.

'It would', Timaeus acknowledges, 'be a hard task to discover the maker and father of this universe of ours, and if we did find him it would be impossible to speak of him to everyone.'[11]

[8] Plato, *Timaeus and Critias* (2008), 6.
[9] Plato, *Timaeus and Critias*, 15.
[10] Plato, *Timaeus and Critias*, 29.
[11] Plato, *Timaeus and Critias*, 17.

Nevertheless although he does not pretend to be 'altogether internally consistent in every respect and precise',[12] starting from the premise that this Craftsman is good and brings order out of chaos, Timaeus is able to construct what he descibes as an *eikos logos*—a likely account.

The most striking feature of this 'likely account' is that 'the maker and father' of the universe is pictured as a geometer, selecting the most beautiful geometry to fashion every aspect of the world. This is most readily seen in the great circular movements of the heavenly bodies, but it is also apparent, Timaeus argues, in the fundamental structure of matter. 'The first thing the god did', he suggests, 'when he came to organise the universe, was to use shapes and numbers and assign them definite forms'.[13] From the most beautiful triangles he formed solid figures (the so-called Platonic solids—tetrahedron, cube, octahedron and icosahedron—he doesn't include the dodecahedron though he knew about it) which he imagines form the basic atomic structure of the four elements. And when, the Demiurgos having created divine beings, these are then given the task of creating mortal beings, geometry is once again to the fore.

Timaeus states as a firm principle that 'when it comes to treating parts of the body, the body of the universe should be our model'.[14] When, however, expounding the details of this secondary level of creation (describing, for instance, how 'a creature's inner triangles . . . start to lose their grip') he is more tentative: 'Is our account true? We could only be certain if we met with the god's endorsement'.[15]

In *Critias*, the unfinished companion piece to *Timaeus*, Timaeus apologizes if 'we inadvertently struck a false note'.[16] In praying for 'the gift of knowledge', however, he is asking not for a recovery of ancient wisdom, but rather that 'in the future any account we give of the creation of the gods may be accurate'.[17]

In his last book, *The Laws* (which Diogenes says were copied out from wax tablets after his death), Plato's spokesman talks about having learned about the rational movements of the planets 'when I was no youngster'.[18] Whereas at one time astronomy might have seemed to be leading people towards atheism, now it was possible to come to the aid of ancient custom with reason (logos). Those previous thinkers 'who had the hardihood to stick their neck out and assert it was reason that imposed regularity and order on the heavens'[19] had been

[12] Plato, *Timaeus and Critias*, 18.
[13] Plato, *Timaeus and Critias*, 46.
[14] Plato, *Timaeus and Critias*, 93.
[15] Plato, *Timaeus and Critias*, 72.
[16] Plato, *Timaeus and Critias*, 103.
[17] Plato, *Timaeus and Critias*, 103.
[18] Plato, *The Laws* (1970), 316.
[19] Plato, *The Laws*, 527.

vindicated by the 'remarkably accurate predictions' about the movements of the planets that it was now possible to make.

Hence he argues religious understanding, ethical thinking, and scientific study needed to be seen as an integrated intellectual enterprise. To 'attain a truly religious outlook' which recognizes 'that reason is the supreme power among the heavenly bodies' and to be able 'to frame consistent rules of moral action', one must make the necessary mathematical studies and 'survey with the eye of the philosopher what they have in common'.[20]

In the long term closing the gap in this way between religion and astronomy turned out to be a hazardous move (an early example of the kind of cycling chute referred to at the end of Chapter Seven), not least because as Copernicus, Galileo, and Kepler between them subsequently established, the whole model of planets revolving round the Earth in circles was, in fact, mistaken. For the time being, however, it provided a powerful model.

Aristotle's *Arche*

10.3 Aristotle.

The religiously motivated mathematical and astronomical studies which Plato had initiated were broadened into a much wider programme of scientific research by a student at the Academy whom Plato referred to as 'the mind of the school' [Fig. 10.3]. Aristotle, 'the young son of a doctor from the colonies', had travelled to Athens to complete his education. He became Plato's pupil at the age of 17 and remained with him at the Academy for the next 20 years.

Aristotle's father was a physician to the king of Macedon, and we know from 'the Hippocratic Corpus'—a collection of medical treatises from the fifth and fourth centuries BC—that Greek doctors of this time placed a high value on empirical observation. Most of these treatises reject both supernatural and speculative explanations of disease, and concentrate on what can be understood by looking (providing sometimes evidence-based case histories). The young Aristotle seems to have inherited from his father both a lifelong interest in biology and a commitment to this kind of empirical curiosity-driven observation.

While Aristotle's focus and method of study were in this respect different from Plato's, his programme of research was no less theologically motivated. In his lectures on *The Parts of Animals*, Aristotle tells a story of some visitors who wanted to meet the philosopher Heraclitus and entered his house to find him warming himself at the stove. The visitors hesitated, but Heraclitus said, 'Come in; don't be afraid; the gods are even here.' Aristotle comments that 'in the same way we should approach the investigation of every kind of animal without being ashamed since in every one of them is something natural and something beautiful. The absence of chance and the serving of ends'.[21] It is 'the end', he argues, 'for the sake of which a thing has been constructed', which 'belongs to what is beautiful', and it is the possibility of glimpsing the deep structure of reality which links the study of nature to a wider theology.

Aristotle argued that it was 'through wonder that men now begin and originally began to philosophize' and that this leads them from 'obvious perplexities . . . by gradual progression' to 'raising questions about the greater matters'.[22] These greater matters involved thinking about 'first principles and causes'—what he describes as 'first philosophy' or *theologia*. Other subjects might be more related to 'extrinsic advantage' but none other was 'more precious' or more 'divine'.[23] His reflections on this subject led him to propose the existence of what he variously refers to as 'the unmoved mover' or 'the final cause': his version of the *arche*, the principle on which 'depend the sensible universe and the world of nature'.[24]

He identifies this *arche* as 'the God' (*ho theos*), and goes on to argue that 'God is a living being, eternal most good'.[25] God's thoughts must be entirely good, so Aristotle concludes it must be of itself that divine thought thinks (since it is the most excellent of things 'and its thinking is a thinking of thinking'[26]). Aristotle in his writings is sometimes contemptuous of popular religion, but this idea was clearly important to him. Although it might appear as if the realm of the divine was wholly separate from the realm of the human, in fact, according to Aristotle, it is in the realm of thinking that the two are united.

At the conclusion of his lectures on ethics Aristotle argues that 'the end of human life' is happiness. Happiness is an activity which involves exercising the highest kind of virtue of 'the best part of us', and 'the best part of us' is what he describes as *theoria*—contemplation or study. It is this activity which gives us knowledge of 'what is noble or divine', and it is either 'itself actually divine' or at least 'the divinest part of us'.[27]

[21] Aristotle, *The Parts of Animals* (2001), 645a621.
[22] Aristotle, *The Metaphysics* (1933), 13.
[23] Aristotle, *The Metaphysics*, 15.
[24] Aristotle, *The Metaphysics*, 149.
[25] Aristotle, *The Metaphysics*, 151.
[26] Aristotle, *The Metaphysics*, 165.
[27] Aristotle, *Nicomachean Ethics* (1968), 613.

'The gods as we conceive them', he argues, 'enjoy supreme felicity and happiness. But what sort of actions can we attribute to them?' It was, he thought, absurd to think of them acting justly or generously as though they made contracts or had a currency. 'It follows that the activity of God which is transcendent in blessedness, is the activity of contemplation; and therefore among human activities that which is most akin to the divine activity of contemplation will be the greatest source of happiness'.[28] This was a principle he acted upon. A large part of Aristotle's own life seems to have been devoted to this kind of contemplation.

Over 500 species of animals, fishes, and insects are referred to in Aristotle's works. In many cases his minute descriptions (which were much admired by Darwin) were the result of his own meticulous dissections. After the death of Plato, Aristotle left Athens and spent some time wandering in the Greek Islands, and his work includes detailed descriptions of the marine animals in the lagoon of Pyrra on Lesbos where he lived for three years. For some years after this he was employed by Phillip of Macedon as tutor to the young Alexander, and when Alexander succeeded to the throne, Aristotle returned to Athens and (probably with the financial support of Alexander the Great) established his own teaching institution at the Lyceum—a precinct sacred to Apollo and the Muses outside the city.

In the Lyceum Aristotle established a kind of collegiate life devoted to the pursuit of *theoria* as he understood it. In a series of logical treatises (known collectively as *The Organon*) in which he discusses the means by which it is possible to gain reliable knowledge, Aristotle mentions the importance of reviewing what has been commonly accepted to be true about a subject in order to help formulate the particular problems which it presents. He also refers to the importance of gathering data and *phainomena*. This was characteristic both of his own work and of that of the scholars he gathered around him.

Thus, a history of geometry and astronomy was produced by Eudemus, a history of medicine by Meno, and a history of physics by Theophrastus. Theophrastus, who had come with Aristotle to Athens and ultimately succeeded him as head of the Lyceum, went on to produce two botanical treatises (which complemented Aristotle's *Inquiry Concerning Animals*), and several treatises, including *On Fire* and *On Stones*. Strato,

[28] Aristotle, *Nicomachean Ethics*, 628.

who succeeded Theophrastus, continued Aristotle's work on dynamics, but disagreed with Aristotle about the existence of vacuum.

Much greater freedom of thought in relation to the founder was allowed by the Lyceum than was permitted towards the founders in other schools. Aristotle seems to have brought with him to Athens a collection of scrolls and items of natural history. These became the basis of a substantial library and museum at the heart of the Lyceum from which the scholars were able to work and to carve out their individual paths. The combination of encouragement to question and provision of resources proved in Aristotle's school to be conducive to the pursuit of curiosity.

This heroic period of work was short-lived. In 322 BC, 12 years after Aristotle's return to Athens, the sudden death of Alexander the Great was followed by a revolt against Macedon in which the Athenians took the lead. As a protégé of the Macedonians, Aristotle was a marked man, and like Socrates was prosecuted for impiety. Unlike Socrates he strategically withdrew to his mother's property at Chalcis, stating that 'he would not let Athens sin twice against philosophy'. He died within a year.

None of the various schools of Greek philosophy that arose after Aristotle's death provided such a direct theological slipstream for the study of nature. The two major schools of thought that developed in the next generation, Stoicism and Epicureanism, were both philosophies that were concerned not so much with knowledge but with what they called 'peace of mind'—*apatheia*.

The Art of Living

Zeno of Citium was a Cyprian merchant who, according to the story told by Diogenes Laertius, wandered into an Athenian bookshop after surviving a shipwreck. Finding a book about Socrates and wanting to meet such a man, he was directed to a cynic philosopher called Crates, and attracted by his ethical teaching became his pupil. When Zeno began teaching independently, his pupils were initially known as 'Zenoians' but later named 'Stoics' after the portico in the Agora where Zeno taught: the Stoa Poikile. None of his writings survive, but their ethical purpose is indicated by Diogenes' story that after his death, the Athenians erected a statue in his honour with an inscription that descibed his conduct as 'a pattern for imitation in perfect consistency with his teaching'.

The Stoics defined philosophy as what they described as *askesis*, the practice or exercise of an expertise in how to live. As one of the later Roman Stoics, the ex-slave Epictetus, put it, 'philosophy does not promise to secure anything external to man; otherwise it would be admitting something beyond its subject matter'. In the same way that the material of the carpenter was wood, 'so the subject matter of the art of living is each man's own life'.[29] How to live in the face of adversity ('sick yet happy, in peril yet happy, dying yet happy, in exile and happy, in disgrace and happy'[30]) was one of the Stoics most distinctive ideas, and one that made Stoicism of particular interest to Christian thinkers.

Stoic thought was rooted in a distinctive conception of both theology and physics. Chysippus, the third leader of the Stoic school and the one who did most to develop Stoic thinking, 'calls the world itself a god, and also the all-pervading world-soul'.[31] Living in accord with nature, he argued, meant living in accord with the *logos*—the rational spirit animating the universe, conceived as a primordial fire. In defending the Stoic outlook Chrysippus developed distinctive approaches to physics, mathematics, and logic, writing, it was reputed, more than 700 books. Nevertheless despite the breadth of this philosophy, the actual study of nature lay 'beyond its subject matter'.

The Epicureans were equally concerned with 'peace of mind'. Unlike the Stoics they found within this a reason for studying nature. Recognizing that natural phenomena had physical causes was a means of ridding oneself of superstitious worries about the gods. The followers of Epicurus thus constituted the first philosophical movement to conceive the study of nature not as an expression of piety, but as a means of marginalizing religious anxieties.

Philosophy as Therapy

Epicurus was an Athenian, raised on the island of Samos, and seems to have studied with a philosopher called Nauctanius, a follower of the Abderan philosopher Democritus [Fig. 10.4]. Democritus had taught that there were only two things in the universe, 'atoms and the void'—minute indivisible particles and

[29] Epictetus, *Discourses* (2014), 33.
[30] Epictetus, *Discourses*, 120.
[31] Cicero, *De Natura Deorum*, I, xv, 39, trans. H. Rackam (1933), 41.

the space through which they moved. Everything else was constructed from, and could be explained by reference to, these
two things (Richard Feynman remarked that if, in some cataclysmic event, all scientific knowledge were to be destroyed
and one had to choose a single observation to pass on to the
next generation that would contain the most information in
the fewest words, it would be that *all things are made of atoms*[32]).

For Epicurus the most important implication of this doctrine was that everything could be explained without reference
to the gods. Plagues, lightning bolts, and earthquakes were the
product of natural causes rather than divine anger. The gods
lived in untroubled bliss and did not interfere in any way with
human affairs. Even the soul was merely a fragile atomic film
which would not survive the dissolution of the body. Hence
there was no need to fear judgement after death.

10.4 Epicurus.

Banishing such religious anxiety was, according to Epicurus,
the only reason to study nature. There was 'no other goal to be
achieved by the knowledge of meteorological phenomena . . .
than freedom from disturbance and a secure conviction, just
as with the rest [of physics]'.[33] Indeed 'if our suspicions about
heavenly phenomena . . . did not trouble us at all . . . then we
would have no need of natural science'.[34] By this token Epicurus' anxieties must have been considerable. Although only a
handful of his writings survive, as letters quoted by a biographer, a library in a villa at Herculaneum contains the charred
remains of a series of lectures on nature that fill 37 scrolls. These
are the likely source of the Roman poet Lucretius' great Epicurean poem *De Rerum Natura* (On The Nature of Things).

Beyond his insistence on the doctrine of atoms and the void
Epicurus was, however, sceptical about the possibility of discovering true physical explanations. Celestial phenomena, he
argued, admitted of many possible explanations. The waxing
and waning of the Moon could, for instance, have a variety of
different causes and to privilege a particular explanation shows
'a failure to understand what is possible for a man to understand and what is not, for this reason desiring to understand
what cannot be understood'.[35]

'In this, and many similar passages', Geoffrey Lloyd has argued,

the unscientific, indeed anti-scientific, aspects of Epicureanism become evident. He believes that research is futile
if it does not contribute to peace of mind. He rules out

[32] R. P. Feynman, R. P. Leighton, and M. Sands, *The Feynman Lectures on Physics* (1965), 1–2.
[33] Epicurus, *The Epicurus Reader* (1994), 19.
[34] Epicurus, *The Epicurus Reader*, 33.
[35] Epicurus, *The Epicurus Reader*, 22.

further inquiry into which of several explanations of a phenomena is correct and abuses those who attempted such investigations first for being dogmatists . . . and secondly for indulging in superstition and mythology. Yet those very astronomers whom Epicurus dismisses with contempt had long ago given the correct explanation for the phases of the moon.[36]

The anti-scientific aspect of Epicurean philosophy had something in common with the sceptical philosophy of the later Academy. Even the Lyceum, after the death of Theophrastus and Strato, seems to have produced little work of any interest. The golden age of Aristotle's Lyceum was not, however, forgotten.

The Hellenistic World

The conquests of Alexander the Great exported Greek philosophy throughout the known world. When Demetrius of Phalerum, a pupil of Theophrastus, was forced to leave Athens, he had ended up in 307 BC at the court of Ptolemy Soter in the newly founded Egyptian city of Alexandria. There, according to the historian Strabo, he persuaded Ptolemy to found a library and museum on the model of Aristotle's original Lyceum. This state-funded institution was to become, over the succeeding generations, a magnet for scholars from all over the Greek-speaking world and the centre of Hellenistic science.

Although the museum over the course of its long history was not absolutely associated with any one philosophical school, what Burkert describes as 'the cosmos religion and star religion' first articulated in the fifth century BC became, 'in the Hellenistic age, the dominant form of enlightened piety'.[37] Ptolemy, the great Alexandrian astronomer of the second century AD, states in the first book of his *Almagest* that 'this study makes those who follow it lovers of this divine beauty and instils and as it were makes natural the same condition in their soul'.[38] His words are a direct echo of the 'up-in-the-air kind of thought' that had been parodied by Aristophanes some 700 years earlier.

Neither was this kind of theological motivation limited to astronomy. Galen, the great second-century AD doctor who had studied at Alexandria, described his anatomical book on

[36] G. E. R. Lloyd, *Greek Science after Aristotle* (1973), 25–6.
[37] W. Burkert, *Greek Religion* (1985), 329.
[38] G. E. R. Lloyd, *Greek Science after Aristotle*, 115.

The Use of Parts as 'a sacred book which I compose as a true hymn to him who created us', and claims that 'true piety' consists not in sacrificing but in 'discovering . . . and then showing to the rest of mankind his wisdom, his power and his goodness'.[39] Galen follows in a direct line of philosophical and religious tradition that had begun with Aristotle.

The aspects of divinity referred to by Ptolemy and Galen as motivating an interest in the natural world—'divine beauty', 'wisdom . . . power and . . . goodness'—had been continually cited throughout these 700 years. Swimming in this powerful slipstream Greek natural philosophy produced some impressive results, including the accurate calculation of the circumference of the Earth and the distance to the Sun by Eratosthenes, the chief librarian at Alexandria in the third century BC.

The spread of Greek philosophy, however, brought it into contact with a religious tradition that had very different roots. This meeting between Hellenistic science and the three Abrahamic religions—first Judaism, then Christianity, and later Islam—would, in time, radically reframe the way in which the relationship between divine order and the physical world was conceived.

[39] G. E. R. Lloyd, *Greek Science after Aristotle*, 151.

PART III
Encounters in Alexandria

CHAPTER ELEVEN

The Two Students

In 2005 excavations in Alexandria near the intersection of Nabi Daniel and Harriya Streets (a district called Kom el-Dikka) began to uncover a large complex of 20 lecture theatres from the fifth or sixth century AD [Fig. 11.1]. There is to date no inscriptional proof, but it seems almost certain that this was the site of a famous school where at the turn of the fifth and sixth centuries AD two men of different religions, one Christian, the other pagan, had begun their studies.

Both young men had enrolled in the school where in the previous century the philosopher Horapollo (whose father had taught at the famous state museum) had been a dominant figure, and both began to attend the Friday morning seminars of Ammonius, the professor of Platonic philosophy. The atmosphere in these seminars was sometimes fractious. Debates between Christians and pagans could become heated. The argument, however, that developed in later life between these two students was remarkable less for its violence than for its persistence: one way or another, it was destined to continue for more than a thousand years.

11.1 The lecture theatre at Kom el-Dikka.

What made this disagreement distinctive was both the identity of the disputants and the subject of their dispute: it was an argument between two professional philosophers who were arguing not so much about the character of God as about the nature of the world. When the two students took their seats in the seminars, however, they were entering into an argument between pagans and followers of the Abrahamic faiths that had already by this time been going on for centuries, as people throughout the Mediterranean world wrestled with radically different ideas about the character of God and the purposes of human life.

Abraham and the *Arche*

How did two not dissimilar religious perspectives come to have such different views of the natural world?

At first sight the philosophic cosmos religion that had come down from the time of Plato and Aristotle might seem to have much in common with the central themes of Abrahamic monotheism. The philosophers' dismissal of popular mythology (with its explanations of natural phenomena and its stories of the gods' immoral behaviour), and their emphasis on 'The One' and 'The Good', were all ideas that apparently chimed with similar concepts in Judaism, Christianity, and later Islam. However, while these separate traditions of thought might have arrived at seemingly similar conclusions, their journeys had begun from such dissimilar points of origin that any convergence was fraught with difficulties.

While the philosophy of the Greeks and the religion of the Jews had both involved revolutionary new ways of thinking, these revolutions had been of very different kinds. Heraclitus' remark that God was 'both willing and unwilling to be called by the name of Zeus' captures not only the ambiguity of the philosophers' critique of traditional religion but also the sense that such criticism came as it were from the outside. Among the pre-Socratic philosophers only Pythagoras had gone as far as founding his own cult, and these Pythagorean communities (which had been subject to continual persecution) had not survived.

After the death of Socrates most philosophers had been circumspect in their criticisms of traditional religion. In his final book Plato had argued that 'no one with intelligence will undertake to tamper with . . . ancient traditions' which 'have made sacred . . . images, altars and temples',[1] and by and large no one did. The existential core of traditional religion with its prayers, processions, sacrifices, dances, and contests remained untouched. Throughout the Hellenistic world (and later the Roman Empire) ancient images, altars, and temples continued to be maintained by their attendant priests, sacristans, and seers, while new ones continued to be dedicated to the traditional Olympian gods (or their Roman equivalents).

Abrahamic monotheism, by contrast, had from its first beginning involved a radical reshaping of the existential core of religious life. The *Shema*, the creedal proclamation of Israel, which begins 'Hear O Israel the LORD our God is one LORD', is immediately followed by the command to 'love the LORD your God with all your heart and all your soul and all your mind'[2] and a warning against following 'the gods of the people around you'.[3] The struggles which followed from these injunctions (which were later echoed in both Christianity and Islam) became central to the narrative of the religious life of the Jewish nation, and never more so than at times of crisis.

These different journeys had, among other things, given rise to very different conceptions of the relationship between God and the natural world. The prophetic writings from the time of the Babylonian exile (when Judaism had been violently separated from its temple and its cultic setting) had emphasized the radical distinction between the gods of the nations, who could be identified with the sun, moon, and stars, and the transcendent God of Israel, who though passionately involved with his people stood outside the whole realm of nature. In time this new configuration of religious ideas would create a powerful slipstream in which all kinds of penultimate curiosity could swim.

The followers of the various Greek philosophical schools, on the other hand, who had more sought to reinterpret traditional religion than to challenge it, had partly done so by putting a great emphasis on the eternity of the cosmos and the divinity of the heavens.

A Visible God

Because the movements of the planets followed rational mathematical formulae which do not allow change (and over cen-

[1] W. Burkert, *Greek Religion* (1985), 334; Plato, *The Laws* (1970).
[2] Deuteronomy 6:4.
[3] Deuteronomy 6:14.

turies of observation had been seen not to change) it followed, Plato argued, that the cosmos was created after the model of the 'perfect living being' to be imperishable and itself a rational being with a soul and mind.[4] The stars likewise were 'visible, created gods'.[5] Aristotle went further and argued that the cosmos was ungenerated without beginning or end. It would be 'a "dreadful godlessness", *atheotes*, even to regard as possible the destruction of "so great a visible god" '.[6]

The change and decay in earthly existence was understood by arguing that the cosmos divided into two realms, a lower earthly sphere and a higher heavenly sphere. The boundary was the moon, which, though it followed a mathematical orbit, had a changing light (borrowed from the sun as Anaxagoras had argued). Below it in the sublunar sphere, objects move in straight lines and are governed by chance. Above it in the translunar realm the stars and planets are not made from the four earthly elements of earth, water, fire, and air; but from a fifth divine element that is governed by reason and moves in perfect geometrical circles.

This synthesis of religion and philosophy could accommodate the traditional myths by a judicious use of allegory. The Stoic philosophers, in particular, worked this out in some detail, identifying Zeus with the sky, Apollo with the sun, and Artemis with the moon.

As time went by, however, such philosophical approaches began to take on more of the characteristics of a free-standing religion. In the third century AD an Alexandrian philosopher, Ammonius Saccas, and his pupil, Plotinus, established a form of Platonic teaching (now known as Neoplatonism) which presented the study of philosophy as a spiritual curriculum designed to lead the soul back to God. By the turn of the fifth and sixth centuries this curriculum had been broadened in some Neoplatonic schools to include quasi-magical spiritual practices known as theurgy ('divine working'), which sought to aid this process.

While Christianity had by this time become the official religion in the Roman Empire, the Christian emperors for a long time depended on pagans for running the empire. In the fourth century the philosopher Themistius was a leading administrator to six emperors in succession, five of them Christian, and the philosophical schools remained the principle centres of education for Christians as well as pagans.

[4] Plato, *Timaeus and Critias* (2008), 21–2.
[5] Plato, *Timaeus and Critias*, 29.
[6] W. Burkert, *Greek Religion*, 330.

Many of the most strenuous arguments of the time took place within Christianity (between duophysites and monophysites—those who believed Christ had two natures, human and divine, and those who believed he had one), but within these school settings it is not perhaps surprising that relations between Christian and pagan students were often tense. During their time at the Alexandrian school the two new pupils (if their careers overlapped there) had little to do with each other. In later life this changed, though not for the better. In a famous book, one (the pagan) violently attacked the writings of the other (the Christian).

What made this conflict so distinctive was not only the territory over which it was fought but also the weapons that were deployed. While the thinking of each man was grounded in their own religious viewpoint, both shared the assumption that the truths of religion and the truths discovered by reason and observation were part of a single seamless fabric. Hence their dispute was conducted not by pitting one claim to revelation against another, but through a series of closely reasoned philosophical arguments. While the echoes of these arguments continued far into the future, their origin lay in a shared past: in the different responses of the two men to the ethos of the school in which both had been students.

The Alexandrian School

The names of the two students were Simplicius the Cicilian and John Philoponus. Whether or not buildings at Kom el-Dikka were the actual site of their school, the size of this massive complex is an indication of the intellectual status of the city in which they were studying. Alexandria with its immense library and state-funded museum had been the intellectual hub of classical culture for more than 800 years and its pre-eminence as the Oxford or Harvard of its day was underwritten by a longstanding generosity towards scholars.

When a previous professor of Platonic philosophy had died, the city authorities continued to pay a salary to his widow Aedesia. With this backing, Aedesia had taken her two sons to Athens to study with the philosopher Proclus (to whom she had once been engaged). Alexandria's generosity was subsequently rewarded. After the completion of his studies her eldest son, Ammonius (the future tutor of Simplicius and Philoponus), returned to the city and succeeded to his father's academic chair.

Academic posts were not generally passed down within families. In Neoplatonic schools it was though common for appointments to pass from master to pupil in a kind of apostolic succession which reflected the religious character of the curriculum. This curriculum was conceived as a kind of religious 'ascent', which began with the study of Aristotle and led up to the higher reaches of Plato. In the Friday morning seminars Ammonius emphasized to his students that the interpretations he passed on were those of 'our divine teacher Proclus, successor to the chair of Plato',[7] and made clear to them that the purpose of their initial Aristotelian studies was to 'ascend to the common principle (*archê*) of all things and to be aware that this is the one, goodness itself, incorporeal, indivisible, uncircumscribed, infinite'.[8]

When Simplicius and John Philoponus joined the seminars of Ammonius they would have entered into a relationship that was more like that of master and disciple than that of teacher and pupil. Pupils addressed their teachers as 'parents' and in some schools would take an oath of loyalty and wear a special scholarly robe (how such a professor–student relationship would be regarded today is an interesting question[9]). In Alexandria the elevated position accorded to the professor was built into the seminar rooms. In the 20 lecture theatres at Kom el-Dikka the pupils sat on rows of steeped benches that ran around the wall in a horseshoe shape, while at the rounded end the lecturer dominated proceedings from an elevated seat or throne [Fig. 11.2].

This elevated position corresponded to the method of teaching. One of Proclus' works was entitled *Sunanagnôsis*—'reading of a text with a master'—and pupils were expected to closely annotate, and in some cases later publish, the master's interpretations of classic texts as they went through them. A pupil named Asclepius of Tralles published a commentary on Aristotle, *From the voice of Ammonius*, while John Philoponus published four books of *School annotations from the seminars of Ammonius son of Hermeias*. These publications could suggest that the young Philoponus may have had some kind of favoured status within the school. If so, it was despite the background suggested by his name.

[7] H. Ammonius, *Ammonius on Aristotle* (1996), 11.

[8] H. Ammonius, *On Aristotle Categories* (1991), 14.

[9] 'In our present Zeitgeist, dominated as it is by a libertarian, atomistic kind of liberalism, the idea that any individual has responsibility for the moral formation of any other—and especially an adult for a late adolescent who is not her own child—is not only implausible but positively suspect.' N. Biggar, 'What Are Universities For? A Christian View', in *Theology and Human Flourishing*, ed. M. Highton, C. Rowland, and J. Law (2011), 238–50.

11.2 The lecturer's podium at Kom el-Dikka.

John the Grammarian

Philoponus, meaning 'lover of work', was a nickname given to several industrious philosophers, but the *Philoponoi* were also a group of lay Christian workers, associated with the Enaton Monastery near Alexandria, who lived together in a guild called a *philoponeion* [Fig. 11.3]. It is unclear from which of

11.3 The Enaton Monastery.

these backgrounds his name derived, but the Christian name John (*Johannes*) is less equivocal. It suggests that this Philoponus came from a Christian family.

He seems to have been appointed to a post of Grammarian (*Grammatikos*) within the school. He wrote two books on the subject. He may then have switched to studying philosophy. Simplicius sneers that when Philoponus came 'to us' from (the works of) the grammarian Herodian he was not in a position to criticize Aristotle's physics, and that in philosophy he was a late learner. His early works had shown few signs of any explicit dissent from the religious assumptions of his teacher, and this made it all the more shocking to his fellow student when in AD 529 'John the Grammarian' published *Against Proclus On the Eternity of the World*: a book which not only attacked Ammonius' teacher, 'the divine Proclus', but mounted a full-scale assault on one of the most fundamental tenets of Aristotelian philosophy.

CHAPTER TWELVE

The Divided City

Alexandria, by this stage of its history, was not unfamiliar with the clash of religious ideas.

Shortly after Alexander the Great founded this new Greek capital of Egypt around 322 BC, his general, Ptolemy, had imported 120,000 captive Jews from the Greek conquest of Judea. Alexander's architect, Deinocrates, laid out the new city on a grid of parallel streets with attendant subterranean drainage canals, and these grids had been divided up into five districts named after letters of the alphabet: the Greeks in Alpha and Beta, the Jews and other immigrants in the north-east Gamma and Delta districts, and the native Egyptians in Epsilon. Multicultural in its origins, by the time the Romans assumed control, Alexandria had a larger Jewish population than Jerusalem and had become the second largest and the most culturally diverse city in the ancient world.

The city's diversity was a frequent cause of friction and Alexandria became famous for its riots. One of the most serious of these took place in AD 38, when the Jewish philosopher Philo led a deputation to Rome to complain (unsuccessfully) to Emperor Caligula of attacks on the Jewish population by the pagan Alexandrians and the placing of statues of the Emperor in Jewish synagogues. In AD 202 persecutions of Christians under Emperor Severus took place with 'special frequency at Alexandria', according to the Christian historian Eusebius. The 'Didascalium' or Catechetical School was forced to close and the head of it, Clement of Alexandria, reported that Christians were being daily 'burned, impaled, beheaded'.[1] The persecution in Egypt became still worse under Emperor Diocletian when, following a decree in AD 303, churches were demolished, sacred books burned, and Christians, not in government employ, enslaved.

All harassment ceased, however, after Emperor Constantine's edict of toleration in AD 313, and went into reverse in AD 391 when acting on the orders of Emperor Theodosius I,

---

[1] Clement of Alexandria, *The Writings of Clement of Alexandria Volume II*, ed. A. Roberts and J. Donaldson (1869), 70.

the patriarch Theophilus oversaw the destruction of all pagan temples in Alexandria, including the famous temple of Serapis. Twenty-four years later the philosopher Hypatia, the first female head of the Platonic school, was barbarically murdered by a fanatical Christian mob.

Although Hypatia is described as sitting on a high seat (like that at Kom el-Dikka) when she was dragged away and killed, her murder is unlikely to have had anything to do with her teaching. She had become embroiled in a power struggle between the patriarch and the imperial prefect, and it was these local politics that seem to have been responsible for a death which a Christian historian of the time described as bringing 'disgrace upon the church of Alexandria'.[2]

A more positive aspect of Alexandria's diversity, however, lay in the mutual discovery of different cultures.

The Sages of the Nations

In the third century BC Ptolemy II Philadelphus had enlisted seventy-two Jewish scholars to translate the books of the Torah into Greek for inclusion in the Library of Alexandria. This translation, known as the Septuagint, was completed over the following two centuries as the remaining books of the Hebrew bible were added. As educated Greeks became able to read the Hebrew scriptures, so too educated Alexandrian Jews became familiar with Greek literature and philosophy.

The Jewish philosopher Philo, mentioned earlier in connection with the riots, was also the most significant figure to attempt to create an intellectual bridge between the two cultures. When referring to 'God's word' the Septuagint translators had employed the Greek term *logos*, which Heraclitus and other Greek philosophers had used to refer to the rational principle of the universe. Hence, Philo (who was an exact contemporary of Jesus) argued that 'the shadow of God is his *logos* which he used as like an instrument when he was creating the world'.[3]

In discussing the Genesis account of Creation Philo argues that while some of the details are 'symbolical rather than strictly accurate',[4] these are not merely 'fabulous inventions' but means of making ideas visible: 'shadowing forth some allegorical truth'.[5] The goodness of Creation of which Genesis speaks is evident in such things as 'the laws and ordinances' seen in the movement of the planets, 'which God has appointed to

[2] D. C. Lindberg and R. L. Numbers, *God and Nature* (1986).
[3] M. Hillar, *From Logos to Trinity* (2012), 57.
[4] Philo, *The Works of Philo* (1993), 1–52.
[5] Philo, *The Works of Philo*, 157.

be unalterable for ever' and 'accomplished in every instance and every country'.[6] It is by trying to understand the divinely instituted 'causes by which everything is regulated', he argues, 'that philosophy has arisen, from which no greater good has entered into human life'.[7]

This kind of positive attitude to philosophy is echoed in a Jewish commentary on *Lamentations* from around the fifth century AD which affirms that 'there is wisdom among the nations of the world',[8] and in the Babylonian Talmud (compiled between the third and fifth century AD), which affirms 'the science of cycles and planets' as a gift of God and specifies a blessing to be spoken 'on seeing one of the sages of the nations of the world'.[9]

The River of Truth

A similar kind of bridge building is found in much early Christian thought. The first sentences of John's Gospel begin by translating the language of Genesis into the terminology of Greek philosophy:

> In the beginning was the *logos*
> and the *logos* was with God
> and the *logos* was God
> . . . through him all things were made

before proceeding to the dramatic claim that 'the *logos* became flesh and dwelt among us full of grace and truth'. Similarly Justin, an early Christian convert who had studied both Stoic and Platonic philosophy (and after his conversion continued to wear a philosopher's stole until his martyrdom in AD 165), argued that Jesus 'is the *logos* of whom every race of men were partakers: and those who lived with the *logos* are Christians, even though they have been thought atheists; as among the Greeks, Socrates and Heraclitus and men like them'.[10]

Within both the church and the synagogue there were suspicions of the syncretistic direction in which this kind of thought might lead. Philo's bridge-building was not accepted by some in the Jewish community whom he referred to as 'sophists of literalness',[11] while in the third century AD there were those in the Christian community in Africa who felt a similar discomfort. Tertullian, a writer from Carthage, famously asked, 'What

[6] Philo, *The Works of Philo*, 53.
[7] Philo, *The Works of Philo*, 53.
[8] Quoted in J. Sacks, *The Great Partnership* (2011), 351.
[9] J. Sacks, *The Great Partnership*, 351.
[10] A. Cleveland Coxe, *The Ante-Nicene Fathers* (1885), 178.
[11] Philo, *The Works of Philo*, 391.

can there be in common between Athens and Jerusalem, between the Academy and the Church?'[12] 'What is there in common between the philosopher and the Christian, the pupil of
Hellas and the pupil of heaven?'[13] His answer was that 'we have
no need of curiosity since Jesus Christ nor of enquiry since the
evangelist'.[14]

Clement of Alexandria, Tertullian's exact contemporary, gave
a very different answer to that same question. Clement argued
that 'philosophy is characterised by investigation into truth and
the nature of things',[15] and that while 'the way of truth is one',
into it, 'as into a perennial river, streams flow from all sides'.[16]
Hence as all truth proceeds from God, 'philosophy is in a sense
a work of divine providence'.[17] Rather as St Paul described the
Torah as a schoolmaster to bring the Jews to Christ, so philosophy might be a schoolmaster to bring 'the Hellenic mind'
to Christ.

In future centuries this principle was to be a critical factor
in determining the strength of the slipstream in which penultimate curiosity could swim, and as head of the Didascalium in
Alexandria, Clement acted on it. Alongside theology, the syllabus included mathematics, medicine, the humanities, and all
the Greek philosophers except Epicurus. Clement's own *Pro-
trepticus*, or 'Exhortation to the Greeks', followed the tradition
established by Justin, using an encyclopaedic knowledge of
Greek literature and philosophy to both point out the absurd-
ities of Greek mythology and draw attention to those places
where Greek philosophers come close to Christian doctrines.

When the school reopened after the Severan and Diocletian
persecutions, this broad tradition continued under a series of
remarkable heads, including Origen and Didymus the blind.
One of Origen's pupils reported that 'he required us to study
philosophy by reading all the existing writings of the ancients,
both philosophers and religious poets.... For us there was
nothing forbidden, nothing hidden, nothing inaccessible. We
were allowed to learn every doctrine, non-Greek and Greek,
both spiritual and secular, both divine and human; with the utmost freedom we went into everything and examined it thoroughly.'[18] Didymus, despite losing his sight at the age of 4, was
reputed to be the most learned man of his day, mastering all the
Greek sciences. Fifteen hundred years before Louis Braille he
reputedly devised a method of wooden lettering that allowed
blind students at the school to read and write.

[12] Tertullian, *Quinti Septimi Florentis Tertulliani Opera* (1890), Book 7.
[13] Tertullian, *Quinti Septimi Florentis Tertulliani Opera*, Apologeticum 46.
[14] Tertullian, *Quinti Septimi Florentis Tertulliani Opera*, De Praesciptionibus haereticorum Book 7.
[15] A. Cleveland Coxe, *The Ante-Nicene Fathers*, 366.
[16] A. Cleveland Coxe, *The Ante-Nicene Fathers*, Book 5.
[17] A. Cleveland Coxe, *The Ante-Nicene Fathers*, 349.
[18] M. L. Clarke, *Higher Education in the Ancient World* (1971), 126–7.

Although as late as AD 398 the Council of Carthage was still prohibiting the reading of pagan books, even in North Africa it was the Alexandrian view that ultimately prevailed. When Augustine became Bishop of Hippo in AD 396 he argued that it was 'disgraceful and dangerous' for infidels to hear a Christian 'talking nonsense' on topics like 'the motion and orbits of the stars ... their size and relative positions, about the predictable eclipses of the sun and the moon, the cycles of the years and seas ... the kinds of animals shrubs, stones and so forth'.[19] Instead he argued that the truths discovered by pagan philosophers should be appropriated rather than ignored: 'if those who are called philosophers, especially Platonists, have said things which are well accommodated to our faith, they should not be feared; rather what they have should be taken from them as unjust possessors and accommodated to our use'.[20]

Philosophy and Legends

This Alexandrian accommodation sometimes led to a conscious fudging of issues. Synesisus was a scholar who had studied with Hypatia in Alexandria and continued to the end of his life to address her in letters as 'mother, sister, teacher'. When he was being pressed to become Bishop of Ptolemais, he confessed to a crisis of conscience. He could not renounce his philosophical belief in the eternity of the world or the pre-existence of the rational soul, nor accept that the resurrection was anything other than a 'sacred and mysterious allegory'. However, he argued that 'the ordinary man' had little in common with philosophy, and while 'Divine truth should remain hidden ... the vulgar need another system'. Consequently he thought he might accept the office on condition that he could 'prosecute philosophy at home and spread legends abroad'.[21] Synesius seems to have got away with this, but pagan philosophers who made some accommodation to their Christian pupils faced criticism from both sides.

Damascius, a pagan colleague of Ammonius, accused the latter of having compromised with the Christian authorities for financial gain. A Christian called Zacharias Scholasticus, on the other hand, complained that Ammonius' *phrontisterion* was being frequented by young men whom Ammonius was converting from Christianity to Hellenism, 'for he is clever in ruining their souls by removing them from God and the

[19] Augustine, *Ancient Christian Writers* (1982), 41–4.
[20] Augustine, *On Christian Doctrine* (1958), 75.
[21] Synesius, *The Letters of Synesius of Cyrene* (1926).

truth'.[22] Zacharias had himself attended Ammonius' seminars
and describes heated classroom debates. In one of these 'many
of those present in the class at the time . . . leaned towards our
arguments or more correctly towards Christianity'. In another
debate, Ammonius 'asked his students to leave the auditorium
so that they would not be persuaded by the argument and convinced to be Christians again'.[23]

What side did the young Philoponus take in these disputes?
In one of his early books of 'school annotations' he draws particular attention to the fact that Ammonius thought that Aristotle believed in a creator God, but in these books Philoponus
also seems to accept the pre-existence of a rational soul and to
endorse Aristotle's idea of a fifth 'eternal' element. It was not
until the publication of *Against Proclus* in AD 529, by which
time Ammonius had retired or died, that Philoponus began to
articulate his own distinctive Christian philosophy.

The School of Athens

Philoponus' fellow pupil Simplicius had left Alexandria some
years before this took place. Over 40 years earlier, in AD 485,
a riot had taken place at the school. In the course of this a
Christian, who had spoken slightingly of his teacher, had been
beaten up by a group of pagan students and rescued by some
student *philoponoi*. The backlash that followed had forced Ammonius' colleague Damascius to leave Alexandria and to go to
Athens, where around AD 520 he became head of the Athenian Platonic Academy. After leaving Alexandria, Simplicius had
gone to study with Damascius in Athens.

In contrast to the uneasy accommodations that took place in
Alexandria, the Academy in Athens had a history of forthright
opposition to Christianity.

When Ammonius' teacher Proclus first arrived in Athens, he
earned the admiration of the resident philosophers (including
presumably Plutarch, the then head of the school) by taking off
his shoes and worshipping the moon goddess 'In plain sight of
them'.[24] According to his pupil Marinus, Proclus would worship the sun 'at dawn, noon and dusk'.[25] He was also an advocate
of 'theurgy'. He practised 'Chaldean prayer meetings' alongside 'the art of moving divine tops' and wrote a booklet about
witnessing 'apparitions of Hecate under a luminous form'.[26]
His advocacy of philosophical religion was not limited to this

[22] R. Sorabji, *Aristotle Transformed*
(1990), 240.
[23] R. Sorabji, *Aristotle Transformed*, 240.
[24] Marinus of Samaria, *The Life of
Proclus or Concerning Happiness*, trans.
K. S. Guthrie (1925), 26.
[25] Marinus of Samaria, *The Life of Proclus or Concerning Happiness*, 39.
[26] Marinus of Samaria, *The Life of Proclus or Concerning Happiness*, 45.

devotional practice. Among his philosophical writings were texts directed against the claims of Christianity.

Proclus' book of 18 arguments, *On the Eternity of the World*, does not explicitly mention Christianity but pointedly argues that anyone who denies the eternity of the world (as Christians did) is being 'irreverent in the extreme'.[27] On at least one occasion he was forced to leave Athens. Finally in AD 529 Emperor Justinian closed down the school in Athens altogether.

Seven philosophers, including Damascius and Simplicius, went into exile, but not without hope of return. In their writings the philosophers express the conviction that Christianity would be no more than a passing phase. They seem to have acted on this assumption. Excavations of the 'philosopher's house' at Athens in 1975 discovered statues which must have adorned the building, carefully hidden in a well, concealed in anticipation perhaps of better times to come [Fig. 12.1]. They were never reclaimed.

12.1 Statue found in the well of the philosopher's house in Athens.

[27] Philoponus, *Against Proclus: 6–8* (2005), 2.

CHAPTER THIRTEEN

Industrious Jack

The appearance of John Philoponus' *Against Proclus* in the same year as the closure of the school in Athens may not have been coincidental.

Unfortunately, both the prologue and the epilogue of the book are missing, so that we don't have Philoponus' own account of what prompted him to write. Various tenth-century Muslim writers suggested either that he only pretended to disagree with Aristotle 'so as to not suffer the fate of Socrates', or that he was paid by Christians to make these attacks on pagan philosophers. There is, however, nothing in the text as it stands that would give any reason to think that its author was insincere. Its method nevertheless is highly distinctive.

Proceeding from Premises

Although the book is addressed to Christians, Philoponus quotes scripture rarely and only by way of illustration. Instead he suggests that the best kind of refutations are those 'which have proceeded from premises . . . accepted beforehand by one's opponents'[1] and this is the technique he adopts. As he works through each of Proclus' 18 arguments Philoponus merits his nickname, exhaustively demonstrating, through close philosophical analysis, that in no cases do Proclus' conclusions follow from his premises. There is only one premise that he cannot accept, even for the sake of argument, and that is the dogma that Plato and Aristotle never disagree.

Neoplatonic philosophers tended to treat Plato and Aristotle's writings as sacred texts which must always be harmonized (though both Proclus and his teacher Syrianus often reject Aristotle when he explicity disagrees with Plato). Philoponus, by contrast, constantly quotes Plato's exhortation to 'pay little heed to Socrates and more to the truth',[2] and Aristotle's remark that although Plato is dear to him 'it is pious to value the truth more highly'.[3] Philoponus does not generally trade insults but

---

[1] Philoponus, *Against Proclus: 6–8* (2005), 118.
[2] Philoponus, *Against Proclus: 6–8*, 103.
[3] Philoponus, *Against Proclus: 6–8*, 35–6.

he does accuse Proclus of being disingenuous in glossing over such obvious divergences between the two philosophers as Plato's belief that the world had a beginning and Aristotle's assertion that it did not.

Starting from his opponent's premises did not, however, prevent Philoponus from challenging some of Proclus' and Aristotle's basic assumptions.

The fundamental argument of Proclus' book is the assertion that the creation must be like the creator: if the creator was infinite and everlasting so also must be the creation. It is not enough for Philoponus to suggest that there is something arbitrary in this assertion. Taking Aristotle's own ideas as his point of departure, Philoponus reminds us that Aristotle himself had argued that there could never be an actual infinite number of things (because there could always be one more) and that you could never pass through an infinite number of things (because it would never be possible to arrive at the end). But 'if the world had no beginning', argues Philoponus, 'and the number of men living before, say Socrates, was infinite', then given that we have somehow arrived at Socrates 'the infinite has become traversable, which is impossible'.[4] What is more if 'those living from Socrates to the present time have been added to it there will be something greater than the infinite', which also 'is impossible'.[5] This was a contradiction at the heart of Aristotle's definition of an eternal universe that had gone unnoticed for 800 years.

Physical Explanation

Although *Against Proclus* was his first openly controversial book, it was not the first time that Philoponus had shown himself willing to challenge Aristotle's ideas about the nature of the universe. In a commentary on Aristotle's *Physics*, a part of which was written around AD 517,[6] Philoponus had already challenged the whole Aristotelian theory of motion. Aristotle's answer to the question of why a javelin continued to move after it had been thrown, or why an arrow continued to fly after it had been shot, was that this was a peculiar property of air. Assuming that something must impart motion and that mover and moved must be in contact, he had argued that the air displaced in front of a projectile rushes round in a kind of vortex and pushes it forward from behind.

Philoponus thought this absurd. If it were true, he said, an army would not need to touch their javelins but could balance

[4] Philoponus, *Against Proclus: 6–8*, 24.
[5] Philoponus, *Against Proclus: 6–8*, 24.
[6] In a forthcoming book (2016) Richard Sorabji argues that this date cannot apply to the whole text.

them on a parapet and set them in motion with a thousand bellows. In practice this method does not work. Instead he suggested that some sort of 'incorporeal motive *enérgeia*' (which in the Middle Ages came to be called 'impetus') must be imparted by the thrower directly into the javelin.

On this hypothesis the air slowed down movement rather than causing it, and there was no reason why movement could not take place in a vacuum. Aristotle had argued that a vacuum cannot exist, but in the same commentary in a section called 'a corollary on the void' Philoponus argues from the evidence of a kind of pipette called a *clepsydra* ('or "snatchers" as people call them here'[7]) that the void has a definite force. In the same section he contradicts Aristotle's assertion that a boulder falls faster than a stone with the observation (later repeated by Galileo) that 'two unequal weights dropped from a given height strike the ground at almost the same time'.[8]

Throughout *Against Proclus*, Philoponus shows a preference for what he calls 'the most scientific (*phusikos*, meaning here physical rather than metaphysical) explanation of these things'.[9] Thus where Aristotle had thought that the spheres were moved ultimately 'by desire', Philoponus compares their rotations to the 'mechanism' (*mêchanêma*) by which water inside a centrifuge is 'whirled around in a circle'.[10] He replaces Aristotle's abstract notion of 'prime matter' with his own more concrete concept of 'three-dimensional extension'.

He also contests Aristotle's idea that while the earth is composed of four elements, the heavens are made of a fifth 'divine' element—'the aether'—which always moves in a circle. Instead he argues that 'the heaven is composed of the same elements that the sublunar living things are constituted of'.[11] This last argument was taken much further in the sequel to *Against Proclus on the Eternity of the World* written a few years later: *Against Aristotle on the Eternity of the World*. This was a book that though immensely influential has only survived in the fragments quoted from it by Philoponus' opponents, most notably by his fellow student Simplicius.

Simplicius in Exile

A Greek historian called Agathias recorded the names of the seven philosophers who left the Academy in Athens: 'Damascius the Syrian, Simplicius the Cicilian, Eulamias the Phrygian, Pro-

[7] Philoponus, *On Aristotle Physics 4.6–9* (2012), 12.
[8] Philoponus, *Corollaries on Place and Void* (1991), 59.
[9] Philoponus, *Against Proclus: 6–8*, 23.
[10] Philoponus, *Against Proclus: 12–18*, 30.
[11] Philoponus, *Against Proclus: 12–18*, 48.

scarius the Lydian, Hermias and Diogenes both from Phoe-
nicia, Priscanius the Lydian . . . the most noble flower of our
times'.[12] Agathias goes on to say that these seven left the Byzantine Empire altogether, and found their way to the court of
the Persian philosopher-king Chosroes.

In Persia, however, the philosophers soon found that life
under a barbarian monarchy did not suit them. When Justinian made a treaty with Chosroes in AD 532, the philosophers
managed to secure a coda which stated that 'they could return to their accustomed haunts and pass their lives without
fear among themselves'.[13] This they seem to have done, though
their actual movements are unknown. One attractive theory
argues that they settled in a Neoplatonic school in Harrân,
near the Persian border (where a tenth-century Muslim trav-
eller found an inscription from Plato on the door knocker of a
building that had been a school or academy).

Wherever he ended up, Simplicius certainly had access to a
library and was able to write, and it was in this period that he
produced his great commentaries on Aristotle in which he included the attack on his fellow pupil.

Simplicius announces his forthcoming assault in the prologue
to his commentary on Aristotle's *De Caelo* (On the Heavens).
His attack, he claims, is not motivated by any personal hostility
to a man whom he tells us three times 'to my knowledge I have
never seen'[14] (which is not quite the same as never having done
so). Philoponus, in fact, is never named. He is referred to only
as 'the grammarian' or 'this individual' (as well as by a variety of
insulting epithets: 'novice', 'raven', 'jackdaw who crows against
the Divine bird of Zeus'). Proclus had regarded Christianity as
a religion 'for the multitude and vulgar people (*hoi polloi*)',[15]
and in a similar vein Simplicius argues that Philoponus' books
address 'the uneducated . . . who always take pleasure in unusual things'.[16] Even reading them is like falling 'into Augeas'
dung'.[17] His own purpose, by contrast, is to maintain Aristotle's
'reverential conception of the universe'.[18]

Simplicius' religious motivation is evident throughout his
commentary, which ends with a prayer to the 'creator of the
whole universe'. He is particularly shocked by Philoponus'
apparent lack of reverence in suggesting that the heavens are
composed of the same material as the 'sublunar sphere'.

Philoponus had pointed out that the transparency and brilliance of the heavens were also properties of some terrestrial

12 Agathias, *The Histories* (1975), 80.
13 Agathias, *The Histories*, 80.
14 Philoponus, *Against Aristotle* (1987), 39.
15 R. Sorabji, *Philoponus and the Rejection of Aristotelian Science* (1987), 60.
16 Philoponus, *Against Aristotle*, 39.
17 Philoponus, *Against Aristotle*, 86.
18 Philoponus, *Against Aristotle*, 40.

bodies. Transparency could be found in 'air, water, glass and certain stones', brilliance in 'fire … fireflies, heads or scales of fish and other similar things'.[19] In response, Simplicius uses Philoponus' own method (using the premises of one's opponents) by quoting from the Bible. 'Even this David, whom he honours so much, teaches the contrary', argues Simplicius, 'for he says that "Heaven declares the glory of God" … but not the "fire-flies" and the "scale of fish" '.[20]

Agathias records that after their return from Persia, Simplicius and the other six philosophers 'spent the rest of their lives in the most agreeable and pleasant manner'.[21] Philoponus' fortunes in Alexandria seem to have been more mixed. *Against Proclus* and *Against Aristotle* were followed by a (now lost) treatise *On the Contingency of the World*. His attempt in this trilogy to establish a distinct Christian approach to philosophy in the Alexandrian School seems though to have had few followers.

He remained as *grammatikos* but was never appointed to a chair of philosophy. Ammonius was succeeded as *scholarch* (head of the institution) first by the mathematician Eutochius and then by the philosopher Olympiodorus, who robustly maintained the pagan tradition of the school. Philoponus, who at some point wrote the oldest surviving treatise on the astrolabe [Fig. 13.1], went on to write two commentaries on an introduction to arithmetic by Nicomachus, but after his trilogy produced no further controversial texts. When he did return to the themes of these earlier works it was in the context of a book of theology.

13.1 The Anti-Kythera Mechanism: a geared astronomical calculating machine from the first century.

---

[19] Philoponus, *Against Aristotle*, 74.
[20] Philoponus, *Against Aristotle*, 75.
[21] Agathias, *The Histories*, 81.

CHAPTER FOURTEEN

The Creation of the World

De Opificio Mundi—'On the Creation of the World'—is a commentary on the Genesis account of Creation. In his preface to the book Philoponus explains how he came to write it.

The work is dedicated to his friend Bishop Sergius (later the patriarch of Antioch) and to his pupil Athanasius (a nephew of the Empress Theodora). People, he writes, have for some time been putting gentle pressure on him not merely to refute the pagan eternalists but positively to defend the Mosaic account of Creation, but 'you', Philoponus says to Sergius, 'pressed on me heavily, urging me, and almost forcing me to contribute my best effort to the cause'.[1] What did this mean?

After the closure of the school in Athens it is not impossible that Philoponus was under pressure from fellow Christians to justify the continued existence of the Alexandrian School. Certainly throughout his commentary he gives the impression of someone fighting on two fronts. Thus, while one thrust of the book is to refute the criticisms of Genesis made by his pagan philosophical colleagues, a larger part is directed against the kind of interpretations of Genesis put forward by some fellow Christians.

The crux of his dispute with the former concerned the nature and purpose of God's relationship with the physical world. The heart of his argument with the latter turned on the nature and purpose of God's Holy Scriptures.

Were the Genesis stories addressing the same kind of questions as the pagan myths, which described Poseidon shaking the earth and Helios driving his chariot around the sky, or were they pointing to a meaning and agency beyond the horizon of the visible world? Back in the first century the Jewish philosopher Philo had described Jewish opponents who looked to the Scriptures for scientific information as 'sophists of literalism'. Five centuries later Philoponus takes as his opponent a contemporary whom he never names but who can almost certainly be identified as an Alexandrian monk known as Cosmas Indicopleustes.

[1] R. Sorabji, *Aristotle Transformed* (1990), 258–59.

Christian Topographia

Indicopleustes (meaning 'who travelled to India') had been a merchant in a former life and had translated his voyages into some of the earliest world maps. These were incorporated into his lavishly illustrated *Christian Topographia*, which, as the title implies, was very much more than a simple atlas. Thus, the first section of the book is headed 'Against those who while wishing to profess Christianity think and imagine like the pagans that the heaven is spherical'.[2] Indicopleustes rejects what he regards as 'the pagan' notion of a round earth within a spherical heaven, arguing instead that the earth is flat and that the heavens arise above it like a box (divided into an upper and lower story) surmounted by the canopy of heaven that he likens to the vaulted ceiling of a bathroom [Fig. 14.1].

Following a theologian called Theodosius of Mopsuestia, Indicopleustes argues that the tabernacle made by Moses in the wilderness was in fact 'a pattern of the whole world'.[3] Taking the most literal possible reading of scriptural texts (as, for instance, Isaiah referring to the earth as 'God's footstool'), Indicopleustes rejects the theories of all those outside the church as 'fictitious, fabulous sophistries', and describes a Christian 'of . . . great learning' (who sounds like Philoponus) as having been 'blinded by his craving for distinction' in taking a different view. 'Wishing to speak against the Pagans' he had 'agreed with them that heaven is a sphere . . . always revolving',[4] failing to recognize that it is the angels who 'move the luminaries and the stars'.[5]

14.1 A page from Cosmas Indicopleustes' *Christian Topographica* showing the vaulted heavens above a flat earth.

[2] C. Indicopleustes, *Kosma Aigyptiou Monachou Christianikē Topographia* (1897), 7.
[3] C. Indicopleustes, *Kosma Aigyptiou Monachou Christianikē Topographia*, 5.
[4] C. Indicopleustes, *Kosma Aigyptiou Monachou Christianikē Topographia*, 274.
[5] C. Indicopleustes, *Kosma Aigyptiou Monachou Christianikē Topographia*, 323.

Kinetiké Dunamis

Philoponus introduces his commentary by trying to establish what kind of questions Genesis is and is not addressing. His conclusion is that the purpose (*skopos*) of the book is not to provide a scientific cosmogony, but 'to teach the knowledge of God to benighted Egyptians superstitiously worshipping the sun, moon and stars',[6] and this is a theme he returns to several times in the course of the book.

The pagan philosophical colleagues who have attacked Philoponus 'hold the opinion that Moses devoted himself to Physics without agreement with phenomena',[7] but in his own philosophical work he has shown (through 'syllogisms of all kinds and labyrinths hard to negotiate for the reader'[8]) that the world did indeed have a beginning. This is indeed asserted in Genesis (where 'in the beginning', he argues, refers to the beginning of time and space), but beyond this assertion 'the fact of God's creation is revealed, but not how it all came about'.[9] The metaphors of Scripture are intended to lift us close to God and cannot be pressed for mechanical information.

Hence there is a place for scientific investigation. Following the principle established by Clement he argues that 'anyone honouring what is true by whomever it may be found, honours Christ the truth',[10] which leads him to the slogan 'let nothing in any manner get in the way of truth'.[11]

In practice this means that if you want to know whether the earth is flat or round and whether the heavens are box-like or spherical you need to make some observations. Thus, he recommends the reader to take their stand in a high place immediately after sunset on a clear but moonless night. They should make a note of the stars that appear near the eastern horizon, near the western horizon, and in the middle of the sky. They should then return shortly before dawn, repeat the observations and note what has changed.

While in his understanding of the heavens Philoponus was a disciple of Ptolemy, the greatest Alexandrian astronomer, he was not an uncritical one. Ptolemy had been a convinced astrologer (his *Tetrabiblos*—a four-part study of astrology—was almost as famous as his astronomical *Algamest*), but Philoponus ends the fourth book of his commentary by arguing that an important vindication of the truth of Christianity lay in its requiring a renunciation of astrology.

[6] Philoponus, *Joannis Philoponi De Opificio Mundi Libri* (1897), i, 1; iii, 42; iv, 17; R. Sorabji, *Philoponus and the Rejection of Aristotelian Science* (1987), 51.

[7] Philoponus, *Joannis Philoponi De Opificio Mundi Libri*, i; B. Elweskiöld, *John Philoponus against Cosmas Indicopleustes* (2005), 17.

[8] Philoponus, *Joannis Philoponi De Opificio Mundi Libri*, i, 7–9.

[9] Philoponus, *Joannis Philoponi De Opificio Mundi Libri*, iii, 13; R. Sorabji, *Philoponus and the Rejection of Aristotelian Science*, 51.

[10] Philoponus, *Joannis Philoponi De Opificio Mundi Libri*, iii, 3.

[11] Philoponus, *Joannis Philoponi De Opificio Mundi Libri*, iii, 17.

In all his writings Philoponus demonstrates a bias towards what Ptolemy had described as explaining phenomena by the 'simplest hypothesis possible'. Thus, Indicopleustes' suggestion that angels move the stars is dismissed out of hand. Did they pull or push? The Scriptures, he points out, made no such claims. But conventional pagan theories as he had pointed out in his philosophical books were equally unsatisfactory. Aristotle had, in effect, split dynamics into a series of unrelated realms. Thus, while the movements of projectiles were explained by the behaviour of air, the movements of the heavens were explained as being like animals which moved 'by desire'. Each of the four elements moved up and down according to their inner nature, while the fifth element moved in a circle of divine perfection.

In his earlier writings on vacuum Philoponus had defended the use of thought experiments, called 'hypotheses', to explore conceptual scenarios.[12] Now in his commentary on creation he proposes another kind of thought experiment. In view of God's omnipotence, instead of thinking of angels pulling and pushing, it was better to ask, 'could the sun, moon and stars not be given by God their creator a certain kinetic force (*kinetiké dunamis*) in the same way that light and heavy things were given their trend to move?'[13] In a few lines he then extends this theory to every remaining category of motion from the movement of the elements to that of the animals. If, in other words, the theory of impetus that he had propounded in his philosophical books was extended to encompass the heavens, all Aristotle's various theories of motion could be replaced by a single unified theory.

John Labour's Lost

After completing his commentary on Genesis, in further theological writings Philoponus attempted to bring a similar philosophical clarity to two of the most vexed theological issues of the time: the nature of Christ and the nature of the Trinity. This was dangerous territory. With respect to the former he argued that Christ had a single nature—a position known as monophysitism (as opposed to the dyophysitism which held that Christ had two natures, human and divine). With regard to the latter he argued for an understanding of the Trinity that its critics described as 'tritheism'.

[12] Christopher Martin, 'Non-reductive arguments from impossible hypotheses in Boethius and Philoponus', *Oxford Studies in Ancient Philosophy* 17 (1999), 279–302.

[13] Philoponus, *Joannis Philoponi De Opificio Mundi Libri*, i, 28–9.

His belief was that the majority of those who contended about such issues were 'only opposed in words to each other', while he held it 'to be a feature of the piety of lovers of truth that each of them can introduce matters which unite the separation caused by ... controversial language'.[14] In this he was disappointed. Both of the positions he argued for were later condemned, and in 680, a century or so after his death, the Third Council of Constantinople placed Philoponus' writings under a formal anathema. In a pun on his name they described him as *mataiponos* 'man whose labour is lost'.

The pun was prophetic. Not only were Philoponus' theological writings ignored in the succeeding generations, but the odium that they brought to his name nearly consigned all the rest of his writings to oblivion. But not quite. Through indirect sources his idea of impressed force and his arguments against the eternity of the world both continued to exercise an influence.

More importantly for our theme was a repetition of the pattern that he illustrated.

As the tides of history ebbed and flowed in different parts of the world in the following centuries, the knowledge of Greek philosophy was recurrently swept away and recovered. When the latter occurred, the Abrahamic religions tended to alternate between a complete acceptance and an outright rejection of this inheritance. And in the midst of this alternation a kind of integration began once again to take place.

As had occurred at Alexandria, it was discovered that the tools provided by philosophy could be used to criticize philosophical assumptions, and to separate what was seen as the baby of Greek science from what now seemed to be the bathwater of Greek religious ideas. In this process, the rejection of the idea that the heavens were divine sat alongside a continued conviction that the world was nevertheless divinely ordered. The result was the emergence of a new stream of penultimate questions and the striking re-discovery of Philoponus' concept of a unified dynamics: his idea that there was a God-given law which governed both the heavens and the earth.

Thus, while at the end of the sixth century the whole culture that had produced both Simplicius and Philoponus was about to be swept away, the argument that these two former classmates had initiated was (in a series of different guises) just about to begin.

[14] U. M. Lang, *John Philoponus and the Controversies over Chalcedon in the Sixth Century* (2001).

PART IV
The Long Argument

CHAPTER FIFTEEN

The House of Wisdom

The siege which swept away the Greek Christian culture of Alexandria lasted six months. The army of Caliph Umar had arrived in front of the great walls of the city in March. By September the head of the Coptic Church was suing for peace. By the following year the Muslim conquest of Egypt was complete.

In the Christian calendar this was AD 642. According to the new calendar that the caliph had just instituted, it was the year 20. Umar took as year 0 the *Hijra*—the Prophet Muhammad's flight to Medina. In the ten years between the *Hijra* and his death, Muhammad had united the whole of the Arabian peninsular. In the ten years following his death (632 according to the Christian calendar) the Arabian tribes, united under the banner of Islam, had overrun the majority of the Persian, Sassanian, and Byzantine empires, rewriting the political map of the world.

Within the next 50 years this political earthquake had begun to effect a profound cultural transformation. Around 690 Caliph Abd-al-Malik decreed that in this vast multicultural and multilinguistic empire only Arabic should be used in official documents. Within a generation Greek, which had been the lingua franca of the Middle East for a thousand years, had been supplanted not only as the medium of daily discourse but as the language in which every kind of theological, philosophical, and scientific debate would now be conducted.

Sixty years later the Abbasids, descendents of the Prophet's uncle Abbas, who were based in Iraq, overthrew the Umayyad Caliphate. In 762 they transferred their capital from the ancient site of Damascus to their newly built cosmopolitan city of Baghdad. Situated at the geographical centre of the empire the new city was open to influences from every point of the compass. The stage had been set for a novel and radically international phase in the struggle to integrate different ways of thinking.

As Far as China

The Arab caliphs' encounters with the ideas of the ancient civilizations they now ruled paralleled Muhammad's own encounters with Judaism and Christianity over a century earlier. These were people, the Prophet said, whom Muslims should regard as 'those who were given the book before you'.[1] Nevertheless 'the Torah and the Gospel which is revealed to you by your Lord'[2] could be accommodated within the new revelation because he asserts (in a repeated phrase) 'to Allah belongs all that the heavens and the earth contain. He has knowledge over all things'.[3]

It is in the light of this overarching perspective that in several *hadith*—sayings of the Prophet recorded by his companions—Muslims are enjoined to seek knowledge. Zirr bin Hubaish recorded, 'I heard the messenger of Allah say "There is no one who goes out of his house in order to seek knowledge, but the angels lower their wings in approval of his action." '[4] One saying with multiple attestations has it that 'seeking knowledge is an obligation upon every Muslim',[5] and another enjoins believers to 'seek knowledge even as far as China'.[6] Within this slipstream all sorts of penultimate curiosities could potentially swim, and while the authenticity of this last saying began to be questioned in the twelfth century, in eighth-century Baghdad it was followed to the letter.

The Round City

In the walls of the round city that stood at the centre of Baghdad were giant gates at each quadrant, through which roads ran to the four quarters of the great Abbasid Empire. From its earliest days the ruling families of the city demonstrated a remarkable openness to the ideas that travelled along these routes. The *majalis* or salons instituted by the Barmakid family that advised successive caliphs encouraged philosophical and religious debates. In these debates Christians, Jews, and Zoroastrians participated alongside Muslim scholars, and in this milieu Greek, Persian, and Sanskrit texts first began to be translated into Arabic.

The importation of Chinese papermaking technology transformed this situation. The establishment in Baghdad of a paper *wiraqah* or 'manufactory' in 794, together with the development

---

[1] Koran, 5. 67.
[2] Koran, 5. 67.
[3] Koran, 4. 125.
[4] *Sunan ibn Majah*, Chapter 1, *The Book of Sunnah*, Hadith 226.
[5] Al-Tirmidhi, Hadith 74 (related by Ibn 'Adiyy, al-Bayhaqi, and al-Tabarani).
[6] Anas by al-Bayhaqi in *Shu`ab al-Iman* and *al-Madkhal*, Ibn 'Abd al-Barr in *Jami' Bayan al-'Ilm Fadlih*, and al-Khatib al-Baghdadi through three chains at the opening of his *al-Rihla fi Talab al-Hadith*.

of new rapid Arabic calligraphy, had an effect not unlike the development of the Greek alphabet. Libraries were built, a book trade developed, and by the beginning of the ninth century the translation of foreign texts had become a state-sponsored project [Fig. 15.1].

It was not only literature that was translated. A Persian scholar, al-Khwarzimi, wrote a book on *The Hindu Art of Reckoning*, introducing Indian numerals, the use of zero, and decimal points, followed by *al-Kitab al-mukhtasar fi hisab al-jabr wa'l-muqabala* (The Compendious Book on Calculating by Completing and Balancing), which introduced the idea of algebra to solve linear and quadratic equations.

Al-Khwarzimi, according to a contemporary account, 'was employed full time . . . in the service of al-Ma'mun'[7] in the latter's famous Bayt al-Hikma: The House of Wisdom.

15.1 Scholars at an Abbasid library: illustration to the 'Maqamat' of Abu Muhammad al-Qasim Hariri (1054–1121).

The Houses of Wisdom

When the Abbasid caliph al-Mansur built Baghdad in the heart of a Persian population and near their ancient capital of Ctesiphon, he co-opted and adapted different elements of this surrounding culture. In Sassanian Iran, books of Persian law, cast into poetry, were deposited in storehouses (*hazā'in*). These were the *buyūt al-hikma*, 'the houses of wisdom'.

The Sassanian empire of Persia had lasted for some four hundred years (226–663). Its state religion was Zoroastrianism and it saw itself as the successor of the mighty Achaemenid empire that had been conquered by Alexander the Great. Alexander's conquest had destroyed much of the literature of Zoroastrianism, including portions of the Avesta, their sacred text. The Sassanian emperors had set themselves the task of collecting, recording, and editing the historical remains of this culture. Central to this enterprise was the task of translating back into Persian the scattered fragments of their literature.

It was an article of faith for Zoroastrianism that Zoroaster himself was the origin not only of religious but also of scientific and philosophical truth. Hence a Sassanian religious text, the Denkard, argues that 'since the root of all knowledge is the teaching of the Religion . . . should he, who speaks wisely, present [his knowledge] to men all over the world . . . his utterance, then, ought to be considered an exposition of the Avesta, even though he has not had it from any revelation of the Avesta'.[8]

[7] D. Gutas, *Greek Thought, Arabic Culture* (1998), 58.
[8] D. Gutas, *Greek Thought, Arabic Culture*, 37.

This principle and this practise could both be adapted to the ideology of the new Islamic empire. Following the construction of his capital al-Mansur invited delegations of scholars from India to come and share their knowledge and, according to a tenth-century historian, 'was the first caliph to have books translated from foreign languages into Arabic'.[9] His successor, Harun al-Rashid, expanded the work of translating foreign texts and it was extended still further by al-Rashid's son, the caliph al-Ma'mun.

Much of the actual work of translation was done by Christians, and at the head of one of the two teams of al-Ma'mun's translators was Hunayn ibn Ishaq, a Nestorian Christian. The Nestorians were dyophysite Christians who, having been denounced as heretical in 451, had, like the Neoplatonist philosophers, relocated to Persia. 'The Sheik of translators', Hunayn was responsible for translating some 116 works into Arabic. To further this project al-Ma'mun dispatched expeditions of scholars as far afield as Constantinople and Egypt to seek out texts.

The role of the House of Wisdom in this project is unclear. It was certainly a repository for documents and as an official bureau of state it may have helped to institutionalize the work of translation. When, however, al-Ma'mun appointed mathematicians and astronomers to its staff it seems to have taken on a different kind of role: he commissioned them to produce a new map of the world, to check the measurement of the circumference of the Earth, and to establish an astronomical observatory.

The Structured Heaven

The Qur'an poses a question to unbelievers: 'Have they not looked at the heaven above them—how we structured it and adorned it and [how] it has no rifts?'[10] Beyond this general encouragement of sky-watching the Qur'anic instructions with regard to both the timing and the direction of prayer provided Muslims with a specific motivation for astronomical study.

The five daily times of prayer commanded to believers required accurate timekeeping, and before the era of clocks this meant accurate astronomical observation. During the years in Medina the *qibla*—the direction of prayer—had changed from facing towards Jerusalem to facing towards Mecca. The instruction to 'turn towards the holy mosque, wherever you be face towards it'[11] henceforward was to be a defining mark of the new religious identity. In the vast empire, establishing

[9] D. Gutas, *Greek Thought, Arabic Culture*, 30.
[10] 'Surah 50: Qaf', Saheeh International, https://archive.org/details/Quran-SaheehInternationalTranslationEnglish, accessed 7 August 2015.
[11] Koran, 2. 144.

this direction involved tricky problems of spherical geometry and accurate observation of stellar reference points. In mosques across the Islamic world the *Muwaqqits* or official timekeepers needed to be competent astronomers. In ninth-century Baghdad such competency was taken to a new level.

Under the direction of Caliph al-Mansur a Persian trained astronomer, Ibrahim al-Fazari, had translated the Indian astronomical text by Brahmagupta known as the *Sindhind*. Al-Mansur's successor, the caliph Harun al-Rashid, instructed al-Fazari to make the first known Arabic astrolabe [Fig. 15.2]. Under al-Rashid's patronage astronomical observations began to be made at Shamasiyyu near Baghdad. His son, al-Ma'mun, commissioned astronomers from the House of Wisdom to build astronomical instruments and to establish Shamasiyyu as a fully fledged observatory. Al-Ma'mun ordered the construction of a similar observatory on Mount Qasioun at Damascus. This was followed over the centuries by a succession of ever larger and more spectacular observatories with giant sextants and quadrants, throughout the Islamic world [Fig. 15.3].

15.2 An early Arabic astrolabe.

15.3 The astronomer Takiuddin at his observatory at Galata 1581.

Catalysed by the translation of Ptolemy's *Algamest*, al-Khwarizmi produced the first major written work of Muslim astronomy, the *Zij al-Sindhind*. Although the whole translation project had taken in texts from all around the world, it was the translation of Greek texts into Arabic that produced the most profound stimulus and challenge to Islamic thought. The injunction to 'seek knowledge' meant that they could not be ignored; but how could they be integrated with the Qur'anic revelation? In the struggle towards such integration the argument that had started in Alexandria between Simplicius and Philoponus began to be revisited.

CHAPTER SIXTEEN

The Dream of Aristotle

The Bookseller of Baghdad

When a tenth-century Baghdad bookseller, Ibn al-Nadim, compiled an index 'of the books of all nations Arab and non Arab alike which are extant in the Arabic language' his text mentions some 70 translators. His index is divided into ten discourses, six on Islamic subjects and four on secular ones. When he comes to the latter, he gives as a reason why 'books on philosophy and other ancient sciences became plentiful in this country': a dream in which Caliph al-Ma'mun saw a man with 'broad forehead, joined eyebrows, bloodshot eyes ... sitting on his bed'. When asked who he was the man replied, ' "I am Aristotle" ' [Fig. 16.1].[1]

In a later account of the dream Aristotle assured the caliph that there was no conflict between reason and revelation. He urged him to devote the resources of the state to translating Greek philosophy into Arabic because 'knowledge has no boundaries; wisdom has no race or nationality. To block out ideas is to block out the kingdom of God'.[2]

Probably the earliest account of the dream, however, has al-Ma'mun asking Aristotle, 'O philosopher, what is the best speech?', to which Aristotle replies, 'whatever is correct according to personal judgement'.[3] In this telling, the import of the dream is not about embracing wisdom beyond the boundaries of the empire, but refers rather to a vexed issue within Islamic thinking.

Kalam

As Arabic became the medium for every kind of philosophical and theological debate so it began to develop its own specialized vocabulary to accommodate this new task. The word kalam, for instance, meant simply word or speech, but as the Qur'an was designated kalam Allah—the word of God—so

[1] M. Ibn al-Nadīm, The Fihrist of al-Nadīm (1970), 583.
[2] E. Masood, Science and Islam (2009), 57.
[3] D. Gutas, Greek Thought, Arabic Culture (1998), 97.

16.1 Aristotle teaching Alexander from the Kitāb al-hayawan.

'ilm al kalam was 'the science of the word' and the term kalam came to describe the discipline of seeking theological principles through dialectic (in later Islamic thought a scholar of kalam was known as a *mutakallim*). In its early stages kalam in this sense may have been essentially a response to debates with Christians and others, but before long these kind of reflections opened up internal arguments within Islamic thought itself.

One of the first of these debates appears to have started in eighth-century Basra, when a scholar called Wasil ibn Ata is said to have withdrawn himself from the instruction circle of his teacher over the issue of the status of a Muslim who sinned. This gave rise to the Mu'tazila school of kalam (the word *i'tazala* means to separate or withdraw, which may be the etymology of the name). This school argued for the superiority of *'aqli* (reasoned faith) to *naqli* (traditional faith), and maintained that reason should be the final arbiter in distinguishing right from wrong.

Though regarded as heretical by some traditionalists it was this Mu'tazilite theology that was adopted by al-Ma'mun (who in his latter years imposed it by force). After some two hundred years in which few traces of philosophical writing survive, the Nestorian Patriarch Timothy is said to have translated Aristotle's Topics for the caliph al-Mansur (the significance of which may have been that it includes useful advice on how to argue a case). From this the appetite for philosophical and scientific texts rapidly grew, and during al-Ma'mun's caliphate, Islamic religion began to seriously engage with the legacy of Greek philosophy.

Whatever the provenance of Ibn al-Nadim's version of the dream story, it does accurately describe the attitude of one of the leaders of the caliph's two translation groups. This was the man whom the bookseller describes as 'the philosopher of the Arabs', Abu Yūsuf Ya'qub al-Kindī [Fig. 16.2].

16.2 Al-Kindī.

The Philosopher of the Arabs

Al-Kindī was an extraordinary polymath who is reported by Ibn al-Nadim to have written some 300 books, the majority of which appear to have covered a vast range of scientific topics including 11 on arithmetic, 23 on geometry, 8 on spherics, 19 on astronomy, 13 on meteorology, and 22 on medicine. Alongside al-Khwarzimi he championed the introduction of Indian numerals into Arabic and pioneered the study of cryptography. His most famous book, however, was the first Arabic philosophical text: *On First Philosophy*.

Aristotle is the hero of this book. Al-Kindī quotes his remark that 'we must thank the fathers of those who brought some truth, since they [the fathers] were a cause of their [sons'] being',[4] and echoing Clement the first section argues that 'we must not be ashamed to admire the truth or to acquire it, from wherever it comes. Even if it should come from far-flung nations and foreign peoples ... all are ennobled by it'.[5]

[4] al-Kindī, *The Philosophical Works of al-Kindī* (2012), 12.
[5] al-Kindī, *The Philosophical Works of al-Kindī*, 12.

In a treatise that gives an overview of the entire Aristotelian corpus he argues that the purpose of Aristotle's metaphysics was 'to affirm the oneness [*tawhīd*] of God ... to explain his beautiful names and that he is the agent cause of the universe'.[6] This did not prevent al-Kindī from voicing occasional disagreements. In the second section of *On First Philosophy* he attacks Aristotle's idea of the eternity of the world, and in doing so reproduces an argument of Philoponus' *Against Aristotle*, which he appears to have known (though without mentioning Philoponus or drawing attention to the fact that these were not the views of Aristotle).

The Baghdad Aristotelians

Although in charge of one translation group al-Kindī did not himself read Greek. Hunayn ibn Ishaq, the leader of the other group, was succeeded by his son Ishaq ibn Hunayn and in the next generation it was these Nestorian Christians who pushed forward the project of rendering the entire Aristotelian corpus into Arabic. Among these was Abu Bishr Matta ibn Yunus, who received an Aristotelian education at the monastery of Mar Mari and, though not a Greek speaker, translated Aristotle from Syriac versions. Among Matta's pupils was the man who became known to Islamic intellectuals as 'the second teacher' (Aristotle being the first): Abu Nasr al-Farabi.

Like al-Kindī, al-Farabi was a prolific author, producing some 117 known works on subjects that varied from scientific, philosophical, and sociological topics to a book on music. He differed from his predecessor in recognizing that the subject of Aristotle's metaphysics was being in itself rather than God. God appeared in this perspective as the principle of absolute being, and al-Farabi followed the Neoplatonists in arguing that this First Cause, overflowing in a series of cascading emanations of necessity, produced an eternal universe.

Al-Farabi was aware that this view had been criticized in antiquity. In the introduction to his book *Against John the Grammarian* (Yahya al-Nahwi) he suggests that 'one may suspect that [Philoponus'] intention ... in refuting Aristotle is either to defend the opinions laid down in his own religion ... or to remove from himself the [suspicion] that he disagrees with the position held by the people of his religion and approved by

---

[6] al-Kindī, *The Philosophical Works of al-Kindī*, 295.

their rulers'.[7] A similar charge was repeated a generation later by the great Ibn Sina, known to the west as Avicenna.

The Flying Man

By the time that Avicenna was born (around AD 980) in a village near Bukhara, the Abbasid Empire had begun to fragment into a number of competing centres of power, and much of his life was spent moving between different patrons. Avicenna was a child prodigy who had memorized the Qur'an by heart at the age of 10, learned Indian mathematics from an Indian greengrocer, studied medicine ('no hard and thorny science', according to his autobiography), qualified as a doctor at the age of 18, and found fame in this role when he cured the Samanid Emir from a dangerous sickness. The multivolume *al-Qanun fi'l-tibb* (The Canon of Medicine), which he went on to compile, systematizing all the medical knowledge of the time, remained a standard textbook for the next 700 years, ensuring that this fame would last [Fig. 16.3].

As a teenager he had struggled to understand Aristotle's *Metaphysics*, reading it many times and praying at the mosque for illumination. In his autobiography he tells us that light finally dawned when he bought al-Farabi's little commentary at a market stall for three dirhams, and in thanksgiving gave alms to the poor.

In a text called *Kitab al-Shifa* (The Book of Healing) (which is not, in fact, about medicine) he developed al-Farabi's insight

16.3 Medical information about the skull, lungs, stomach, and heart, from *Canon of Medicine* (*al-Qanun fi'l-tibb*), by Avicenna (Ibn Sina).

[7] M. Mahdi, 'Alfarabi against Philoponus', *Journal of Near Eastern Studies* 26 (1967), 233–60.

that the subject of *Metaphysics* was 'being', distinguishing between 'contingent being' (which could exist or not exist without contradiction, and whose existence was due to something other than itself) and 'necessary being' or 'necessary of existence through itself'—wajib al-wujud bi-dhatihi—whose existence was due to itself. Since the idea of an infinite regress of causes was impossible, necessary being (which had to be one, and could have neither counterpart nor opposite) must exist and pointed towards the existence of God. He devised a famous 'flying man' thought experiment, in which he argued that a man suspended in air and deprived of all sensory experience would still affirm his own existence. He argued that this pointed to the soul as being something whose essence did not depend on the body.

Avicenna's concept of 'necessary being' seemed (like the ideas of al-Farabi) to chime more closely with Aristotle's concept of an eternal universe than with the Qur'anic picture of voluntary creation. He was not, however, uncritical of Aristotle, preferring the idea of *'mayl* (imparted impetus) to Aristotle's theory of motion, and advancing the notion of *tajriba* (experimentation) over Aristotelian induction, which he claimed 'does not lead to the absolute universal and certain premises it purports to provide'.[8] It was, however, his dependence on Aristotelian physics that was questioned by a polymath with a range and brilliance that were equal to Avicenna's own.

Questions and Answers

Al-As'ila wa-l-Ajwiba (Questions and Answers) was a book in which a Persian mathematician and astronomer, Abu Rahyan al-Biruni, published his correspondence with Avicenna. It consisted of 18 questions which al-Biruni had posed to 'The wise one', Avicenna's replies, al-Biruni's further set of questions objecting to these replies, and the responses that Abu Said al-Ma'sumi, a star pupil of Avicenna, then gave to these objections.

Al-Biruni was born and lived the first part of his life in Khwarazm (the modern Khiva) in Persia, where he studied mathematics, astronomy, and physics. Among his achievements was an accurate estimate of the Earth's circumference. In later life he extended this range (his most famous book *India* has been described as the first anthropological study), but he always considered himself more of an empirical investigator than a

[8] J. McGinnis, 'Scientific Methodologies in Medieval Islam', *Journal of the History of Philosophy* 41 (2005), 307–27.

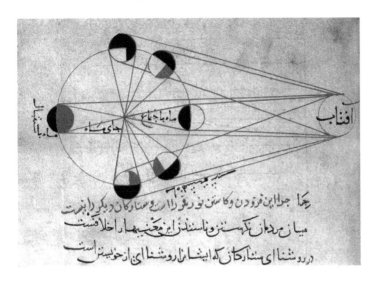

16.4 Different phases of the Moon, from *Kitab al-Tafhim* by Abu Rayhan al-Biruni.

philosopher [Fig. 16.4]. Hence he argued that the metaphysical axioms on which philosophers built their physical theories did not constitute valid evidence for the mathematical astronomer. Having performed numerous experiments ('I broke as many flasks as would be enough to hold the water of the [river] Jayun'[9]), he concluded there is no valid evidence that rules out the possibility of a vacuum. Likewise he argued that there is no inherent physical reason why planetary orbits should be circular rather than elliptical.

Al-Biruni's first 10 questions all related to Aristotle's *De Caelo*. The second question he asks is why Aristotle gave such weight to traditional ideas about the eternity of the world since 'anyone who is not stubborn and does not insist on falsehood would agree that this is not a known fact'. Though more diverse in their range, his 18 questions may deliberately echo Proclus' 18 arguments for the eternity of the world and Philoponus' 18 refutations. Avicenna remarks that 'it is as if you had taken this objection from John Philoponus who was opposed to Aristotle simply because he himself was a Christian'. He then goes on to point out that whoever read Philoponus' other books would discover his agreements with Aristotle.[10]

Al-Biruni questions this suggestion: 'I think that you, O wise man, have not seen [Philoponus'] book on the Response to Pericles [*sic*] in which [the latter claimed] that the world is eternal, nor his book on what Aristotle embellished, nor his commentaries on Aristotle'. In fact, he argues, 'John Philoponus

[9] R. Berjak and M. Iqbal, 'Ibn Sina–al-Biruni Correspondence (5)', *Islam & Science* 3 (2005), 57–63.

[10] R. Berjak and M. Iqbal, 'Ibn Sina–al-Biruni Correspondence (2)', *Islam & Science* 1 (2003), 253–60.

is far from [deserving] to be called mischievous; Aristotle, the embellisher of his own infidelity, is more deserving of this description'.[11]

This more critical attitude to Greek philosophy was shared by a remarkable contemporary of Avicenna and al-Biruni who lived most of his life in Egypt, Ibn al-Haytham.

The Mad Philosopher

Ibn al-Haytham's most famous work was his *Kitab al-Manazir* (The Book of Optics). *Thesaurus Opticus*, as it became known in the West, exercised an influence that reached all the way from Roger Bacon and De Witelo in the thirteenth century, to Kepler and Descartes in the sixteenth and seventeenth centuries. In the book Ibn al-Haytham combined the geometrical tradition of Euclid, who thought that visual rays came out of the eye in straight perpendicular lines (the so-called 'extramission theory'), with the medical tradition of Galen, and Aristotle's idea that light travelled into the eye (the theory of 'intromission') [Fig. 16.5]. His influence derived as much from the means by which he arrived at his conclusions as from their content.

16.5 Diagram of the eyes and related nerves, MS illustration from *Kitab al-Manazir* (Book of Optics) by Ibn al-Haytham.

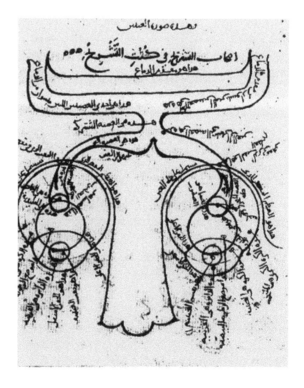

[11] R. Berjak and M. Iqbal, 'Ibn Sina–al-Biruni Correspondence (5)'.

Ibn al-Haytham was born in Basra. According to a thirteenth-century story, he travelled to Egypt to persuade the Fatimid caliph al-Hakim to build a dam to regulate the flooding of the Nile. After surveying a site just short of Aswan, he realized the impracticality of his proposal. To avoid punishment from the unstable al-Hakim he feigned madness and was kept under house arrest from 1011 until the caliph's death in 1021.

If this story is true, it might help to explain Ibn al-Haytham's wariness of intellectual hubris. In a brief autobiography he describes how after examining the claims of many religious sects in his youth he had been won over by mathematics and Aristotelian philosophy. Intellectual enquiry as he understood it was central to the religious life: 'I constantly sought knowledge and truth, and it became my belief that for gaining access to the effulgence and to closeness to God there is no better way than that of searching for knowledge and truth'.[12] An early (now lost) essay was entitled 'All Matters Secular and Religious Are the Fruits of the Philosophical Sciences'.

'Finding the truth', however, as he wrote in later life, 'is rough. God has not preserved the scientist from error'.[13] Instead of proceeding from first principles and metaphysical axioms on the classical model, *Thesaurus Optics* is composed throughout of systematically described experiments combined with geometrical proofs. It was this combination of experimentation and mathematics that later exercised such a profound influence in the West, providing, as it did for Ibn al-Haytham himself, a perspective from which to criticize the writings of the Greeks.

In a book called *Al-Shukuk 'ala Batlamyus* (Doubts about Ptolemy) he suggests that 'truths are immersed in uncertainties' and that scientific authorities were not immune from error. Therefore, he argues,

the seeker after truth is not one who studies the writings of the ancients and following his natural disposition puts his trust in them, but rather the one who suspects his faith in them and questions what he gathers from them, the one who submits to argument and demonstration, and not to the sayings of a human being whose nature is fraught with all kinds of imperfection and deficiency. Thus the duty of the man who investigates the writings of the scientists . . . is to make himself an enemy of all that

[12] *Global History of Philosophy*, vol. 2, ed. John C. Plott (2000), 465.
[13] A. I. Sabra, 'Ibn al-Haytham: Brief life of an Arab Mathematician: Died circa 1040', *Harvard Magazine* (2003), 54–55.

he reads and, applying his mind to the core and margins of its content attack it from every side. He should also suspect himself as he performs his critical examination of it, so that he may avoid falling into either prejudice or leniency.[14]

(A similar observation was made in the twentieth century by the Nobel laureate Richard Feynman, 'The first principle is that you must not fool yourself—and you are the easiest person to fool.'[15])

Some 50 years after Ibn al-Haytham's death, this kind of suspicion was to take a more radical turn.

[14] A. I. Sabra, 'Ibn al-Haytham'.
[15] R. P. Feynman, *Surely You're Joking, Mr Feynman!* (1985), 343.

CHAPTER SEVENTEEN

Al-Ghazali's Pilgrimage

In 1095 the head of the Nizamiyyah college in Baghdad, Abu Hamid Muhammad ibn Muhammad al-Ghazali, resigned his teaching posts, gave away his goods, and set out on a pilgrimage, a quest for religious certainty, during which he vowed never again to serve the political authorities or teach in state-sponsored schools. The year before he had published his *Tahâfut al-Falâsifa* (The Incoherence of the Philosophers) in which he analyses a series of philosophical doctrines. Although he implicitly accepts many of them, he nevertheless seeks to show that the doctrines of the eternity of the world and denial of miracles do not satisfy the philosophers' own criterion of rational demonstration.

The Islamic assimilation of Greek philosophy—*falsafa*—reached a high point in the tenth and twelfth centuries. Despite the support of ruling families this assimilation had never been uncontroversial. In the last years of his life Caliph al-Ma'mun had established the *Mihna* (Inquisition), which interrogated and imprisoned scholars who refused to accept the rationalist Mu'tazilite theology that the caliph favoured. The legacy of this inquisition was a deep hostility towards Mu'tazilite theology. Many theologians opposed it in private, but already in 912 a forty-year-old Mu'tazilite theologian had publicly rejected the Mu'tazilite doctrines of his teacher, al-Dhubbai, and embraced a more traditional orthodoxy.

Al-Ash'ari

According to some later accounts Abu al-Hasan al-Ash'ari had acted in response to three visions he had experienced during Ramadan in which the Prophet had called him to adhere to tradition. Because the orthodox disapproved of any form of kalam (rational argument) al-Ash'ari was ready to abandon this also, but (according to the story) in the third vision he was told to adhere to tradition but not abandon kalam.

Whatever the truth of these stories, this was the path that al-Ash'ari pursued. In the handful of texts that survive (from between the 100 and 300 that he is reported to have written) he uses philosophical arguments to defend the reality of the Qur'anic picture of God. Contesting the idea of an impersonal absolute whose only attribute was essential being, which Mu'tazilite theologians argued lay behind it, he argued that while the anthropomorphic expressions in the Qur'an which refer to God's hand and face were (as the Mu'tazilites said) not corporeal and literal, they nevertheless pointed to real attributes of God whose precise nature was unknown.

At the same time he argued equally forcefully against the literalist Zahirite school. In a small treatise *Istihsan al-Khaud* he writes that

> a section of the people made capital of their ignorance; discussions of rational matters became a heavy burden for them and therefore they became inclined to blind faith and blind following (*taqlid*). They condemned those who tried to rationalize the principles of religion as 'innovators'. They considered discussion about motion rest, body, accident, colour, space, atom, the leaping of atoms and the attributes of God to be an innovation and therefore a sin.[1]

Al-Ash'ari's response is that since the Prophet discusses none of these things, condemning such discussion is itself an 'innovation'.

The Incoherence

Al-Ghazali's treatise, written some two hundred years later, adopts a similar strategy to this Ash'arite theology. Using the tools of philosophy he mounts a full-scale assault on the hubris of the philosophers' claims to have established the nature of God and the universe by demonstrative proofs that are inherently superior to prophetic revelation.

In a preface to the book al-Ghazali makes it clear that his examination will not obscure the philosophers' 'solid achievements'. With regard to matters like lunar eclipses 'he who thinks it is his religious duty to disbelieve such things is', he argues, 'really unjust to religion and weakens its cause. For these things have been established by astronomical and mathematical evidence which leaves no room for doubt'.[2] Far from helping religion 'if you tell a man who has ... sifted the data that these things are contrary to religion you will shake his faith in religion not in these things'. Theologians should not commit themselves in these matters: 'what we are interested in is the product of God's creation, whatever the manner of the action may be'[3,4].

[1] Abū al-Ḥasan ʿAlī ibn Ismāʿīl al-Ashʿarī, *Risālat Istiḥsān al-khawḍ fī ʿilm al-Kalām* (1925).

[2] al-Ghazzālī, *al-Ghazali's Tahafut al-Falasifah* (1963), 6.

[3] al-Ghazzālī, *al-Ghazali's Tahafut al-Falasifah*, 8.

[4] Al-Ghazali's argument here precisely parallels Augustine's, who seven centuries earlier had written, 'Usually, even a non-Christian knows something about the earth, the heavens, and the other elements of this world, about the motion and orbit of the stars and even their size and relative positions, about the predictable eclipses of the sun and moon, the cycles of the years and seasons, about the kinds of animals, shrubs, stones, and so forth, and this knowledge he holds to as being certain from reason and experience. Now, it is a disgraceful and dangerous thing for an infidel to hear a Christian, presumably giving the meaning of Holy Scripture, talking nonsense on these topics; and we should take all means to prevent such an embarrassing situation, in which people show up vast ignorance in a Christian and laugh it to scorn.' Augustine's *The Literal Meaning of Genesis: An Unfinished Book*, trans. J. H. Taylor, S.J. (1982), Book 1, Chapter 19, Section 39.

Al-Ghazali argued that some twenty of the philosophers' propositions concerning the nature of God and the universe rested on unproven premises. Of these the first that he examines (and treats in most detail, repeating some of the arguments in *Against Proclus* and *Against Aristotle*) is the question of the eternity of the world. While insisting that true demonstrations cannot conflict with revelation, here and elsewhere he points out that such demonstrations only apply to quite restricted phenomena. The Qu'ranic statement that 'you will not find any change in God's habit' implies that God acts in an orderly manner (in his later writings he compares the workings of the universe to those of a water clock) does not provide grounds, he argues, for denying God's freedom or thinking that all his actions are necessary.

Instead he suggests that even in an orderly world miracles are possible, either by God's direct intervention or by natural processes not understood: 'in the treasury of things that are enacted by [God's] power there are wondrous and strange things that one hasn't come across'.[5]

In his autobiography *The Deliverance from Error* al-Ghazali describes his wanderings after leaving Baghdad, and how he eventually found the certainty he was seeking in the mystical experience of God pursued by the Sufis (he eventually founded a Sufi convent in his home town of Tûs) [Fig. 17.1]. Yet despite his departure from public life the influence of his criticisms of *falsafa* in the eastern empire was immense. The principal voice raised in defence of the classical Aristotelian positions came not from Baghdad, but from the new centres of Islamic scholarship that had begun to develop in the West.

17.1 Last page of al-Ghazali's autobiography, *The Deliverance from Error*, in MS Istanbul, Shehid Ali Pasha 1712, dated AH 509 (AD 1115–16).

A Palm Tree in Rusafa

The coup that had brought the Abbasids to power in AD 750 had involved the massacre of the entire Umayyad family. A single teenager, Abd al-Rahman, survived. After a dramatic escape which involved swimming the Euphrates and trekking through Egypt and north Africa pursued by Abbasid soldiers, he finally reached Spain (where some 40 years earlier invading Berber tribes from Africa had established Islamic rule). There he succeeded in making himself emir and established his capital in the city of Cordoba.

Al-Rahman's ambition was to recreate the Umayyad court in Andalusia. To that end when he built the Rusafa palace

5 al-Ghazzālī, *The Incoherence of the Philosophers* (Tahāfut Al-Falāsifah): A Parallel English-Arabic Text (2000), 222.

outside the town he imported Syrian plants to fill the gardens, including a palm tree (and in a famous poem compared himself to this exiled plant). More critically he established the basis of one of the largest libraries in the world that made Cordoba a centre of culture and scholarship. This cultural programme continued through the following centuries and by the beginning of the eleventh century the intellectual centre of gravity had begun to shift from Baghdad to Islamic Spain.

In 1090 a Berber tribe from Morocco called the Almoravids conquered Cordoba, ending the reign of the Umayyad caliphs. The Almoravids espoused a comparatively puritan interpretation of Islam but this did not prevent the intellectual culture established by the Umayyads from flourishing as never before. Ibn Bajja, who served as vizier to an Almoravid governor, was known as a poet, musician, and astronomer, and also as the first Andalusian philosopher. He was later referred to by Latin writers as Avempace. Among the many commentaries Avempace wrote on Aristotle was one on his *Physics* in which he refers to the commentary of *Yahwa al Nawi* (John the Grammarian) and once again puts forward a theory of impetus based on Philoponus' writings. According to some accounts, Avempace was murdered with a poisoned eggplant. Many of his writings were subsequently lost. His commentary on the *Physics* was preserved by a man who may at one time have been his pupil: Ibn Rushd, known to the West as Averroës.

A judge and doctor as well as a philosopher, Averroës was a more classical Aristotelian than Avempace. He defended Aristotle's theory of motion and responded to al-Ghazali's *The Incoherence of the Philosophers* with a book entitled *The Incoherence of the Incoherence* (*Tahâfut al-Tahâfut*), in which he insisted that Aristotle had demonstrated the eternity of the world so that on al-Ghazali's own principles it must be considered a religious as well as a philosophical doctrine. Because 'the truth cannot contradict the truth' and philosophers had the greatest insight in to what the truth was, their interpretations of sacred texts had, he argued, the greatest validity.

Something not dissimilar was argued by a contemporary of Averroës, who like him also undertook the task of integrating Greek philosophy with a scriptural tradition. This was a man known to the Muslims as Musa ibn Maymun, to his own community as 'Rambam' (an acronym of 'Rabbi Mosheh ben Maimon'), and to posterity as Moses Maimonides.

Rambam

Maimonides, like Averroës, had grown up in Cordoba. In 1130 a Berber tribe called the Almohads seized Andalusia from the Almoravids. The Almohads (reversing a previous policy of toleration) announced a policy of 'no synagogue, no church' and gave the Jews the choice of conversion or exile. Maimonides' family chose exile. After a spell in North Africa they ended up in Egypt where he became both court physician to the sultan Saladin and chief rabbi of the Jewish community.

This ability to bridge cultures was reflected in his major works [Fig. 17.2]. Thus while the *Mishneh Torah* is an encyclopaedic codification of Jewish law and ethics, his *Dalalatul Ha'irin* (The Guide for the Perplexed) is addressed to those religious persons who while adhering to the Torah 'have studied philosophy and are embarrassed by the contradictions between the teachings of philosophy and the literal sense of the Torah'.[6]

In the introduction he argues that God could only communicate the profundity of Creation 'using allegories, metaphors and imagery' and much of Book One is taken up with showing, by close textual analysis, that the anthropomorphism in the biblical text (speaking of 'God's hand' and so on) was never intended to be taken literally. On the one hand, he argues that 'we should examine the scriptural texts by the intellect, after having acquired a knowledge of demonstrative science'.[7] On the other, he maintains that we can only say what God is not,

17.2 A manuscript of Maimonides'.

[6] M. Maimonides, *The Guide for the Perplexed* (1956), Introduction 2.

[7] M. Maimonides, *The Guide for the Perplexed*, Book II, 29.

and that the purpose of theological language is not to define God but 'to conduct the mind towards the utmost reach that man may attain in the apprehension of him'.[8]

This ability to bridge cultures which Maimonides demonstrated in Egypt was paralleled by Averroës back in Spain.

A New Universe

The latter part of Averroës' career had been spent as court physician to the Almohad caliph Yusuf. When Yusuf complained on one occasion about the difficulty of understanding Aristotle, it inspired Averroës to embark on three series of commentaries (each of a different level of difficulty) which covered the entire Aristotelian corpus. Although it used to be thought that Averroës' work signalled the decline of Islamic philosophy, more recent scholarship points to an explosion of philosophical activity in the twelfth–thirteenth centuries, often in the context of kalam, and a revival of interest in Greek philosophy in the subsequent Safivid period.[9]

Although Averroës' commentaries had much less influence in the Islamic world than the writings of al-Ghazali or Avicenna, in the Latin West they were to provide a critical introduction to a new universe.

A hundred years earlier in 1085 the city of Toledo, whose library rivalled that of Cordoba, had fallen to the Christian king Alfonso VI. Almost immediately Greek texts, which had long ago been translated into Arabic, began for the first time to be translated from Arabic into Latin, as Western Christian scholars set out to repair a cultural deficit that had lasted for more than 500 years.

[8] M. Maimonides, *The Guide for the Perplexed*, Book I, 58.
[9] P. Adamson, *Classical Philosophy: A History of Philosophy without Any Gaps*, vol. 1 (2014). Volume 2 is due for publication in 2016; http://www.historyofphilosophy.net, accessed 7 August 2015.

CHAPTER EIGHTEEN

A Tale of Two Cities

The origins of the loss of Greek literature in the Christian West go back to a bitter winter in AD 406. In that year the Rhine had frozen and German tribes poured across the river into the territories of the Roman Empire. Four years later Rome itself was sacked by the Goths. In 455 it was sacked again by the Vandals. In 476 the young Romulus Augustulus, the last Emperor of Rome, was forced to abdicate by the barbarian king Odvocar. For centuries educated Romans had been fluent in Greek, but the barbarian tribes that now carved up the Western Empire spoke their own languages. Latin survived as the language of the Church, but the knowledge of Greek quickly faded.

At the beginning of the sixth century the Christian philosopher Boethius (a Roman contemporary of John Philoponus) had embarked on the project of translating the entire works of Aristotle into Latin [Fig. 18.1]. He had produced a number of short treatises on arithmetic and music, and had just completed the translation of some of Aristotle's logical treatises, when he was arrested by the Ostrogoth king Theodric (who suspected him of plotting with Constantinople). In prison Boethius just managed, before his execution, to write his masterpiece *The Consolation of Philosophy*, but no more. The translation project died with him, and for the next 500 years the few texts that he had produced constituted almost the sum total of all that the Western Christian world knew of Greek thought.

18.1 A medieval image of Boethius from Boethius' *De institutione musica*.

The Poverty of the Latins

The fall of Toledo in 1085 rapidly changed this situation. The city had a substantial population of Arabic-speaking Christians who collaborated with Latin-speaking scholars to produce translations, at first of mainly mathematical and astronomical texts. In the next century the city began to attract scholars from all over Europe. The most famous of these was Gerard of Cremona, who, it was reported, made his way to Toledo where 'seeing the abundance of books in Arabic on every subject and pitying the poverty he experienced among the Latins concerning these subjects'[1] he set himself to learn Arabic and to translate. Gerard went on to translate scores of texts, including many of the works of Aristotle, as well as those of al-Kindī, al-Farabi, and Avicenna.

At the beginning of the next century some of these texts formed the core curriculum of the teaching institutions that had begun to develop.

The education of European nobles had, up to this point, mostly taken place in royal courts. Those destined for the church or administrative posts had been educated in cathedral or monastic schools, while professions like lawyers or doctors had taught their own apprentices. In 1158, however (12 years after Gerard had first arrived in Toledo), Emperor Barbarossa granted the law school in Bologna the right to become a self-governing corporation. Other schools followed suit in seeking this kind of self-governing status. In 1200 the University of Paris was formed from the monastic schools on the left bank and the cathedral school at Notre Dame, when the King of France bestowed on them a similar recognition. Within the next two decades universities emerged at Oxford and at Cambridge and were given royal charters (thus preparing for the education of both authors). Over the course of the thirteenth century similar institutions were established all over Europe.

These new self-governing universities became a magnet for young men who would previously have found their education elsewhere. Enjoying a measure of independence from both church and state, they had (almost inadvertently) planted the seed of a new concept: an idea of education that stood at a slight remove from a direct instrumental purpose and open to the pursuit of a more free-ranging curiosity. They provided a context within which the ideas found in the texts translated by Gerard and others could begin to be explored.

The rapid influx of young men and novel ideas, however, produced both social and intellectual tensions in the cities and towns where these new universities had been established that could on occasion explode into violence.

Oxford and Paris

In December 1209 the city of Oxford experienced the first of what was to be a recurrent erruption of social tensions

[1] C. Burnett, 'The Coherence of the Arabic–Latin Translation Program in Toledo in the Twelfth Century', *Science in Context* 14 (2001), 249–88.

between town and gown. After a furious row, a student who
had been living in a house-share with two others murdered
his mistress (a local Oxford girl) and fled from the city. Faced
with an angry mob of locals, the magistrates seized the two
remaining students and summarily hanged them. In protest at
this high-handed action (the students should have been subject to church law), seventy masters and hundreds of students
promptly decamped, leaving the city to cope with the economic consequences.

In Paris the same year, intellectual tensions turned lethal
when the exhumed corpse of a university lecturer and ten of
his living followers were burned at the gates of the city. A few
years earlier this same teacher, whose name was Amalric, had
been attracting excited crowds to his lectures on Aristotle in
the theology faculty. In these he had advanced the novel idea
that *omnia sunt deus* ('all things are god'). Amalric's heretical
ideas had been condemned in 1204, but even after his death
his teachings continued to be promulgated, and a year after the
burnings the Bishop of Paris decreed that 'neither the works of
Aristotle on natural philosophy, nor their commentaries are to
be read at Paris in public or private'.[2]

A result of the rapid spread of the new teaching institutions
was that when problems arose in one place people could move
elsewhere. Thus, many of the scholars who left Oxford settled
in the small fenland town of Cambridge where a new university was initiated. When the teaching of Aristotle was banned in
Paris, the new University of Toulouse produced a flyer adver-
tising its own lectures on Aristotle's *Libri Naturales* 'which has
been prohibited in Paris'.[3]

This sort of competition between institutions provided a
strong motive for a university to resolve its own conflicts. The
ban on Aristotle, though ratified by the Pope, was soon recognized to be absurd. In 1231 Pope Gregory wrote, pardoning
those who had ignored it, acknowledging that 'the books on
nature prohibited in Paris ... are said to contain both useful
and useless matter', and commanding that they be examined
'subtly and with prudence', so that 'with the suspect material
removed the rest may be studied without delay and without
fault'.[4]

In Oxford meanwhile a settlement was arrived at with the
city in 1214, which established the university with its own
charter and its own chancellor. The first person to fill that role

[2] J. Hannam, *God's Philosophers* (2009), 79.

[3] S. C. Easton, *Roger Bacon and His Search for a Universal Science* (1952).

[4] J. Hannam, *God's Philosophers*, 81.

was a formidable man who immediately set himself to engage with both social and intellectual tensions.

Bob Large Head

18.2 Thirteenth-century portrait of Robert Grosseteste as Bishop of Lincoln.

It is not clear whether Robert Grosseteste held the actual title of chancellor (in the sources he is referred to only as *magister scholarium*—master of scholars) but he does seem to have performed the same role as later chancellors, being involved in discipline and administration and contributing to the early statutes and constitution of the university. He subsequently became Bishop of Lincoln [Fig. 18.2], and throughout his career was active in church politics and an energetic reformer (in his eighties he travelled to the papal court to accuse the Pope of leading the church astray). This was only one side of his life. In addition to his public duties he was also a prolific author.

Publishing some 120 works, ranging from French poetry to theology, science and estate management—this in a little book known as *Les Reules Seynt Robert*—Grosseteste, whose name in Norman French means 'large head', was an English equivalent to the great Arab polymaths of the previous century. His most important contributions to learning sprang out of a deep engagement with the newly available Greek and Arab texts. At the age of 60 he taught himself Greek and went on to produce his own translation of Aristotle's *Nichomachean Ethics* and his *De Caelo* together with Simplicius' commentary on it.

Long before this, while still at Oxford, Grosseteste had produced the first Western Latin commentary on Aristotle's *Posterior Analytics*. Aristotle's book was concerned with how we arrive at valid scientific knowledge—the kind of knowledge of a thing we possess 'when we know its cause'. In a process which Grosseteste refers to as 'resolution and composition', Aristotle suggests that from primary sense observations we arrive at universal principles, and that by valid demonstrations and syllogisms based on these principles we can advance to predictions and further knowledge.

Commenting on this Grosseteste argues that because the human mind is darkened we do not advance automatically from sense impressions to universal principles. In childhood we begin to correlate repeated experiences, but then as reason is awakened we need to investigate and test which of these correlations are valid. As an example he takes Avicenna's observation that when people ate the plant scammony (convolvulus) a discharge of red bile (through diarrhoea and vomiting) would follow. To discover whether there was a cause underlying this correlation, rather than just coincidence, Grosseteste suggests conducting what we would call a controlled experiment: removing everything else that was known to produce a discharge of red bile, and then feeding someone scammony to see what happens. By this (rather unpleasant) means it would be possible to arrive 'from sense at an observational (*experimentale*) universal principle'.[5]

Grosseteste does not say more about this method of arriving at the truth, and does not himself seem to have been an experimenter. He was, however, deeply interested in physical processes, and in particular he was fascinated by light. His commentary includes a long quotation from an earlier commentary describing the twinkling effect of stars.[6]

His interest in light continues in a commentary on Aristotle's *Physics* and Grosseteste's own *On Light*, in which he tries to give a physical content to Aristotle's abstract notion of 'prime matter' and 'first form'. What, he asks, is there in the physical world that instantaneously propagates itself in three dimensions from a single point? The only apparent answer was 'light'. A single candle lit in a darkened church in the instant of lighting sends it beams into every corner. Hence he concludes light could be described both as 'prime matter' and as 'first form': the fundamental physical reality from which everything else in the universe is made.

Theologically this seemed to be consistent with the biblical first words of Creation, 'let there be light', which Grosseteste expounded in his *Hexaëmeron*, a commentary on the six days of the Genesis creation story. While some of his contemporaries glossed over the contradiction between Christian doctrine and Aristotle's idea of the eternity of the world, in the *Hexaëmeron* Grosseteste robustly rejected the harmonizations of those who 'pointlessly try to make a catholic of Aristotle . . . making heretics of themselves'.[7]

Grosseteste does not follow the traditional argument against the eternity of the world: that infinity cannot be increased. He is generally credited with being the first person in the Latin West to recognize the possibility of so-called 'unequal infinites': while the sum of all even numbers is infinite and the sum of all odd and even numbers is also infinite, the latter must

---

[5] R. Grosseteste, *Commentarius in Posteriorum Analyticorum Libros* (1981), 215.

[6] Grosseteste's quotation probably comes from a commentary put together under the patronage of a Byzantine princess called Anna Comnena at the beginning of the 12th century. In Byzantium the tradition of philosophical thinking had largely withered in the centuries following the loss of Alexandria. Anna had defied this trend and despite the suspicion of the court had gathered around her scholars who could teach philosophy. From one of these, a man called Michael of Ephesus, she commissioned commentaries on the works of Aristotle. Michael (who complained that he ruined his eyesight in the process) incorporated into his commentaries material from wherever he could find it. Among the fragments he assembled were some pages of Philoponus which are (though he was unaware of it) what Grosseteste is quoting.

[7] R. Grosseteste, *Hexaëmeron* (1982), 61.

still be larger than the former because 'it exceeds it by the sum of all the odd numbers'.[8] Rather he argues that Aristotle's idea of a beginningless world rests on a failure to adequately comprehend God's non-temporal being—his transcendence over time.

Since light was a part of the physical world which God had made, as well as its theological importance it could also be physically understood, and in a series of books, *On Light, On the Rainbow, On the Heat of the Sun, On Colour*, this is what Grosseteste attempted to do.

Aristotle had thought that a rainbow is formed by the reflection of light from drops of water. Grosseteste corrected this misunderstanding, using concepts which he had learned from a book by al-Kindī on using a lense to focus the Sun's rays for burning [Fig. 18.3]. He attempted to make this quantitative, but without success not least because he thought that refraction was caused by a whole cloud acting as a single lens.[9] He was though able to show the potential of lenses and describes how 'it may be possible for us to read the smallest letters at incredible distances, or to count sand or seed or any sort of minute objects'.[10] The method by which he suggested these physical phenomena could be studied was to have a profound and lasting impact.

It had been recognized both by Euclid and later by al-Haytham that the way that light was reflected and refracted could be described by mathematics and geometry. For Grosseteste it followed that mathematics was a central aspect of God's creation. He frequently quotes the passage in the Book of Wisdom that addresses God as the one who has 'ordered all things in measure and number and weight',[11] and argued that whereas all human measurements are relative—comparing one finite thing with another finite thing—for God to whom the infinite is finite and who can embrace infinity in his mind, absolute measurement is possible [Fig. 18.4].

This implied that mathematics and geometry were relevant not only to light but to all physical phenomena. In a concise but significant study of colour, for instance, Grosseteste first argues (in a way that parallels the modern geometry of three-dimensional colour mapping) that the infinite variety of colours can be mapped on three independent axes. He then concludes that what can be understood about colour in this way 'by reason' can also by *experimento* be made visible through

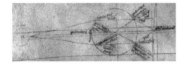

18.3 Diagram illustrating Grosseteste's *De Natura Locata*.

[8] L. Baur, *Die Philosophischen Werke des Robert Grosseteste* (1912), 52–3.
[9] G. Dinkova-Bruun et al., *The Dimensions of Colour: Robert Grosseteste's De Colore* (2013), 31–2.
[10] L. Baur, *Die Philosophischen Werke des Robert Grosseteste*, 75.
[11] Wisdom of Solomon 11:20.

18.4 God measuring the world with a compass, *Bible moralisée* (*c.*1250).

the 'skillful manipulation' of 'those who thoroughly understand the depths and principles of natural science and optics'.[12]

Aristotle had argued that natural philosophy and mathematics were intrinsically separate, but Grosseteste maintained that 'the consideration of lines, angles and figures is of the greatest utility since it is impossible for natural philosophy to be known without them'.[13] Consequently, he argues, 'the diligent investigator of natural phenomena can give the causes of all natural effects . . . by the rules, roots and foundations given from the power of geometry'.[14]

Grosseteste in this sense was a slipstreamer. In his book about him, James McEvoy writes, 'Grosseteste's interest in natural questions was both genuine and deep. Part of his motivation for inquiry undoubtedly lay in his religious faith.'[15] He saw Scripture and nature as having a single author, so that he fully expected consistency between the two. It was as if he considered

[12] G. Dinkova-Bruun et al., *The Dimensions of Colour: Robert Grosseteste's De Colore*, 19.

[13] L. Baur, *Die Philosophischen Werke des Robert Grosseteste*, 59–60.

[14] L. Baur, *Die Philosophischen Werke des Robert Grosseteste*, 65.

[15] J. J. McEvoy, *Robert Grosseteste* (2000), 80.

that his knowledge of science was a key to understanding the meaning of the Word of God.

Grosseteste bequeathed his books to the Franciscan convent library in Oxford, and it was this institution, which he himself had done much to promote, that was responsible for the preservation and subsequent influence of his work.

Friars and Friaries

At the beginning of the thirteenth century two young men, an Italian, Francesco Bernadone, and a Spaniard, Dominic de Guzman (subsequently canonized as Saint Francis and Saint Dominic), had been given permission by the Pope to found orders of preaching friars. Instead of living isolated lives of prayer in what had become sometimes very wealthy rural monasteries, these friars were to preach in towns and cities, living on alms and instructing ordinary people how to live Christian lives.

A group of Dominicans arrived in Oxford in 1221 and started living in a small community near St Aldates. They were joined three years later by a group of Franciscan brothers who established a small house near St Ebbes. In later years both communities established two enormous adjacent abbeys outside the city walls—around the site now occupied by the main Oxford police station.

Although Francis' dedication to poverty had included a rejection of learning with its temptation to intellectual pride, his successors recognized that the friars could not be effective without scholarship. Robert Grosseteste became the first lecturer to the small Franciscan community and it is possible (though perhaps unlikely) that Grosseteste's lectures here or elsewhere in the university were attended by a very young student called Roger Bacon. Bacon subsequently wrote that 'no one really knew the sciences except the Lord Robert ... he knew mathematics and perspective ... and ... was sufficiently well acquainted with languages to be able to understand the saints and philosophers and wise men of antiquity'. In later life Bacon himself joined the Franciscans, probably under the influence of Grosseteste's friend Adam Marsh, who became the leader of the Oxford Franciscans. That, however, lay in the future. After completing his degree Bacon moved to Paris, where the ban on teaching Aristotle had just been removed.

The Dumb Ox

Roger Bacon arrived in Paris around 1240, where as someone familiar with the works of Aristotle he is likely to have been welcomed with open arms. Paris had now become the Mecca of those wishing to study the new learning. Some five years later a quiet, rotund young Dominican friar (known to his fellow students as 'the Dumb Ox') also arrived in the city to pursue his own Aristotelian studies.

The son of an Italian nobleman, Landulf of Aquino, Thomas Aquinas had escaped from his family (who wanted him to pursue a more lucrative ecclesiastical career) to join the Dominicans. The Dominicans, sensing his potential, had sent him to Paris to study with Albert the Great (as he had already become known), whom Thomas then followed to Cologne [Fig. 18.5].

Albert was the most learned of all the Dominican teachers of the time. He had a comprehensive knowledge of the Aristotelian writings and the Arab commentators, and wrote on topics in natural philosophy that ranged from a study of falcons to mineralogy and chemistry. He is even credited with the discovery of arsenic. Although not uncritical of Aristotle, warning that since he was not a god 'he has certainly made mistakes just like the rest of us'[16] and recognizing that 'it is the task of natural science not simply to accept what we are told but to enquire into the causes of things',[17] he nevertheless recognized that what was of value in the Aristotelian corpus needed somehow to be integrated with the Christian revelation. This was a task he passed on to his star pupil. Almost since first meeting him he had become aware that Thomas had exceptional gifts. 'We call this lad a dumb ox', he remarked on one occasion, 'but I tell you the whole world will hear his bellowing'.

The project which Aquinas took over was not without precedent. Aquinas quotes with approval the works of 'Rabbi Moses', and follows his example of using a 'via negativa' when speaking of the ultimate nature of God. With respect to the natural world, however, Aquinas argues that there is much that can be known. God has created man to be exactly the 'rational animal' seeking his own highest good that Aristotle had described, and the divine reason implanted within us is sufficient for knowing those things that 'pertains to natural knowledge'.[18]

To go beyond this, to knowledge of grace and salvation, required the supernatural revelation transmitted through the

18.5 Albertus Magnus by Tommaso da Modena (fresco, 1352, Treviso, Italy).

[16] O. Pedersen, *The First Universities* (1997), 281.
[17] M. S. Albertus, *Book of Minerals* (1967), 69.
[18] T. Aquinas, *Summa Theologica*, I–II Q109, 119.

Scriptures. However, the 'natural revelation' that we receive through our God-given reason could, he believed, lead us to a basic knowledge of God's existence. Thus, in his *quinque viae*, or five ways, he develops both Aristotle's idea of 'the unmoved mover' and Avicenna's concept of 'necessary being', to show how a rational reflection on such things as motion, causation, existence, and order leads us to posit the existence of something that is not moved by something else and which has no cause: a necessary being that is the source of the world's order 'and this we call God'.

Challenges to Aquinas' project of integrating Aristotle's philosophy with Christian teaching came from two directions: on one side from some Franciscans who questioned the value of philosophy, and on another from those who advocated a strict Averroist interpretation of Aristotle. It was to meet this latter challenge that Aquinas was called back to Paris in 1268 to serve as regent master and to confront the rise of what was known as 'Latin Averroism' within the university.

The most prominent proponent of this teaching at Paris was a professor called Siger of Brabant. In his lectures and treatises Siger argued in favour of what Averroës had identified as central Aristotelian positions: that the laws of nature were necessary and could not be altered by God, that the world was deterministic and eternal, and that all these teachings were philosophically correct even if faith taught otherwise. Despite his great respect for Averroës (whom he referred to as 'the Commentator') Aquinas vigorously disputed this interpretation.

In a replay of the sixth-century Alexandrian exchanges, Aquinas replied to Siger's treatise *On the Eternity of the World* with his own treatise of the same title in which he argued that it was not possible to prove philosophically either that the world was eternal or that it had a beginning in time. Aquinas left Paris in 1272 and died two years later (not long after having an experience in prayer after which all that he had written seemed 'like straw'). Even before he left, the argument had taken on an official dimension when the Bishop of Paris, Etienne Tempier, had weighed into the debate.

The Paris Condemnations

In 1270 Tempier condemned thirteen propositions, among which was the claim that 'the world is eternal' and that 'human acts are not ruled by the providence of God'. In 1277 this was expanded to a list of 219 condemned propositions that ranged from the statement that 'theology is based on fables' to the claim that 'God cannot make more than three dimensions exist simultaneously'. Tempier had a fundamental objection to any lack of consistency which implied that 'there were two contrasting truths ... as if against the truth of sacred Scripture, there is truth in the sayings of condemned pagans'. More positively he wanted to affirm God's freedom to act as he chose: what Philoponus' lost book had described as 'the contingency of the world'.

Tempier's list included some 20 propositions that had been affirmed by Aquinas, and as a result the suspicion of heresy hung over Aquinas' name for some years after his death. In 1323, however (after years of vigorous campaigning by the Dominicans), Aquinas was pronounced a saint by Pope John XXII and the list of condemnations was hastily modified.

The rapid seesawing of Aquinas' posthumous reputation gives some indication of an underlying social and intellectual turbulence that was to have an even more direct effect on the career of his one-time contemporary in Paris, Roger Bacon.

CHAPTER NINETEEN

Imposed Silence

Thomas Aquinas had thought highly enough of Albertus Magnus to have followed him to Cologne. Roger Bacon, after teaching in Paris alongside the great German theologian, was rather less impressed. Although admitting that Albert was a man of infinite patience who had amassed a great store of information, he saw four faults in his work: 'the first is a boundless puerile vanity; the second is ineffable falsity; the third is a superfluity of bulk, and the fourth is his ignorance of the most useful and beautiful parts of philosophy'[1] (Bacon was not a man to mince words).

His belligerence was not without basis. He had arrived on the medieval intellectual stage at a moment when free-ranging curiosity was under threat. How far should open-minded inquiry be allowed? How far could ultimate questions be pursued without constraint?

'The most beautiful parts of philosophy' in Roger Bacon's view were those that he may have learned from a mysterious French soldier known as Pierre de Maricourt or Petrus Peregrinus (Peter the Pilgrim).

Peter the Pilgrim

Almost the only thing that is known about this remarkable man is that on 8 August 1269 he sent a letter to one Sygerus de Foucaucourt, setting out a treatise on magnetism.

The treatise contains the first extant account of magnetic polarity and describes how to make a pivoted compass 'by which you will be able to direct your steps . . . to any place in the world'. As remarkable as Peter's discoveries was the method by which he had arrived at them.

Peter begins by telling Sygerus that 'the investigator in this subject . . . must be very diligent in the use of his own hands'. In the first part of the letter he gives detailed instructions for carrying out a series of experiments which will provide

[1] B. Clegg, *The First Scientist* (2003), 120.

demonstrative proofs of his argument, 'break off a small piece of the needle or iron . . . then put it on the spot that was found to be the pole' [Fig. 19.1].

Whether Bacon actually knew Peter has been disputed, but the text of Bacon's most famous work describes a man who sounds remarkably like him.[2] It declares that 'the business of experimenting no one in Europe really understands save only Master Peter', and describes how this 'master of experiment' gains knowledge 'of matters of nature . . . through experiment', even trying out 'the illusions and wiles of conjurors . . . so that nothing may escape him which ought to be known . . . that he may perceive how far to reprove all that is false and magical'.[3]

On his return to Oxford Bacon seems to have set himself to following this path. In a later work he describes how for 20 years 'after abandoning the usual methods' he had laboured in the study of wisdom, spending more than £2,000 (perhaps the equivalent today of £1 million) 'on secret books and various experiments and mathematical tables'.[4]

Bacon's enthusiasm was fuelled in part by his sense of the extraordinary potentiality of such studies. He raises the possibility of mechanically powered ships, 'a car . . . which will move . . . without the help of a living creature', a mechanically driven device for flying and 'devices . . . whereby . . . a man may walk on the bottom of the sea or of a river'.[5] In his later works he expresses a great enthusiasm about how the discoveries of natural philosophy might forward the spread of the Gospel, and this desire to put his discoveries to the service of the Church may have contributed to his decision to join the Oxford Franciscans as a Friar Minor. In doing so it may be that Bacon had hoped to follow in Robert Grosseteste's footsteps by becoming the chief lecturer to the Oxford Franciscans.

His career was not to work out that way. Within a few years of joining the order he seems to have been moved to the Franciscan convent in Paris where, according to his own account, 'My superiors and my brothers . . . kept me under close guard and would not permit anyone to come to me, fearing that my writings would be divulged to others than the chief pontiff and themselves'.[6] The details of what happened are unclear, but it seems that Bacon had become caught up in the same kind of crossfire from the perceived need to control ideas that caused the rapid seesawing of Aquinas' reputation.

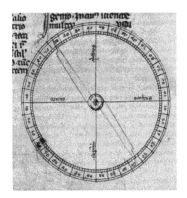

19.1 Pivoting compass needle in a fourteenth-century hand copy of Peter Peregrinus' *Epistola de magnete* (1269).

---

2 It has however been argued that this is a marginal gloss added to Bacon's text, cf. E. Grant, 'Peter Peregrinus', in *Dictionary of Scientific Biography* (1975), 10: 532.'

3 B. Clegg, *The First Scientist*, 33.

4 B. Clegg, *The First Scientist*, 37.

5 B. Clegg, *The First Scientist*, 42.

6 B. Clegg, *The First Scientist*, 74.

The motivations behind these controls were sometimes practical. Then as now inflammatory ideas could turn into issues of
public order with frightening speed. On a trip to Paris in 1251,
shortly before taking his monastic vows, Bacon had witnessed
the march of the so-called Pastoreaux rebels through the city.
Although the Pastoreaux were mainly shepherds, their leader,
known as 'the master of Hungary', claimed to have a document from the Virgin Mary that gave them the right to reclaim
the Holy Land. Bacon recalls seeing this man 'with my own
eyes carrying openly in his hand something as though it was
a sacred object', and while 'he went with bare feet ... was always surrounded by a host of armed men'.[7] This last detail was
significant. As they moved through France the Pastoreaux not
only set up their own church with a pope, bishops, and clergy,
but (before being routed outside Bourges) had begun attacking
and killing clerics, university students, and Jews [Fig. 19.2].

Mobs like this could be corralled by soldiers, but how could
the religious ideas that inflamed them be policed? The Church's
control of doctrine, and the related question of whether it
could employ secular force in doing so, would affect not only
Bacon's life but the whole future relationship of science and

19.2 Crusade of the Pastoreaux.

De la meute des pastoureaux.

religion. By the thirteenth century it had already been an issue that had dogged Christianity for almost 900 years: ever since the conversion of the emperor Constantine in AD 312.

The Hounds of God

The conversion of the Roman emperor in that year had brought to an end the persecution of Christians and had placed before Christian leaders the prospect of themselves exercising influence and power. The question that then arose was how should the Church respond to those who challenged or questioned its ideas whether from without or within? Two possible models presented themselves.

The Hebrew Scriptures recorded that when the Israelites conquered Canaan they had been commanded to eliminate all trace of Canaanite religion. The 'detestable ways' of nations who 'burn their sons and daughters in the fire as sacrifices to their gods'[8] were to be eradicated by the destruction of whole populations. The same draconian punishments that were penalties for social crimes were also prescribed for those who returned to such religious practices. Thus, the death penalty for sexual adultery was to be paralleled by the stoning of those who forsook the Lord and worshipped other gods.

These commands might stand as indelible markers of the consequences of leaving the path of truth, but they had not been taken by the prophets of Israel as God's last word or as the final expression of his nature. The books which contain these prescriptions also contain the injunctions to 'love your neighbour as yourself'[9] and to 'love the LORD your God with all your heart, and all your soul and all your strength'.[10] Within them the deity is referred to in terms that suggested an ultimate loving parent: 'the eternal God is your refuge and underneath are the everlasting arms'.[11] These are all expressions which hint at a divine mercy and compassion that can override legal prescriptions. Among the prophets of Israel such hints were dramatically developed.

Thus, with respect to the sins of the chosen people, the prophet Hosea's forgiveness of his own wife's adultery was enacted as an image of God's forgiveness of the spiritual adultery of the nation. Similarly with respect to the false gods of other nations the psalmist affirms that even such enemies of Israel as Babylon, Egypt, Philistia, and Tyre are 'born in Zion',[12] while

[8] Deuteronomy 12:31.
[9] Leviticus 19:18.
[10] Deuteronomy 6:5.
[11] Deuteronomy 33:27.
[12] Psalm 87:4.

the prophet Isaiah looks forward to a time when God will bless the nations of the earth, 'saying, Blessed be Egypt my people, Assyria my handiwork and Israel my inheritance.'[13]

These prophetic themes in the Hebrew Scriptures became central and explicit in the Christian writers of the New Testament, who take it as their fundamental commission that 'beginning at Jerusalem' the message of mercy and the forgiveness of sins is to be preached 'to all nations'.[14] The teaching within the Gospels of non-retaliation and overcoming evil with good translated into a conviction that the purposes of God were to be accomplished not by violent subjugation but by self-sacrificial preaching. Within the Church itself moral discipline could only be enforced by exclusion from fellowship. No compulsion was possible. For the first three centuries of Christianity there is evidence for this as a model of understanding and practice.

Tertullian, for instance (AD 160–225), writes to the proconsul of Carthage, 'that it is a fundamental human right, a privilege of nature (*humani iuris et naturalis potestatis*) that every man worships according to his convictions. One man's religion neither harms nor helps another man. It is assuredly no part of religion to compel religion ... to which free will and not force should lead us'.[15] Similarly Lactantius, writing around AD 308, argues that 'There is no occasion for violence and injury, for religion cannot be imposed by force; the matter must be carried on by words rather than by blows, that the will may be affected ... Torture and piety are widely different; nor is it possible for truth to be united with violence or justice with cruelty'.[16]

The centrality of religion to human life, as Lactantius saw it, made no difference: 'It is true that nothing is so important as religion, and one must defend it at any cost ... but', he insists, 'religion is to be defended not by putting to death, but by dying; not by cruelty but by patient endurance'. 'If you wish to defend religion by bloodshed, by tortures', he concludes, 'it will no longer be defended, but polluted and profaned. For nothing is so much a matter of free will as religion'.[17]

This conviction did not immediately disappear with the conversion of the emperor. In the fourth century Hilary of Poitiers (300–68), his disciple Martin of Tours (316–97), and the teacher of St Augustine, Ambrose of Milan (340–97), had all strongly objected to the use of violence against the heretical Bishop Pricillian and his followers. There were those in the fifth century like John Chrysostom (347–407) who still

---

[13] Isaiah 19:25.
[14] Luke 24:47.
[15] Tertullian, *The Writings of Quintus Sept. Flor. Tertullianus* (1882), vol. III, 142.
[16] Lactantius, *The Works of Lactantius* (1871), vol. XXI, 399–400.
[17] Lactantius, *The Works of Lactantius*, 400.

maintained that it was not right 'for Christians to eradicate error by constraint and force but to save humanity by persuasion and reason and gentleness'.[18]

Writing in 396, Augustine of Hippo (354–430) stated, that 'I would have no man brought into the catholic communion against his will'.[19] Attitudes, however, were beginning to change. Some 12 years later he began to justify the use of violent coercion. Another African prelate, Optatus of Mileve, had been the first Christian bishop to invoke the Deuteronomic legislation against a heretical Christian group called the Donatists. Augustine wrote threatening letters to the leaders of the same Christian group, in which he employed all his rhetorical skills to make his case.

Augustine's theological prestige in the following centuries ensured that his arguments provided a ready basis for the use of state violence to ensure religious conformity. In the Western church this had taken on an external dimension in 1095 when Pope Urban II had called for a crusade to liberate Jerusalem. Fifty years later Bernard of Clairveaux had called for the Christian equivalent of a jihad to reverse the Islamic conquests. These external crusades against Islam in turn inspired the so-called 'Albigensian Crusade', undertaken in 1209 against the Cathars, a heretical Christian group within Christendom.

It was largely in the wake of this crusade against heretics that the Pope began to appoint 'inquisitors' or judges to try individual cases of heresy. The newly established Dominican and Franciscan orders seemed to some to have been providentially supplied for just this purpose.

The previous focus of both Francis and Dominic themselves had been on preaching and example rather than coercion. Following a meeting at the siege of Damietta in 1219, the sultan Malik al-Kamil gave Francis permission for his friars to live among the Mohammedans and practice the 'silent preaching' of good deeds. Similarly the strategy that Dominic had developed from his encounter with the Albgensian heretics in 1204 was to encourage Christians to live better and to educate themselves so as to be able to counter heretical arguments. Despite this background, the intellectual orientation of the Dominicans seemed to fit them peculiarly well for the new inquisitorial task. The zeal with which they subsequently performed it earned them a title that was a Latin pun on their name: *Domini canes*—'the hounds of God' [Fig. 19.3].

19.3 Visual punning on *Dominicani/Domini canes* in the Spanish Chapel at St Maria Novella in Florence.

[18] St John Chrysostom, *Apologist* (2001), 400.
[19] R. Forster and P. Marston, *God's Strategy in Human History* (2001), 333.

The Chapter of Narbonne

The Franciscans never proved well suited to the task of inquisition, not least because of their own internal divisions and occasional heretical tendencies. The divisions had been apparent from the early days in a split between the so-called *spirituales* who believed the vow of poverty was absolute and the *relaxati* who maintained that while friars themselves could own nothing, the order as a body could own property like the abbey at Oxford. The heretical tendencies had become apparent when a head of the order, John of Parma, had allowed a Franciscan brother, Gerard de Borgo, to promote a text called *The Everlasting Gospel*. Gerard's tract looked forward to a coming age of universal love when there would be no need for the Church. According to Gerard this time would be ushered in by the Franciscans.

A new minister general of the Franciscans named Bonaventura was appointed in 1257. One of his first concerns was to heal the divisions in the order and to halt the slide into heresy. His means of doing so was to establish a new set of rules for the order at the chapter of Narbonne. These included the decree that no brother could even keep a book without explicit permission, and that no books could be published without being sanctioned by the leadership of the order.

Bacon's removal to Paris may have been unconnected to these wider issues (it was probably the result of some insulting remarks he made about the stupidity of the Franciscans' most senior academic in Oxford, Richard of Cornwall) but the restrictions placed on him while in France seem likely to have been a direct result of the chapter of Narbonne.

This imposed silence lasted some ten years. His release from it came by the direct intervention of the chief pontiff himself.

CHAPTER TWENTY

Experimental Science

On the 22nd of June 1266 Roger Bacon was sent a letter stamped with the papal seal of Clement IV (a contemporary copy of which still exists in the Vatican archive), instructing him to send the Pope 'writings and remedies for current conditions . . . not withstanding any prohibitions from his order'. Bacon had appealed to the then cardinal Guy de Foulques three years earlier and had received an encouraging response. This instruction, coming after Guy's surprise elevation to the papacy, now faced him with the task of summarizing some 30 years of work and thought.

The Great Work

His initial response was the *Opus Majus* (the greater work), but while this was being copied he seems to have added first the *Opus Minus* (the lesser work), and then a third, *Opus Tertium*. At least the first two volumes of this trilogy were sent to the Pope together with two enclosures: a map of the world and a lens.

The significance of these enclosures would not at first have been apparent. The *Opus Majus* begins with a general consideration of the importance of wisdom in such things as the governance of the Church, the conversion of the heathen, and the repression of evil. In asking why wisdom is hard to obtain, Bacon identifies four causes that obstruct it: subjection to unworthy authority, the influence of habit, popular prejudice, and a false conceit of our own wisdom. 'Wise men', he argues, '. . . feel they are more lacking than fools', and for this reason are 'humbly disposed to receive instruction from another'.[1] This, in particular, includes Greek science. After reviewing the history of the Church's relationship with Greek thinkers Bacon goes on in the second part to a more general consideration of the relationship of philosophy and theology.

His starting point (like that of his predecessors) is that since all wisdom comes from God it cannot be inconsistent with

[1] R. Bacon, *The Opus Majus of Roger Bacon* (1928), Part I, Chapter X, 25.

itself: 'truth wherever found is thought to belong to Christ'.[2] There may, however, be differences of focus, so that in some instances scripture may describe a phenomenon and reveal its final cause 'from which the efficient cause can be investigated'.[3] An example of this was the rainbow whose significance is described in Scripture, but whose physical cause 'was not clearly understood by the philosophers'[4] and which, like Robert Grosseteste, he was very interested in trying to work out.

It was very helpful to Bacon's case to argue, as had many Christian authors, that Greek philosophy had ultimately derived (via the Chaldeans and Egyptians) from the Hebrew patriarchs, but this does not lead him to gloss over its distinctive character. His conclusion is rather that 'philosophy is ... the unfolding of the divine wisdom by learning and art'.[5] 'There are', he points out, 'many common rational truths which any wise man would easily accept from another, although he might be ignorant of them himself'.[6] Therefore, Christians ought 'not only to collect the statements of philosophers in regard to divine truths, but should advance far beyond to a point where the power of philosophy as a whole may be complete'.[7] The philosophizing Christian 'can unite many authorities and various reasons and very many opinions' and this must be undertaken not merely 'to complete philosophy' but because 'all truth' is 'divine truth'.[8]

With this programme established, Bacon then proceeded to describe two principal tools which were needed to pursue it. The first of these was a knowledge of languages, the second was a knowledge of mathematics (a concept whose full significance would become apparent in future centuries) [Fig. 20.1] (see Chapter Forty-Nine).

The study of language, Bacon argues in Part III, is critical to give us accurate translations both of the Greek and Arabic philosophical texts and of the texts of Scripture (not to mention such practical things as 'intercourse with foreign nations'). Mathematics, he argues in Part IV, is 'the gate and key'[9] to all other sciences. 'The fundamental principles of demonstration' cannot be made clear 'except in the realm of mathematics ... Therefore of necessity logic depends on mathematics'.[10] The dependence of astronomy on mathematics 'was obvious', as was the study of optics [Fig. 20.2]. 'The things of this world', he insists, 'cannot be made known without a knowledge of mathematics'.[11] It was necessary for astrology (then considered a science) and for such practical matters in the life of the Church as

[2] R. Bacon, *The Opus Majus of Roger Bacon*, Part II, Chapter V, 43.
[3] R. Bacon, *The Opus Majus of Roger Bacon*, Part II, Chapter VIII, 51.
[4] R. Bacon, *The Opus Majus of Roger Bacon*, Part II, Chapter VIII, 51.
[5] R. Bacon, *The Opus Majus of Roger Bacon*, Part II, Chapter XIV, 65.
[6] R. Bacon, *The Opus Majus of Roger Bacon*, Part II, Chapter XIX, 73.
[7] R. Bacon, *The Opus Majus of Roger Bacon*, Part II, Chapter XIX, 73.
[8] R. Bacon, *The Opus Majus of Roger Bacon*, Part II, Chapter XIX, 74.
[9] R. Bacon, *The Opus Majus of Roger Bacon*, Part IV, Chapter I, 116.
[10] R. Bacon, *The Opus Majus of Roger Bacon*, Part IV, Chapter II, 120.
[11] R. Bacon, *The Opus Majus of Roger Bacon*, Part IV, Chapter III, 128.

20.1 Illustration at the beginning of Euclid's *Elementa*, in the translation attributed to Adelard of Bath.

the reform of the calendar. It was also needed for the accurate determination of longitude and latitude that was necessary if the Church was to fulfil its worldwide mission (hence a chapter on geography and the inclusion of a map of the world).

Part V concentrates exclusively on optics, the science with which Bacon was particularly fascinated (hence the inclusion of the lens). He is keen to point out the benefits that might flow from it and argues, like Grosseteste, that it could 'from an incredible distance' allow us to 'see the smallest letters' and 'might cause the sun and moon to descend in appearance'.[12] Not only this but a new kind of 'geometrical figuring' could give such a reality to religious art that 'the evil of the world would be destroyed in a flood of grace'.[13] Giotto's frescoes in the basilica of St Francis, executed 50 years later, might be considered the first installment of this promise.

His focus nonetheless is on the means by which this science can be advanced. Thus, he starts with a detailed examination of the physical structure of the eye, and goes on to describe some of his own experiments with reflection and refraction. This then leads him in Part VI to a more general reflection on the nature of experimental science.

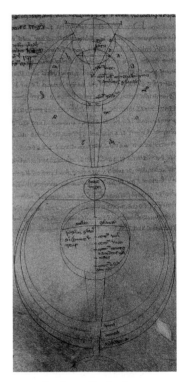

20.2 Optic studies by Roger Bacon from *De Perspectiva*.

[12] R. Bacon, *The Opus Majus of Roger Bacon*, Part V, Chapter IV, 582.
[13] R. Bacon, *The Opus Majus of Roger Bacon* (1897), 219.

Measuring Rainbows

Experience, Bacon suggests, comes in two kinds. There is the internal experience of things spiritual which 'comes from grace', and there is that which 'is gained through our external senses'. We gain our experience of things that are in the heavens 'by instruments made for the purpose', and of things that do not belong in our part of the world 'through other scientists who have had experience of them'.[14] It is with the latter that he is principally concerned. The distinguishing characteristic of what he calls 'the science of experiment'—*De Scientia Experimentali*—is that 'it investigates by experiment the . . . conclusions' of other sciences.[15] To illustrate this he returns to the example of the rainbow.

While the natural philosopher, he argues, comes up with theories or 'judgements' about phenomena like rainbows and haloes, 'experimental science attests them'.[16] He first of all looks for visible objects in which the colours of the rainbow appear in the same order, and finds 'this same peculiarity in crystalline stones correctly shaped and in other transparent stones'[17] as well as when 'the solar rays penetrate drops . . . in water falling from the wheels of a mill and likewise when one sees on a summer morning the drops of dew on the grass in the meadow'.[18] Armed with these 'terrestrial facts' the experimenter looks at 'those phenomena that occur in the heavens' using 'the recquired instrument'.[19]

Bacon then proceeds to describe how using his own measurements (made presumably with an astrolabe) 'the experimenter . . . taking the altitude of the sun and of the rainbow above the horizon will find that the final altitude at which the rainbow can appear above the horizon is 42 degrees and this is the maximum elevation of the rainbow'.[20]

Bacon was the first person to discover this. He went on to argue that each raindrop in the bow worked as a spherical mirror, producing because of their proximity a continuous image. He was aware that his account of how the rainbow was produced was far from satisfactory 'because I have not yet made all the experiments that are necessary'[21] and was convinced that 'experiments on a large scale and by various necessary means are recquired'[22] (this was, in fact, achieved some 50 years later by a Dominican friar, Theodric of Freiburg, who demonstrated using spherical globes filled with water how primary and

[14] R. Bacon, *The Opus Majus of Roger Bacon*, Part VI, Chapter I, 585.
[15] R. Bacon, *The Opus Majus of Roger Bacon*, Part VI, Chapter II, 587.
[16] R. Bacon, *The Opus Majus of Roger Bacon*, Part VI, Chapter II, 588.
[17] R. Bacon, *The Opus Majus of Roger Bacon*, Part VI, Chapter II, 588.
[18] R. Bacon, *The Opus Majus of Roger Bacon*, Part VI, Chapter IV, 589.
[19] R. Bacon, *The Opus Majus of Roger Bacon*, Part VI, Chapter IV, 590.
[20] R. Bacon, *The Opus Majus of Roger Bacon*, Part VI, Chapter IV, 592.
[21] R. Bacon, *The Opus Majus of Roger Bacon*, Part II, Chapter XII, 615.
[22] R. Bacon, *The Opus Majus of Roger Bacon*, Part II, Chapter XII, 615.

secondary rainbows are produced by double refraction within a raindrop). His point is that he is 'proceeding ... not by the method of compiling what has been written on the subject' because 'experiment not reasoning determines the conclusion in these matters' and he ends with 'a plea for the study of science'.[23]

Bacon concludes his discourse on experimental science by arguing that a 'philosophical chancellor' could organize science and the pursuit of technology for the benefit of the Christian world. He dispatched the great work to the Pope in 1267, presumably with the hope that Clement might take on this role. It was not to be. By the end of 1268 news had spread throughout Europe that Clement had died in November.

Bacon's fate after the death of his patron is unclear. He wrote a 'collected study of philosophy' which contains a bitter attack on the corruption of every aspect of church and society. This may explain why in *The Chronicle of the 24 ministers general of the Franciscans* written a century later, he is named as a master of theology imprisoned for 'suspect novelties'.

In 1274 Bonaventura was succeeded as head of the Franciscans by Girolam D'Ascoli, who is known to have imprisoned a group of suspect friars who remained in confinement until released by his more merciful successor, Raymond de Gaufredi, in 1290. There is no direct contemporary evidence that Bacon was imprisoned, but given that he seems to have written nothing further until 1292 when he appears to have been back in the Franciscan abbey in Oxford, it would fit with the facts.

While Bacon's larger plans for the development of what he called experimental science may have failed, two of his ideas found strong echoes among his successors in both Oxford and Paris. The first of these was his emphasis on the role of mathematics as the key to all other sciences. The second was his conviction that the laws of reflection and refraction were a part of what he describes as *leges communes nature*—the universal laws of nature.

[23] R. Bacon, *The Opus Majus of Roger Bacon*, Part II, Chapter XII, 615.

CHAPTER TWENTY–ONE

The Universal Law

In the century that followed Roger Bacon's death in 1294 his successors in both Oxford and Paris began to develop various ways of using the idea of a mathematically universal law to understand the physical world [Fig. 21.1]. In Oxford this was particularly associated with a series of fellows at Merton College.

The Merton Calculators

Thomas Bradwardine, after having been elected to a fellowship at Merton, went on to become a professor of divinity and chancellor of the university, ending his career as (for a few months) Archbishop of Canterbury. Hailed by the Pope as 'Doctor Profondus', Bradwardine argued that anyone who tried to pursue

21.1 Miniature of Richard of Wallingford, Abbot of St Albans, from a fourteenth-century *History of the Abbots of St Albans*.

physics without mathematics 'will never make his way through the portals of wisdom'.[1] In his *Treatise on Proportions* written in 1328 he devised a mathematical formula to establish the relationship between the force applied to an object, the resistance to its motion, and the velocity that results. He also speculated that in a vacuum, objects of different weights would fall at the same speed.

A younger Merton contemporary, Richard Swineshead, was known as 'The Calculator'. In 1350 he produced his *Liber Calculationem* (Book of Calculations) that contained more than 50 mathematical variations of Bradwardine's law, and attempted a mathematical analysis of rates of variation of qualities like heat and speed.

It was, however, a Merton logician, William of Heytesbury, who in his *Rules for Solving Sophismata* (logical puzzles) writes down a formula which actually corresponds to physical reality and has stood the test of time. This was the so-called 'mean speed theorem' or 'Merton rule', according to which, as Heytesbury puts it, 'a moving body will travel in an equal period of time a distance exactly equal to that which it would travel if it were moving continuously at its mean speed'.[2] That is to say, putting aside factors like wind resistance, a cart bowling along at a steady 4 mph would be caught up by a horse smoothly accelerating from a standing start at the moment it reached 8 mph.

These mathematical approaches to physics were soon echoed in Paris where they were combined with an idea of motion that contradicted the traditional theory of Aristotle. In the second half of the twelfth century al-Ghazali's *Maqâsid al-falâsifa* (The Intentions of the Philosophers) was translated into Latin as *Logica et philosophia Algazelis arabis*. Within it was the discussion of the idea of impressed force that Avicenna had derived from Philoponus. In mid-fourteenth-century Paris this idea found a new champion.

The Age of Buridan

Jean Buridan, who became the rector of the University of Paris in 1328 and again in 1340, seems to have been the first person to find a Latin term for the concept of impressed force. What Philoponus had called *kinetiké dunamis* and Avicenna had called *mayl*, Buridan referred to as *impetus*. In accordance with the

[1] J. Hannam, *God's Philosophers* (2009), 176.

[2] J. Hannam, *God's Philosophers*, 180.

21.2 God the geometer creating the world, *Bible moralisée de Tolède* (1240).

doctrine that God had 'ordered all things in measure, number and weight' [Fig. 21.2], he proposed that impetus was a quality whose magnitude was proportional to both weight and speed. Buridan followed Philoponus in recognizing that air resistance would slow down a moving object rather than propelling it forward, but went beyond him in arguing that in a vacuum where there was no resistance an object would continue to move indefinitely.

Neither al-Ghazali nor Avicenna seem to have had any knowledge of Philoponus' book of Christian theology *De Opificio Mundi*, and Buridan is unlikely to have had any other access to it. Nevertheless, starting from the same premises as his Alexandrian predecessor he followed the same train of thought.

Buridan argued that an omnipotent creator could give to the whole of creation a single law of motion. In his commentary on Aristotle's *Physics* Buridan suggests that impetus explains why someone trying to jump a long distance 'drops back a way in order to run faster so that by running he gains an impetus which would then carry him further in the jump'. Contrary to Aristotle, he points out, 'the person so running and jumping does not feel the air moving him but feels the air in front strongly resisting him'. He then, in light of the Genesis account of Creation, immediately goes on to extend this idea to the movement of the planets:

> Since the Bible does not state that appropriate intelligences move the celestial bodies it could be said that it does not appear necessary to posit intelligences of this kind because it could be answered that God when he created the world, moved each of the celestial orbs as he pleased and in moving them impressed in them impetuses which moved them without his having to do any more except as a co-agent in all things which take place.[3] [Fig. 21.3]

This last point was crucial to Buridan's approach to natural science. In a book of questions relating to Aristotle's books on the heavens and the Earth, Buridan argues that 'God is no less the cause of this world and its order than if it were eternal', and connects this with the conviction that 'in natural philosophy one should always consider processes and causal relationships as if they always come about in some natural fashion'.[4]

[3] M. Clagett, *The Science of Mechanics in the Middle Ages* (1959), 536.

[4] J. Buridan, *Iohannis Buridani Quaestiones super Libris Quattuor de Caelo et Mundo* (1942), 164.

21.3 God the geometer creating the world, Gautier de Metz (1425–50).

Many of Jean Buridan's ideas were developed by his most brilliant pupil, Nicholas Oresme [Fig. 21.4]. Coming possibly from a peasant background, Oresme became first tutor and later economic advisor to Charles V, ending his career as Bishop of Lisieux. Among his other gifts Oresme was a formidable mathematician who devised the idea of using a graph to represent motion. By plotting speed on the vertical axis (which he called *latitudo*) and time on the base line (*longitudo*), he was able to draw a line that represented acceleration. By then calculating the size of the figures underneath, he provided a geometrical proof of the Merton rule [Fig. 21.5].

21.4 Nicolas Oresme, miniature from *Traité de la Sphère*.

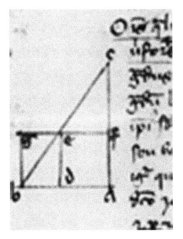

21.5 Diagram from Nicolas Oresme's *De Configuationibus Qualitatum*, showing a geometrical proof of the 'Merton rule'.

5 J. Hannam, *God's Philosophers*, 187.
6 N. Oresme, *Le Livre du ciel et du Monde*, ed. Menut, A. D., and De-nomy, A. J., trans Menut, A. D. (1968), 289.

Like Grosseteste, Oresme regarded mathematical order as a fundamental aspect of the way that God had created the world, but stressed that God was free to create the world however he chose. In a work called *Livre du Ciel et du Monde* he argued that Aristotle's idea, that the Earth remained stationary while the planets and stars moved around it, could not be proved. It was equally possible and more economical to think that it was the Earth that was moving. In the end he concluded that the Bible suggested a stationary Earth, but he recognized that many biblical passages that seem to favour this view simply conform 'to the normal use of popular speech . . . which are not to be taken literally'.[5]

Following Buridan, Oresme argues that 'when God created the heavens, he put into them the motive qualities and powers just as he put weight resistance against these motive powers in earthly things'. 'The situation', he suggests, 'is much like that of a man making a clock and letting it run down and continue its motion by itself'.[6]

In statements like these leaders of the Church in both Oxford and Paris seemed to be rapidly edging towards something not unlike the mechanical philosophy of the seventeenth century. The obstacle that derailed all such developments was introduced not by the Inquisition but by a black rat.

Rattus rattus

Bubonic plague (carried both by fleas on the black rat, *Rattus rattus*, and by infected people) arrived in Genoa and Venice in spring 1348, brought by merchants fleeing from the siege of Kaffa (where the besieging Tartars had catapulted plague-infected corpses into the town). Once the plague had arrived in Europe it spread with terrifying speed. Ships from Genoa took it to Marseilles. By summer it was in Paris and by September it had reached England. Within 50 years between a third and a half of the population of Europe had perished.

Thomas Bradwardine died in the first wave of the plague, Jean Buridan probably in the second. Although university towns were not affected as badly as the villages, the devastation of primary education (which in some areas threatened a complete loss of literacy) had a rapid knock-on effect. By the end of the fourteenth century what a writer from the next century described as 'the age of Buridan' had ended.

The Gun and the Book

Europe's economic and intellectual recovery from this catastrophe was slow and piecemeal. Far from disappearing the plague continued to return in successive generations (the last major outbreak was in the eighteenth century). As, however, populations were gradually rebuilt and economies grew, intellectual life was energized by two events that occurred almost simultaneously in the middle of the fifteenth century.

The first of these was the fall of Byzantium.

In 1453 Sultan Mehmet II, with the aid of a gigantic siege gun constructed by a Hungarian engineer called Urban, finally succeeded (where generations of Moslem armies had failed) in breaching Theodosius' great walls and seizing the city. A side effect of this cataclysmic event was the sudden appearance in Western Europe of numerous Greek-speaking scholars willing both to teach and to translate, rather as in the 1930s Oxford and Cambridge benefited from the influx of Jewish scientists fleeing the Nazi threat. Within a few decades a knowledge of Greek had once again become the mark of an educated person in Europe, and Greek manuscripts began to be hungrily sought and collected. An event that occurred two years after the fall of Byzantium ensured that these texts were not long to remain as manuscripts.

In 1455 Johannes Gutenberg printed the first copies of his great folio Bible. Some years earlier, using a hard metal punch to make a softer copper mould in which type could be cast, Gutenberg had devised a system of making large quantities of movable type. Using an oil-based ink which could be printed onto paper with an adapted agricultural screw press, he had arrived by 1453 at a complete workable system for printing text. The production two years later of his folio Bible demonstrated to the world the extraordinary potential of this technology.

In the decades that followed, print shops using this system sprang up all over Europe. Alongside the other books that began to pour from these presses, special type was cut so that ancient Greek manuscripts could be printed in the original language, and for those without Greek, Latin translations of these hitherto unknown texts were made. Among the new texts that became suddenly available were all the writings of Plato, and others (including the books of Philoponus) which challenged what had become the received Aristotelian approach to natural philosophy.

This fresh stimulus to contrarian curiosity was met by a new formalized version of the Inquistion. 'The Holy College' was established in Rome in response to the rift powered by Gutenberg's print technology that was tearing the religious fabric of Europe apart. The Holy College had the task of investigating any and every kind of suspect novel thinking. A clash—a *chute*—was perhaps inevitable.

The Socratic argument that the pursuit of truth was a religious duty—an argument echoed by Clement, Philoponus, al-Kindī, and Bacon—had been critical to the strength of the slipstream of religious motivation in which penultimate curiosity could travel. The corollary of this, Tertullian's claim that it 'is no part of religion to compel religion' and Lactantius' insistence that 'there is no connection between truth and violence', had, while equally critical, been hard to live by. It was in the clash between the investigation of physical reality and the attempt to police religious doctrine that this principle began to be articulated and dramatized.

On one side of this clash was an amalgamation between Aristotelian science and biblical literalism (backed up by repressive authority). On the other side was the rediscovery of a tradition of Christian criticism both of Aristotelian science and of biblical literalism, backed up by discoveries made possible by advances in optical instrumentation.

The consequences of the temporary victory of the former were to make the study of the heavens into a closed book. The consequences of the ultimate victory of the latter were precisely the opposite.

PART V

The Open Book of Heaven

CHAPTER TWENTY–TWO

Against Aristotle

On a Thursday morning in 1613 a Benedictine monk had just come out of the Medici palace in Pisa (where he had been a guest at a grand breakfast) when he was called back in by the porter. At breakfast the monk, Benedetto Castelli, had replied to some questions put to him by the Grand Duke Cosimo de Medici. The Duke had wanted to be informed about some objects which the Benedictine (who was a professor of mathematics at the University of Pisa) had been observing through his telescope the previous night. These were the four moons of Jupiter which Castelli's teacher, Galileo Galilei, had discovered four years earlier and named the 'Medician stars'. As a loyal pupil Castelli used this opportunity at breakfast to expound to the Medici his teacher's 'proof of the motion of these planets'.

When he came back into the palace Castelli was led to the chambers of Cosimo's mother, the Grand Dowager Duchess Christina, where a group was assembled that included the Duke, his wife and mother, and the professor of logic and philosophy at the University of Pisa, Cosimo Boscaglia. Throughout the breakfast 'Dr Boscaglia had talked to Madame', telling her, 'that the earth's motion was incredible and could not be, particularly since Holy Scripture was obviously contrary to such motion'.[1] This was the objection that the Grand Duchess now put to Father Castelli.

A thousand years earlier, when John Philoponus had tried to articulate a Christian philosophy, he had found himself arguing on two fronts: opposing the Aristotelian dogma of Simplicius on one side and the biblical literalism of Cosmas Indicopleustes on the other. By the beginning of the seventeenth century Aristotelian thinking had become so much a part of Christian theology that these two fronts had in a sense united. A philosopher like Boscaglia now had no hesitation in using biblical literalism as a knock-down argument to support the cosmology of Aristotle and Ptolemy.

[1] G. Galilei, *Discoveries and Opinions of Galileo* (1957), 151.

To counter this, Castelli, the Benedictine monk, found himself obliged 'to play the theologian'. He afterwards thought he had done so with such 'confidence and dignity' that his teacher would be pleased with him. Galileo, however, was not sure that his erstwhile pupil had said all that was necessary. He therefore sent Castelli an extended letter intended for circulation. In it he suggested that Aristotelian philosophers who are wrong about a physical question try to use Scripture like 'an irresistible and terrible weapon' to frighten their opponents. Those who are right, however, can support their argument with 'a thousand experiments and a thousand necessary demonstrations'.[2]

Galileo's recent telescopic discoveries (which Boscaglia did not dispute) came into this category. They were yet more evidence in support of a campaign against Aristotelian ideas that Galileo had begun during his own time (as first student, then teacher) at the University of Pisa. This was a period where in his unpublished notes he refers to the works of Philoponus, whose books had first begun to appear in print some 50 years earlier.

Philoponus Regained

The man primarily responsible for the new appearance of these and other books had been Cardinal Bessarion, whose palazzo in the late 1450s had become an outpost of Greek culture in the centre of Rome. Learned refugees from the Byzantine Empire were employed there to transcribe and make Latin translations from ancient Greek manuscripts, while Bessarion himself (who was born in Trebizond and had been a metropolitan in the Orthodox Church before becoming a Catholic cardinal) attempted to promote a crusade to free Byzantium from the Turks. The cardinal's crusading efforts were a failure, but his cultural endeavours bore considerable fruit. The library of Greek codices that he eventually presented to Venice in 1468, and which formed the nucleus of the Biblioteca Marciana, was the largest yet assembled in Western Europe.

Bessarion had not been the first person in Italy to form a library of Greek manuscripts. In 1397 a Byzantine scholar called Chrysoloras had arrived in Florence and begun teaching Greek to a group of enthusiastic students. One of his pupils, Guarino da Verona, had followed Chrysoloras back to Byzantium to

[2] G. Galilei, *The Essential Galileo* (2008), 106–7.

continue his studies, and had returned with a haul of 50 ancient manuscripts. In 1423 Giovanni Aurispa had brought some 238 classical manuscripts back from Byzantium. Others like Poggio Bracciolini had gone on manuscript-hunting expeditions in monastic libraries in France, Germany, and Switzerland. Bessarion, however, outdid them all. At a reputed cost of 30,000 florins he assembled more than 800 Greek codices, and among these were nearly all the surviving texts of Philoponus. Manuscripts of Philoponus were also listed in the earliest catalogue of the Vatican library made in 1475, as well as in the famous library of Pico della Mirandola. Pico's nephew, Gianfrancesco, refers to the manuscript in his uncle's library in his own most substantial work, talking about 'that Greek author most illustrious in the Peripatetic family'.

During the 1490s both uncle and nephew came under the influence of the fiery preacher Girolamo Savonarola, whose sermons had electrified Florence, and this early experience left Gianfranceso with the conviction that all human philosophy was fruitless. His later career was dedicated to attacking the prevailing Aristotelian tradition and he used Philoponus' argument that motion in vacuum did not involve a contradiction, to show that 'the dogma of Aristotle cannot be true'.[3]

Philoponus' commentary on the *Posterior Analytics* first appeared in print in 1504. When *Against Proclus* was printed in 1535 the editor 'rejoiced at having the good fortune to come across the work and be able to present it to the public'.[4] The commentary on *Physics* appeared the same year, and a Latin translation of it was reprinted nine times between 1546 and 1581 [Fig. 22.1].

These newly discovered writings were not always well received. Defenders of Aristotle generally followed Simplicius (whose writings were also beginning to appear in print) in their rejection of Philoponus' arguments. Although some hailed Philoponus as '*Christianus Philosophus*',[5] others did not. One of Galileo's own teachers at the University of Pisa, Francesco Buonamici, in a vast treatise on motion, accused 'the Alexandrian philosophers' of 'wishing to be seen as Christian when they deal with Aristotle' with the result that they had 'fallen into being simultaneously pseudo-philosophers and pseudo-Christians'.[6]

In Galileo's manuscript notes on motion, written at the same time, Philoponus is much more favourably mentioned. Galileo cites Philoponus a number of times (more often than Plato),

[3] R. Sorabji, *Philoponus and the Rejection of Aristotelian Science* (1987), 219.
[4] R. Sorabji, *Philoponus and the Rejection of Aristotelian Science*, 213.
[5] R. Sorabji, *Philoponus and the Rejection of Aristotelian Science*, 213.
[6] R. Sorabji, *Philoponus and the Rejection of Aristotelian Science*, 223.

22.1 Frontispiece from a Latin translation of Philoponus' *On the Eternity of the World.*

describing him as one of those forced 'by the power of truth' to realize the falsity of Aristotle's views, and in all that he says about 'motion in a void' and 'impressed force' echoes Philoponus' arguments even where he does not directly cite him. Galileo qualifies his approval of Philoponus by including him in a list of those who arrived at truth 'by belief rather than by real proof'.[7] Nevertheless in one instance, he directly replicated the Alexandrian's experimental observation.

The Pisan teacher that Galileo perhaps reacted against most strongly during his student days was called Girolamo Borro. Borro had set his face against what he seems to have regarded as the new-fangled fashion for Platonism with its emphasis on the use of mathematics in the study of nature. He emphasized

[7] G. Galilei, *On Motion, and on Mechanics* (1960), 49.

instead Aristotle's insistence on the importance of observation and experience, but seems to have practiced this with a certain lack of rigour.

Aristotle had stated that objects of the same material but different weights would fall with a speed proportional to their weight. Borro accepted this without testing it directly, but to demonstrate his own theory (that materials that contained more air would fall faster) would repeatedly hurl pieces of wood and iron, that appeared roughly the same weight, out of the upper window of his house (the streets of Pisa could be a hazardous place). His students dutifully reported that the wood arrived first.

Galileo, however, describes in his notes carrying out many times a 'periculum'—meaning a test or experiment—in which he dropped lead and wood from 'a high tower' (his pupil Viviani informs us that this was the leaning tower). In his experience if wood and lead were 'both let fall from a high tower, the lead moves out far in front. This is something I have often tested [*De hoc saepe periculum feci*]'.[8] He goes on though to argue the need for a 'firmer hypothesis'. When later he found such a hypothesis and tried dropping differently weighted objects of the same material, he discovered that, as Philoponus had noted a thousand years earlier, both objects struck the ground at approximately the same time (see Chapter Twenty-Six).

In 1592 Galileo had moved from Pisa to Padua where he was appointed professor of mathematics. It was during his time there (the happiest period of his life he later recalled) that, with the help of a new instrument of discovery, he was able to dramatically extend his attack on Aristotle's picture of the world.

The Starry Message

On a visit from Padua to Venice Galileo heard news 'that a Dutchman had constructed a spyglass by means of which certain objects although at a great distance from the eye of the observer were distinctly seen as if near'.[9] The idea proposed in Oxford by Robert Grosseteste and Roger Bacon, of using the refraction of a lens to achieve this effect, had finally been realized by two Dutch spectacle-makers: Hans Lipperhey and Sacharias Jansen. Galileo immediately set to work grinding his own lenses and succeeded in producing much greater magnification. So much so that he felt justified in announcing himself as 'the inventor' of a spyglass that he attempted to sell to the Venetian state, and began to manufacture in Padua [Fig. 22.2].

22.2 Galileo's telescope.

8 W. A. Wallace, 'Dialectics, Experiments and Mathematics in Galileo', in *Scientific Controversies*, ed. Machamer, P., Pera, M., and Baltas, A. (2000), 106.

9 G. Galilei, *The Essential Galileo*, 49.

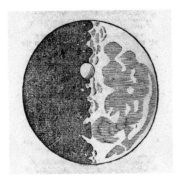

22.3 Galileo's sketches of the Moon.

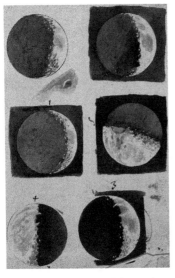

22.4 Five diagrams of the surface of the Moon, during its phases. Aquatint after Galileo Galilei.

When he directed this instrument up at the night sky he immediately started to find evidence that the heavenly bodies were not the perfect and unchanging entities that Aristotle had supposed. Rather they showed every sign of being formed, as *Against Aristotle* had argued, from the same kind of elements as the Earth. The Moon was 'not perfectly smooth, free from in-equalities and exactly spherical (as a large school of philosophers holds with regard to the moon and the other heavenly bodies), but that on the contrary it is full of inequalities, uneven, full of hollows and protuberances, just like the surface of the earth itself, which is varied everywhere by lofty mountains and deep valleys' [Fig. 22.3].[10] A cavity at the centre, in fact, 'produces the same appearance . . . as an area like Bohemia' [Fig. 22.4].[11]

His most dramatic discovery (made in January 1610) re-vealed another similarity between our own world and one of the planets: 'four planets which our sense of sight present to us circling around Jupiter (like the moon around the earth)',[12] and in a series of diagrams he detailed the observations which had forced him to that conclusion [Fig. 22.5]. This was 'The Starry Message', which he used as the title of the book which he rushed into print to describe his discoveries. It concluded by promising that 'the gentle reader may expect more soon'. In 1612 he published further discoveries that continued to under-mine the idea that the heavens were a realm of geometrical perfection. Now Galileo claimed the Sun had spots (*maculae*) which rotated in a circular motion, Saturn had 'ears' (which later turned out to be rings), and Venus had phases like the Moon.

The Starry Message turned Galileo into an international ce-lebrity. The English ambassador in Venice, Sir Henry Wotton, reported home that 'the author runneth a fortune to be either exceeding famous or exceeding ridiculous';[13] it soon became apparent that it was to be the former. When Galileo visited Rome in 1611 he was feted by the establishment. Cardinal Robert Bellarmine had asked the Collegio Romano whether they could confirm Galileo's discoveries, and this (after a new telescope was constructed) they duly did.

Galileo's telescope was set up in Cardinal Bandim's garden on the Quirinal where he could demonstrate his discoveries to an admiring crowd. Meanwhile a series of banquets was held in his honour. One of these was held by the Academia de Lincei (the Academy of Lynxes) that had been founded in

[10] G. Galilei, *The Essential Galileo*, 52.
[11] G. Galilei, *The Essential Galileo*, 56.
[12] G. Galilei, *The Essential Galileo*, 84.
[13] L. P. Smith, *The Life and Letters of Sir Henry Wotton* (1907), vol. I, 486–7.

1603 by the Marquis of Monticelli to forward the study of nature and mathematics. Galileo was elected the sixth member of
this group (where his spyglass was first called a 'telescope') and
thereafter regularly signed his name 'Galileo Galilei Linceo'.

The crowning glory was a public lecture endorsing his discoveries held at the Jesuit Collegio Romano in the presence of
dukes, cardinals, and Galileo himself. Christopher Clavius, the
most famous Jesuit astronomer of the time, called him 'the second Ptolemy' and added an outline of the new discoveries to
his own textbook, noting that 'astronomers ought to consider
how the celestial orbs should be rearranged to model these
phenomena'.[14]

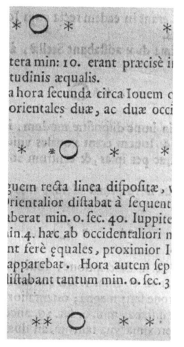

22.5 Galileo's notes of the
'Medicean stars'—the moons of
Jupiter.

[14] J. Hannam, *God's Philosophers*
(2009), 313.

CHAPTER TWENTY–THREE

Free Philosophizing

Some 70 years earlier, in 1543, a Polish canon at Frombork Cathedral, Mikołaj Kopernik—Nicholas Copernicus—had published a book *On the Revolutions of the Heavenly Spheres*, which argued (as the Bishop of Lisieux had suggested) that the Earth and the other planets all revolved around the Sun. The first printed copy of the book was placed in the author's hands on the day that he died, but Copernicus' idea (which was the fruit of a lifetime of astronomical observation and mathematical calculation) had already been widely circulated. In 1533 the Pope's secretary, Johann Widmannstetter, had delivered a series of lectures about Copernicus' theory to Clement VII and two of his cardinals, which they heard with great interest.

Galileo was probably first introduced to Copernicus' idea by Buonamici. His earliest works show no sign that he had been convinced by the idea, but by 1597 he had changed his mind. In a correspondence with a philosopher friend, Jacopo Mazzoni, he was able to show by simple trigonometry that Mazzoni's objections to Copernicus were invalid. Two months later, in letter to a German astronomer, he was claiming to have come over to Copernicus' opinion many years ago, but fearing ridicule had said nothing. When his correspondent urged that they should join together in speaking out, Galileo did not reply, until his telescopic discoveries convinced him to do so. In making public the discoveries and the conclusions they pointed to, he was responding, it seemed to him, to the evidence which 'divine grace' had placed in his hands.

A Beam of Wisdom

Galileo's modern status as a hero of secular thought can make this language seem unexpected, but it was of a piece with his life and thought. He had begun his education as a novice at the monastery of Santa Maria di Vallombrosa where, like Castelli, he might have gone on to become a friar had not his father withdrawn him (in fact he ended his life as a cleric, wearing a tonsure from 1630, and according to his own account, reciting a daily office[1]). As it was, his earliest lecture, 'On the Shape, Location and Size of Dante's Inferno', uses mathematics to explicate a religious poetic text about the invisible world. In a parallel way his later investigations of the visible world are often described in the terminology of religious revelation.

The Starry Message begins by describing 'the spyglass devised by me, through God's grace first enlightening my mind',[2] and in the follow-up book, *Letters on Sunspots*, published three years later by the Lyncean Academy, he develops this theme. In a letter about the book to the Vatican prelate Piero Dini, he writes that 'One should not lose confidence that the divine Goodness sometimes decides to instil a beam of its immense wisdom in humble minds, especially when they are adorned with sincerity and holy zeal.'[3] In *Letters on Sunspots* he argues that while mathematical properties 'run through the divine mind like light in an instant', mathematical demonstrations proved that 'the human mind is a work of God and one of the most excellent'.[4] This was not to deny the frailty and limitations of human wisdom. In the third *Letter on Sunspots* he contrasts the intellectual light 'for which we now search almost like blind men in the impure and material sun', with that which 'shall come to us from the grace of the true pure and immaculate sun together with all other truths in him'.[5] Nevertheless, a trust in 'the divine goodness' should engender, he argues, a basic intellectual optimism.

Looking back later in life to his time in Padua, he described how 'during mass one morning an idea flashed into my mind, and after I had immersed myself in it, I went to confirm for myself and since then . . . have been considering it as worthy of admiration as a marvellous example of how nature works'.[6] This kind of experience was to be expected. It is through 'divine grace', he argues, that we are 'able to philosophise better about other and more controversial qualities of natural substances', while the same grace 'finally by elevating us to the ultimate end of our labours, which is the love of the divine Artificer . . . will keep us steadfast in the hope that we shall learn every other truth in Him, the source of sight and verity.'[7]

Galileo's theologically informed intellectual optimism made him particularly anxious that the Church should not stumble into the cul-de-sac of condemning Copernicanism.

[1] J. L. Heilbron, *Galileo* (2010), 299.
[2] G. Galilei, *The Essential Galileo* (2008), 49.
[3] R. J. Blackwell, *Galileo, Bellarmine, and the Bible* (1991), 212.
[4] P. K. Machamer, *The Cambridge Companion to Galileo* (1998), 187.
[5] P. K. Machamer, *The Cambridge Companion to Galileo*, 187.
[6] P. K. Machamer, *The Cambridge Companion to Galileo*, 198–9.
[7] G. Galilei, *Discoveries and Opinions of Galileo* (1957), 124.

Inquisitions

The risk of this arose almost immediately after the establishment of the Roman Holy Office of the Inquisition, which Pope Paul III had created in 1542 as part of his campaign to combat the new rapidly spreading Lutheran heresy. The following year he was presented with a copy of Copernicus' *De Revolutionibus*. Paul passed it over to his personal theologian, who died before he could review it. It was then passed to a Dominican friar, Giovanni Maria Tolosani. Tolosani, in an appendix to a treatise *On the Truth of Sacred Scripture* written in 1545, denounced Copernicus as a braggart and a fool. For nearly seventy years nevertheless, Copernicus' text had not been placed on the Inquistion's *Index of Prohibited Books*.

The first hint of this possibility had been a report in November 1612 that a Dominican friar in Florence, Niccolò Lorini, had observed that 'the opinion of Ipernicus, or whatever he was called seemed to be contrary to scripture'.[8] Castelli's breakfast with the Medici had taken place the following December. A year later on 20 December 1614 another Dominican friar in Florence, Tommaso Caccini, had used a scriptural lecture to denounce Copernicanism from the pulpit of Santa Maria Novella, and to decry 'mathematics' (meaning probably astrology) as a diabolical art.

When Friar Lorini read Galileo's *Letter to Castelli*, he assumed it was a reply to Friar Caccini's sermon. He forwarded an inaccurate copy of it to Cardinal Sfondrati, secretary of the Roman Inquisition. Enclosed with it was a complaint on behalf of his fellow Dominicans at San Marco that the 'Galileists' were spreading all sorts of 'impertinences' in Florence. Caccini travelled to Rome where he went in person to the Inquisition, repeating his complaint and adding dark hints that Galileo was in correspondence with German heretics.

Galileo, however, was not without clerical supporters. He had received a letter from a Dominican preacher-general in Rome, just before Caccini's visit to the city, apologizing for the nuisance he was being caused, and a Carmelite friar, Antonio Foscarini, had just written a book arguing the compatibility of Copernicanism and Scripture. Meanwhile, using his own contacts Galileo had already begun to take steps to further the Copernican cause. In February he sent an accurate copy of the *Letter to Castelli* to his Vatican friend Piero Dini, with a request

[8] M. Sharratt, *Galileo* (1994), 109.

LETTERA
DEL SIGNOR
GALILEO GALILEI
ACCADEMICO LINCEO
SCRITTA ALLA
GRAN DUCHESSA
DI TOSCANA.
IN CUI

Teologicamente,e con ragioni faldiffime,cavate da'Padri più fen-
titi,fi rifponde alle calunnie di coloro,i quali a tutto potere
fi sforzarono non folo di sbandirne la fua opinione intorno alla conftituzione delle parti dell'Univer-
fo, ma altresì di addurne una perpetua
infamia alla fua perfona.

IN FIORENZA,
MDCCX.

23.1 Title page of Galileo's *Letter to the Grand Duchess Christina.*

that he might show it to Cardinal Bellarmine (the Jesuit theo-logical advisor of the Pope), which he duly did. At the same time with the aid of Castelli, who recruited a helpful Barnabite priest to find appropriate quotations from the doctors of the Church, he set about expanding the *Letter* into a longer treatise to be called *Letter to the Grand Duchess Christina* [Fig. 23.1].

A Letter to the Duchess

In his *Letter to Castelli* Galileo had argued that there was no contradiction between science and Scripture. Holy writ was given to teach us the path to salvation but it was not neces-sary 'to believe that the same God who has furnished us with senses, language and intellect would want to bypass their use and ... the information we can obtain with them'.[9] He then

[9] G. Galilei, *The Essential Galileo*, 106.

argued that the passage in the Book of Joshua which describes God lengthening a day of battle by making the Sun stand still (which was used by opponents of the Copernican theory to show it contradicted Scripture) actually made better sense if Copernicus was right (this was because according to Ptolemy's theory the Sun's own motion was 'contrary to the diurnal turning' from west to east while the primum mobile moved it in the other direction. Hence if the Sun's own motion was stopped, the length of the day according to that theory would become shorter not longer).

This detailed kind of exegesis risked violating the edict that the Council of Trent had issued in 1564, which decreed that Scripture could not be interpreted in any way other than in accordance with the unanimous agreement of the fathers. In 1615 some friendly advice passed on from his fellow Florentine, Cardinal Barberini, warned Galileo not to provoke the theologians by interpreting Scripture. In his letter to the duchess therefore he omits the detailed exegesis and concentrates on showing that an unbiased investigation of Copernicanism would be entirely consistent with the teaching of the fathers.

He begins as before by suggesting that his opponents 'lack confidence to defend themselves as long as they remain in the philosophical field' and have therefore 'decided to try to shield the fallacies of their arguments with the cloak of simulated religiousness and with the authority of the Holy Scriptures'.[10] Despite the aspersions that have been cast, Copernicus was 'not only a Catholic'[11] but a priest and a canon, while for himself his only goal is that 'if these reflections . . . lead someone to advance a useful caution for the Holy Church in her deliberations about the Copernican system, then let it be accepted . . . if not, let my essay be torn up and burned, for I do not intend or pretend to gain from it any advantage that is not pious or Catholic.'[12]

He then proceeds to quote long passages from St Augustine. In one place Augustine had written that 'It is also customary to ask what one should believe about the shape and arrangement of heaven according to our Scriptures. In fact, many argue a great deal about these things, which with greater prudence our authors omitted, which are of no use for eternal life to those who study them'.[13] In another passage he had remarked that 'Some brethren have also advanced a question about the motion of the heaven, namely whether heaven moves or stands

[10] G. Galilei, *The Essential Galileo*, 111.
[11] G. Galilei, *The Essential Galileo*, 112.
[12] G. Galilei, *The Essential Galileo*, 115.
[13] G. Galilei, *The Essential Galileo*, 118.

still', but had replied 'that these things should be examined with very subtle and demanding arguments, to determine truly whether or not it is so; but I do not have the time to undertake and to pursue these investigations', and must concentrate on instructing people 'for their salvation'.[14] Augustine's position was almost exactly the same as 'an ecclesiatical person in a very eminent position' (almost certainly Cardinal Baronio), 'that the intention of the Holy Spirit is to teach how one goes to heaven and not how heaven goes'.[15] How then, Galileo asks, 'can one now say that to hold this rather than that proposition on this topic is so important that one is a principle of faith and the other erroneous? Thus, can an opinion be both heretical and irrelevant to the salvation of souls?'[16]

In the Book of Ecclesiastes (3:13) the Bible declares that 'God has delivered the world to their consideration so that man cannot find out the work which God hath made from beginning to the end'. Consequently Galileo argues, 'one must not, in my opinion, contradict this statement and block the way of freedom of philosophizing about things of the world and of nature, as if they had all already been discovered and disclosed with certainty'.[17] To do so and to try 'to banish the opinion in question from the world' it would be necessary, he argues, 'not only to prohibit Copernicus' book and the writings of the other authors who follow the same doctrine', but even 'to ban all astronomical science completely . . . to forbid men to look toward the heavens'.[18] To do that, however, would be 'to reject hundreds of statements from the Holy Writ, which teach us how the glory and the greatness of the supreme God are marvelously seen in all of His works and by divine grace are read in the open book of the heavens.'[19]

Who could do such a thing? 'No one can doubt', he admits, 'that the Supreme Pontiff always has the absolute power of permitting or condemning'; however, no creature has the power of making things 'be true or false, contrary to what they happen to be by nature and de facto'.[20] 'In brief', he concludes, 'it is inconceivable that a proposition should be declared heretical',[21] while we remain in doubt as to its truth. With this conviction and against advice to let sleeping dogs lie, he set out for Rome to clear his name and to put the case for Copernicanism in person. He succeeded in the former, while the latter campaign culminated in his persuading Cardinal Orsini to put the case directly to the Pope.

[14] G. Galilei, *The Essential Galileo*, 119.
[15] G. Galilei, *The Essential Galileo*, 119.
[16] G. Galilei, *The Essential Galileo*, 119.
[17] G. Galilei, *The Essential Galileo*, 121.
[18] G. Galilei, *The Essential Galileo*, 121.
[19] G. Galilei, *The Essential Galileo*, 128.
[20] G. Galilei, *The Essential Galileo*, 140.
[21] G. Galilei, *The Essential Galileo*, 140.

This was a strategy that backfired. Paul V was inclined to condemn Copernicanism outright but was persuaded to refer the question to the Inquisition, who responded in 1616 by listing Copernicus' book on the *Index Librorum Prohibitorum* as 'suspended until corrected'. Meanwhile Cardinal Bellarmine summoned Galileo, gave him a certificate to this effect, and, according to an unsigned record of the interview, extracted a promise from him not to hold, teach, or defend Copernicanism in any way.

For the next eight years Galileo complied in public with Bellarmine's restriction. Apparent release from this imposed silence came, however (as it had for Roger Bacon 400 years earlier), with the election of a new pope.

CHAPTER TWENTY–FOUR

The Freedom of Intellect

Eight months after his landslide election to the papacy Urban VIII, formerly Maffeo Barberini, welcomed Galileo as an honoured guest to the Vatican. The Barberinis were one of the great merchant families of Galileo's native Florence, to where in 1610 (following his strategic naming of the moons of Jupiter) he had returned as court philosopher and mathematician to the Medici. The new Pope to that extent was already a natural ally. Cardinal Barberini had, in fact, long admired him on other grounds.

In August 1620 he had written a poem, *Adulatio Perniciosa* (Harmful Adulation), praising Galileo's rhetorical skills. Three years later Galileo returned the compliment by dedicating his latest book to the cardinal. *Il Saggiatore* (The Assayer) appeared with the three bees of the Barberini crest on its title page. Even before Galileo received his own copies the supreme pontiff was reported to be much enjoying the book as it was read to him at meals [Fig. 24.1].

The Assayer

The Assayer was Galileo's latest contribution to a debate that had begun with the appearance in August 1618 of the first of three comets. These had been observed by the Jesuits of the Collegio Romano using their Galilean telescope. Orazio Grassi, the professor of mathematics at the college, had published a report in which he had argued that the comet was located between the Moon and the Sun, supporting this with arguments that included the confused suggestion that the more distant an object, the less it was magnified by a telescope. Galileo first heard of this publication in a letter which warned him that some people (though not the Jesuits) were saying it overthrew Copernicus' system. This was enough for him to launch (or at least sponsor) an attack. His one-time pupil Mario Guiducci published a *Discourse on Comets* (which may have been largely the work of

24.1 Title page of Galileo's *Il Saggiatore* (The Assayer).

Galileo himself), which dismissed the Aristotelian contrast between heaven and Earth as illusory and argued that it was impossible to determine the location of comets. Grassi had replied in a book called *The Astronomical Balance* (using the nom de plume of his pupil Lothario Sarsi). Galileo in turn responded with *The Assayer*, which was written in the form of a letter to his friend Virginio Cesarini, attacking the views of this 'Sarsi'.

The Copernican theory was not the subject under discussion in any of these exchanges, but it hovers in the background in all of them. *The Assayer* reads almost as a manifesto of Galileo's approach to natural philosophy. If he could not advocate Copernicus' system directly, he could nevertheless, as one of his biographers has put it, 'teach a method which would inevitably transform all natural philosophy and thus sooner or later establish a true system of the universe and demolish Aristotelian physics'.[1]

The Manifesto

The Assayer begins by rehearsing the ill treatment that Galileo had received as a result of his discoveries. Much of the rest of it is then taken up with showing that Sarsi's assertions are only conjectures and that he has failed to understand the principle of the telescope. Among these specific arguments, however, are several more general statements.

The first of these is an attack on the Jesuit's public commitment to Aristotelianism. In Sarsi, Galileo says, he discerns 'the firm belief that in philosophising one must support oneself upon the opinion of some celebrated author', as though philosophy was a work of fiction in which the truth did not matter. Well, he informs Sarsi, 'that is not how things stand'. Rather he asserts, in a famous restatement of the credo of Grosseteste et al.:

> Philosophy is written in this grand book, the universe, which stands continually open to our gaze. But the book cannot be understood unless one first learns to comprehend the language and read the letters in which it is composed. It is written in the language of mathematics and its characters are triangles, circles and other geometric figures without which it is humanly impossible to understand a single word of it.[2,3]

Where Sarsi seems to think 'that our intellect should be enslaved to that of some other man' and that 'all the host of philosophers may be enclosed within four walls', Galileo is sure that 'they fly and fly alone like eagles'.[4] Later on he returns to this theme in a still more pointed way. There is an easy way, he suggests, 'to find out how much force human authority has upon the facts of nature' and that is to put it to the test. 'I cannot but remain astonished', he writes, 'that Sarsi should persist in trying to prove by witnesses something that I may see for myself by means of experiment at any time'.[5] Sarsi may say that he does not wish to be among those who affront the sages by disbelieving and contradicting them but, counters Galileo, 'I do not

---

[1] M. Sharratt, *Galileo* (1994), 139.

[2] G. Galilei, *Discoveries and Opinions of Galileo* (1957), 238.

[3] It seems to be a recurrent theme for the penultimate curiosity, at least for the more physical aspects, that mathematics and geometry provide essential intellectual skills. To describe it as language suggests that it can be translated. But mathematics, like reason, is a methodology that has to be learned. It provides a way of thinking that can no more be translated into non-mathematical language than a symphony can be translated into a sculpture. See G. A. D. Briggs, 'The Search for Evidence-Based Reality', in *The Science and Religion Dialogue: Past and Future*, ed. M. Welker (2014), 201–15.

[4] G. Galilei, *Discoveries and Opinions of Galileo*, 239.

[5] G. Galilei, *Discoveries and Opinions of Galileo*, 271.

wish to be counted as an ignoramus and an ingrate toward Nature and toward God; for if they have given me my sense and my reason, why should I defer such great gifts to some error of man? Why should I believe blindly and stupidly what I wish to believe and subject the freedom of my intellect to someone who is just as liable to error as I am?'[6]

This finally leads Galileo to consider the differences between so-called 'primary' and 'secondary' qualities: the differences, that is, between those qualities which reside in our senses and those that are essential properties of an object (he gives the example of our experience of tickling and the construction of a feather). Simply naming a quality 'heat' (in the manner of an Aristotelian philosopher) does not, he argues, lead to a fundamental explanation of its nature, which must be found in the deep 'corpuscular' character of physical objects. This, however, he is aware goes far beyond anything he can demonstrate, not wishing 'to engulf myself inadvertently in a boundless sea from which I might never get back to port'.[7]

Urban's enjoyment of this manifesto was encouraging. In the hope that he might persuade the Pope to change the Church's attitude to Copernicanism Galileo travelled to Rome in April 1624. At their first meeting the two men had talked for an hour. This was followed by five more audiences. The last of these concluded with the Pope presenting Galileo with a painting and some medals, promising a pension for his son, and the same day sending an official testimonial to the Grand Duke.

Urban did not reverse the decree of 1616 (though he did later admit to Father Castelli that if he had been Pope at the time it would not have been promulgated). Nor did he give permission to ignore it. Galileo did, however, come away from these meetings with the impression that he was free to discuss the two opposing world systems so long as he did not assert that Copernicanism was true.

Urban had previously suggested to him that no mathematical model could produce a definite answer to the question. Aristotle himself had argued that mathematics could not be a tool in natural philosophy (but only in 'subordinate sciences' such as mechanics), and Aristotelian natural philosophers held mathematics (which did not deal directly with material realities) to be distinct from mathematics and astronomy. Mathematical models could provide predictability—i.e., save the phenomena—but not could provide a true explanation of

[6] G. Galilei, *Discoveries and Opinions of Galileo*, 272.
[7] G. Galilei, *Discoveries and Opinions of Galileo*, 279.

the causes involved.[8] Hence Urban asked that both positions should be described but neither advocated, and that his own view should be stated.

With this encouragement Galileo returned to Florence. Eight years later he published *A Dialogue Concerning the Two Chief World Systems.*

---

[8] P. Harrison, *The Territories of Science and Religion* (2015), 110f.

CHAPTER TWENTY-FIVE

Simplicius Reborn

Galileo's *A Dialogue Concerning the Two Chief World Systems* takes the form of a conversation conducted over the course of four days. A discussion between three men takes place in the Venetian palazzo of one of the three who is called Sagredo. Sagredo (named after one of Galileo's friends) is an intelligent layman who is at first neutral in the debate between the other two. Galileo's own arguments are put into the mouth of one of these two: Salviati (named after another friend) [Fig. 25.1].

The third character, Simplicius, is described as a philosopher 'whose greatest obstacle in apprehending the truth seemed to be the reputation he had acquired by his interpretations of Aristotle'. Galileo tells us that this character (who is recognizably modelled on two of his opponents) 'because of his excessive affection towards the commentaries of Simplicius, I have thought fit to leave . . . under the name of the author he so much revered'.[1]

The discussion on the first day centres on the argument of Aristotle's *De Caelo*, which had been the subject of the historical Simplicius' famous anti-Philoponian commentary. Salviati proposes that they examine Aristotle's thesis that nature consists of two utterly different substances: a celestial substance which is eternal and unchanging, and an elemental which is temporary and destructible. He will propound his own views which can then be criticized by Simplicius, 'that stout defender and champion of Aristotle's doctrines'.

Simplicius does his best in this contest, but runs into difficulties when Salviati introduces Galileo's telescopic discoveries and begins to refer to such obvious evidence of change as the rotating spots on the Sun and the cratered surface of the Moon. Simplicius tries to insist that Aristotle argues from first principles, but Salviati counters that if Aristotle ('who preferred sensible experience to any argument') had been able to use a telescope he would have changed his mind. The 'honest hour' of the day being past, Sagredo suggests moving to a

[1] G. Galilei, *Dialogue Concerning the Two Chief World Systems* (1953), 7.

25.1 Title page of Galileo's *A Dialogue Concerning the Two Chief World Systems.*

gondola to enjoy the cool of the evening and returning to the debate the next day.

Having shown that the fundamental assumption of the Aristotelian system is mistaken, the debate, when it resumes, turns to the question of whether the Sun moves round the Earth or visa versa. The arguments on day two deal with common objections to the Copernican system (such as that if the Earth moved, it would create such a gale that we would be swept off the ground). The arguments on day three then go on to show that Copernicus' system accounts for the movements of the planets in a way that hangs together much better than the alternative.

Finally on day four of the dialogue Galileo introduces the phenomenon of the tides, which he mistakenly thought was a conclusive argument in favour of Copernicanism. Even Simplicius admits that this is a powerful argument but reminds the others of 'a doctrine which I learn't from a most ingenious person and with which it is necessary to set down',[2] namely that God could if he wished 'confer upon the element of water the reciprocal motion that we observe in some other way'.[3] Since they must admit he could do so, they must also admit that 'it would be an extravagant boldnesse for anyone to goe about and limit the Divine Power and wisdom to some particular conjecture of their owne'.[4] All agree with this 'Angelical doctrine' and according to custom the three adjourn to 'spend an hour taking the air in the gondola that waiteth for us'.[5]

Vehemently Suspect

The dialogue was published in February 1632 but by August it had been suspended. When the Tuscan ambassador had an audience with the Pope in the September of that year, Urban 'errupted in anger'. To have placed what amounted to a parody of the Pope's argument on the lips of Simplicius was bad enough, but when the unsigned record of Galileo's meeting with the now-deceased Cardinal Bellarmine came to light, deliberate deception was suspected. Galileo was peremptorily summoned to Rome, and when he protested age and illness, was threatened with being brought there in chains.

At his subsequent trial in front of the Inquisition he stubbornly maintained that he had not taught or held Copernicanism after 1616. The judges refused to believe this obviously untrue assertion and found him 'vehemently suspect of heresy'. His book was to be publicly condemned, while he was compelled to kneel, dressed in penitential garments, and abjure his former beliefs. Although subject to formal imprisonment he was allowed to move first to the Archbishop of Sienna's palace, and later to live under house arrest at his villa in Arcetri outside Florence.

Once back in Arcetri, however, he almost immediately began work on a second dialogue in which Salviati and Sagredo resume their discussion with Simplicius.

[2] T. Salusbury, *Mathematical Collections and Translations* (1661), 424.
[3] T. Salusbury, *Mathematical Collections and Translations*, 424.
[4] T. Salusbury, *Mathematical Collections and Translations*, 424.
[5] T. Salusbury, *Mathematical Collections and Translations*, 424.

CHAPTER TWENTY–SIX

The Creation

The Archbishop of Sienna, Ascanio Piccolomini, encouraged Galileo to resume his studies. Under his solicitous promptings Galileo recovered from his depression, returned to his early thoughts about motion, and began to arrange both old ideas and new in dialogue form [Fig. 26.1].

Discourses and Mathematical Demonstrations Concerning Two New Sciences begins in a resolutely down-to-earth way by considering some of the everyday problems encountered in 'the famous arsenal of Venice'—why beams snap and boats break their back [Fig. 26.2].[1] The study of the strength of materials is the first of the 'two new sciences' and appears at first sight to have nothing to do with Galileo's more controversial works.

Salviati, continuing his role as the spokesperson for Galileo's own thoughts, continues to pursue his anti-Aristotelian agenda. In the *Dialogue* he had begun by rejecting the claim 'that nature consists of two utterly different substances'. Here he starts by asserting that 'Since I assume matter to be unchangeable and always the same, it is clear that we are no less able to treat this constant and invariable property in a rigorous manner than if it belonged to simple and pure mathematics.'[2]

Cosmos and Cannonballs

The first dialogue begins with a series of these mathematical demonstrations before going on to provide experimental refutations of two of the basic principles of Aristotelian physics. In an echo of Philoponus' remarks about 'snatchers' Salviati shows how it is possible to demonstrate that a vacuum exerts a real force: 'two flat pieces of glass . . . exquisitely planed', though they can move horizontally, 'if you try to pull them vertically you will encounter great resistance'.[3] Later it

[1] One of us as the holder of the Professorship of Nanomaterials at Oxford has a special fondness for this aspect of Galileo's work.

[2] G. Galilei, *The Essential Galileo* (2008), 296.

[3] G. Galilei, *Dialogues Concerning Two New Sciences* (1914), 16.

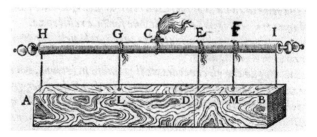

DISCORSI
E
DIMOSTRAZIONI
MATEMATICHE,
intorno à due nuoue scienze
Attenenti alla
MECANICA & i MOVIMENTI LOCALI,
del Signor
GALILEO GALILEI, LINCEO,
Filosofo e Matematico primario del Sereniffimo
Grand Duca di Toscana.

Con vna Appendice del centro di grauità d'alcuni Solidi.

IN LEIDA,
Appreffo gli Elfevirii. M. D. C. XXXVIII.

26.1 Title page of Galileo's *Discourses and Mathematical Demonstrations Concerning Two New Sciences* (1638).

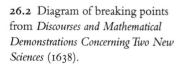

26.2 Diagram of breaking points from *Discourses and Mathematical Demonstrations Concerning Two New Sciences* (1638).

is Sagredo who interrupts a description of Aristotle's account of falling objects:

> Hold Simplicius, for I myself have proved by experience that if a common bullet weighing 100, 200 or more pounds let fall from a height of 200 yards and at the same instant a musket ball that weighs but half a pound, at the arrival of the cannon bullet at the ground the other will not want a hand's breadth of it.[4]

This account belongs to the second of Galileo's new sciences: the study of the motion of objects. Throughout each discussion in his book every mathematical argument is supported by clearly described physical demonstrations. In the third dialogue he uses the mean speed diagram developed by Nicholas Oresme to show that the distance an object falls is proportional to the square of the time elapsed [Fig. 26.3].

Galileo follows this up by describing a controlled experiment in which he rolled 'a smooth and very round bronze ball' down a slope in a groove lined with parchment. To time the descent he allowed water to flow through a spout which was 'weighed after every descent on a very accurate balance'. 'The collected differences and ratios of these weights, gave us', he discovered, 'the differences and ratios of the times, and this with such accuracy that although the operation was repeated many, many times there was no appreciable discrepancy in the result'.[5]

Finally, in the fourth dialogue Galileo turns to the theme of projectile motion.

Rejecting Aristotle's assertion that no object can move unless moved by some other object, he supersedes Philoponus' idea of impetus with the principle of inertia: 'Imagine any particle projected along a horizontal plane without friction ... this particle will move along this same plane with a motion that is uniform and perpetual, provided the plane has no limits'.[6] He then proceeds to show that a projectile like a cannon ball is acted on by two principles: inertia, which keeps it going in straight line, and gravity, which pulls it down, and consequently moves in a kind of curve known as a parabola. These are principles Sagredo suggests which might apply throughout the cosmos—to planets as well as to cannon balls.

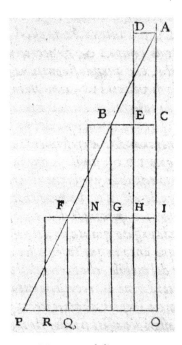

26.3 Mean speed diagram, *Discourses and Mathematical Demonstrations Concerning Two New Sciences* (1638).

---

4 G. Galilei, *Dialogues Concerning Two New Sciences*, 92.
5 G. Galilei, *Dialogues Concerning Two New Sciences*, 213.
6 G. Galilei, *The Essential Galileo*, 357.

The Universal Law

The historical Simplicius had objected on theological grounds to Philoponus' comparison of celestial phenomena to 'fire flies or the scales of fish' with its implication that matter and motion obeyed the same laws throughout the universe: 'how could anyone with a normal mind possibly conceive of such a strange God who ... hands over to nature the generation of the elements one out of another and the rest out of the elements'.[7] To try to understand the universe from this perspective seemed to him 'an extraordinary way for a stupid person to enquire into truth'.[8]

The belief common to the Abrahamic religions that God was 'outside' the physical universe (in a way Simplicius would not admit) had provided its own impetus for Philoponus' argument that the Sun, Moon and stars could 'be given by God their creator a certain kinetic force'.[9] Both Jean Buridan's similar argument that God could impress laws on the celestial orbs which then moved bodies 'without His having to do any more except as a co-agent in all things which take place'[10] and Nicholas Oresme's suggestion that 'when God created the heavens, he put into them the motive qualities and powers ... much like that of a man making a clock and letting it run down and continue its motion by itself'[11] seem to have been arrived at independently. Although direct influence cannot be ruled out, it seems perfectly possible that in each case a trajectory of thought, acted upon by similar principles, described a comparable arc. Similarly, it is not clear whether Galileo (who rarely acknowledged intellectual debts) was helped towards his conclusions by any of his predecessors. What is apparent is that like them he found a congruence between the direction of his scientific thought and his theological understanding.

For once this is put into the mouth of Simplicius, who remarks that 'there must be some great mystery hidden in these true and wonderful results, a mystery related to the creation of the universe ... and also to the seat of the first cause'. To which Salviati replies, 'I have no hesitation in agreeing with you', only to qualify his reply by warning that 'profound considerations of this kind belong to a higher science than ours'. Students of nature must be satisfied 'to belong to that class of less worthy workmen who procure from the quarry the marble out of which, later, the gifted sculptor produces those

[7] S. L. Jaki, *Science and Creation* (1986), 131.
[8] S. L. Jaki, *The Relevance of Physics* (1966), 417.
[9] Philoponus, *Joannis Philoponi De Opificio Mundi Libri* (1897), 28–9.
[10] M. Clagett, *The Science of Mechanics in the Middle Ages* (1959), 536.
[11] N. Oresme, *Le Livre du ciel et du Monde*, ed. Menut, A. D., and Denomy, A. J., trans. Menut, A. D. (1968), 289.
[12] G. Galilei, *Dialogues Concerning Two New Sciences*, 194.

masterpieces'.[12] Even when Sagredo had suggested that what applied to cannonballs might apply to planets Salviati had been cautious, replying that 'he did not wish to speak of it, lest in view of the odium which his many new discoveries had already brought upon him this might be adding fuel to the fire'.[13]

In writing the book Galileo had been walking on eggshells. In order to evade the need for a licence from the Inquisition he had arranged for it to be published by a Dutch Protestant printer. His decision to do so was a straw in the wind. Should penultimate questions about the nature of the physical world fall under the jurisdiction of the Church? Was is necessary or even possible to eliminate the space between 'students of nature' and the study of 'a higher science'? Although a loyal Catholic, Galileo's insistence on 'the freedom of . . . intellect'[14] and his advocacy of the need for 'free philosophising about . . . physical things'[15] seemed, both to opponents (like Caccini) and to some supporters (like his German correspondent), to have a close affinity with the central emphasis of a religious movement that had been rolling through Europe for more than a hundred years.

[13] G. Galilei, *Dialogues Concerning Two New Sciences*, 262.
[14] G. Galilei, *Discoveries and Opinions of Galileo* (1957), 272.
[15] G. Galilei, *The Essential Galileo*, 121.

PART VI

Priests of Nature

CHAPTER TWENTY–SEVEN

A New Era

On a spring evening in 1676 boats began to arrive at the steps of the Dorset Garden Theatre, bringing a fashionable audience, which included King Charles II, to a performance of Thomas Shadwell's new comedy *The Virtuoso* [Fig. 27.2]. While the periwigged playgoers alighting from their craft might have expected broad comedy, Shadwell was interested in stimulating thought about ideas (his fellow playwright John Dryden described him as 'the king of dullness'). Embedded within the comic goings on of his new play was a satirical commentary on an emerging way of thinking, which (like that ridiculed in *The Clouds*) was poised to revolutionize intellectual life.

As the most modern theatre in London, the Dorset Garden with its 'machine house' of theatrical effects was an appropriate venue for such a sharply topical drama [Fig. 27.1]. It was also an ironic one. Built on a site destroyed by the Great Fire ten years earlier, the theatre was part of the programme of reconstruction overseen by the surveyor general Christopher Wren and the city surveyor Robert Hooke, one or other of whom may have

27.1 The Dorset Garden Theatre in a nineteenth-century print based on a 1681 print by Hollar.

27.2 *The Virtuoso* by Thomas
Shadwell.

been involved in its design. Both men were among the leading
'virtuosos' of the day.

The word had originally referred to people with an interest in antiquities, but its meaning had evolved. By the second
half of the seventeenth century it was being applied to a group
of people who were applying themselves to investigate natural phenomena. It was this apparently obsessive preoccupation (which still seemed as eccentric in 1676 as it had to Aristophanes' audience 2000 years earlier) that was the target of
Shadwell's play.

The Lucid Sirloin

The first act opened by introducing the audience to a comic
scenario in which two young men-about-town are trying to
gain access to a pair of high-spirited girls: wards of the repressive but hypocritical Sir Nicholas Gimcrack—the virtuoso of
the title. The second act took the audience inside Sir Nicholas'
house, where he is discovered 'learning to swim upon a table'
and informs the young men that he is 'interested only in the
speculative parts of swimming' not 'the practic', because 'I sel-
dom bring anything to use; 'tis not my way. Knowledge is my
ultimate end'.[1]

This turns out to be true of all his experiments. Towards the
end of the play when the virtuoso's house is surrounded by

[1] T. Shadwell, *The Virtuoso* (1966), Act
II, scene ii, 47.

angry ribbon weavers complaining his mechanical loom will put them out of work, Gimcrack protests that 'I never invented so much as an engine to pare cream cheese with. We virtuosos never find out anything of use, 'tis not our way'.[2] Finally bankrupt and abandoned by all, he decides to 'study for use' and to discover the philosopher's stone.

Shadwell's satire of ideas was calculated to appeal to his royal audience. 'To study for use' was a central theme of what was beginning to be described as 'the experimental philosophy', and it was with this in view that Charles II had granted a charter to the Royal Society in the same year that he had reopened the theatres. Three years later, however, Pepys reported in his diary that the King 'spent an hour or two laughing . . . at Gresham college' (where the society met) 'for spending time only in weighing of ayre and doing nothing else since they sat'.[3]

When Robert Hooke, the society's curator of experiments, came to see the play the following week, he was rather less amused. According to a classical writer called Marinus, at the end of the *The Clouds* when someone in the audience had called out 'who is this Socrates?' the philosopher had stood up and silently bowed. Hooke was more thin-skinned. His diary for the 2nd of June 1676 records 'with Geoffery and Tompion at play. Met Oliver there. Damned dogs. *Vindica mea deus* (God grant me revenge). People almost pointed.'[4]

Hooke had some reason to feel got at. Unlike Aristophanes, Shadwell had done his homework. Gimcrack's experiments with transfusing blood, his nights spent looking through a microscope at 'the fundament of ant', and his collection of bottles of air were all references to actual investigations undertaken respectively by Wren, by Hooke, and by Hooke's erstwhile employer, Robert Boyle. Perhaps the most pointed reference of all is the description in Act V of 'the lucid sirloin of beef'.

In 1672 Boyle had reported in the *Philosophical Transactions of the Royal Society* 'Some observations concerning shining flesh' after his servant had been startled to discover that a leg of veal left in the larder had become luminous. In the play the luminosity is so bright that Sir Nicholas claims that 'I myself have read a Geneva Bible by a leg of pork'.[5] Shadwell's (rather good) joke hit multiple targets. It not only alluded to Boyle's well-known piety and the Puritan sympathies of some members of the Royal Society, but also hinted at a broader association. From the outset, the English virtuosi had linked their novel

[2] T. Shadwell, *The Virtuoso*, Act V, scene iii, 119.
[3] S. Pepys, *Pepys' Diary* (1963), 1663–4.
[4] R. Hooke, *The Diary of Robert Hooke* (1935), 235.
[5] T. Shadwell, *The Virtuoso*, Act V, scene ii, 110.

philosophy to a radical new approach to the Scriptures epitom-
ized by the so-called 'Geneva Bible'.

Reformation Day

This new approach had begun some 250 years earlier. The 31st
of October 1517 has been remembered by Lutheran churches
as 'Reformation Day': the occasion on which the young German theologian Martin Luther nailed 95 'questions for debate'
to the door of Wittenberg's castle church. For Luther himself
a day in July 12 years earlier, when lightning had struck the
ground next to his horse as he was travelling back to university,
had been the turning point of his life. He had sold his books,
left the university, and entered a monastery. The monastic life
had, however, driven him to the edge of despair, and he had
returned, on the advice of his abbot, first to study and then to
teach at Wittenberg University.

It was in the course of lecturing there on St Paul's letter to
the Romans that he had come to an understanding that salvation was a gift from God to be received by faith, not a reward to
be earned by merit. This discovery not only liberated him from
a personal prison of despair, but confirmed his growing conviction that the received interpretations of even the most central teachings of Scripture could not be trusted simply because
they were hallowed by tradition or backed by papal authority.
His own teaching soon reflected this new conviction. In August 1513 he began a series of sermons on the psalms. He arranged for the university printer to provide his students with a
wide-margined text that, instead of being filled with the glosses
and commentaries of past authorities, were left pointedly blank
for them to insert notes and commentaries of their own.

Four years later the sale of so-called 'indulgences'—written
declarations that purported to remit the punishment for sins—
brought Luther into conflict with the church authorities. His
95 theses, all protesting against this practice (which was being
used to finance the building of the new St Peter's), were handed to the Archbishop of Magdeburg and forwarded to Rome.
What transformed this action from a private complaint into
a public issue was not the legendary posting of the theses on
the door of the castle church, but the new medium of print.
In January of 1518 friends of Luther translated a copy of the
Latin theses into German. Within two weeks printed copies of

this version were appearing throughout Germany. Within two months they were being circulated throughout Europe.

When Luther was summoned to a series of meetings and interrogations he appealed over the heads of Church authorities by once again going into print. *To the Christian Nobility of the German Nation* (the first of what was to be a stream of publications) took as its starting point the New Testament idea that every baptized Christian is part of a 'royal priesthood': 'a cobbler, a smith, a peasant, every man has the office and function of his calling, and yet all alike are consecrated priests and bishops'.[6] Consequently, the idea that the Pope alone could interpret the Scriptures was, he argued, without foundation.

Following his final interrogation at the 'Diet of Worms' in 1521, Luther was declared an outlaw. He was concealed by the Elector of Saxony in a room under a staircase in Wartburg Castle at Eisenach. There he set about putting into the hands of cobblers and smiths the means to realize their own priesthood by translating the Bible into German. In doing so he set an example which was rapidly followed.

In 1526 a complete Dutch Bible was published based on Luther's translation. In the same year William Tyndale published his English translation of the New Testament. In 1530 a French translation of the Bible was published, and in 1535 Miles Coverdale published a complete English Bible. Tyndale's translation had been suppressed in England, but after Henry VIII's break with Rome copies of Coverdale's Bibles were placed in all English churches. When Henry's daughter Mary set out to reverse the reformation in England, Coverdale and other scholars fled to Europe where, in Geneva, work was begun on a new English Bible.

'The Geneva Bible', as it was called, was completed in 1560 and published in England in 1575. Printed in roman type with the new Stephanus numbering system, it included maps, illustrations, cross references, and notes: all the means to allow laymen to understand and interpret the text for themselves, in the way that Luther had intended [Fig. 27.3]. It immediately became the most popular English Bible. A century later, 60 years after the publication of the King James Bible, it remained the preferred text of Puritan Christians.

The immediate effects of Luther's writings had been far more political than he had imagined. In 1525 50 members

6 C. W. Eliot, *The Harvard Classics* (1910), vol. 36.

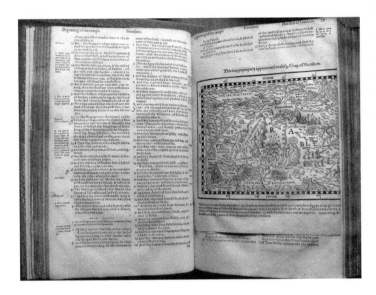

27.3 Geneva Bible (1575).

of the Upper Swabian farmers group drew up an article in which they proclaimed that 'we have been held as villeyne which is pitiful given that Christ redeemed all of us ... the shepherd as well as the highest, no one excluded. Therefore it is devised by the scripture that we are and that we want to be free.'[7] Luther was initially sympathetic, arguing that 'it is unbearable to tax and slave-drive people like this forever'. But after witnessing the atrocities and the burning of libraries in convents and monasteries that were being carried out in his name, Luther (who in later life became increasingly repressive and anti-Semitic) published a tract *Against the Murderous Thieving Hordes of Peasants*.

These disturbances were the beginnings of the civil strife that was to wrack Europe. Individual revolts could be put down, but the ideas that had been unleashed were not so easily suppressed. At roughly the same time that Luther was writing, a parallel movement had begun in Geneva under Ulrich Zwingli, and as similar ideas were seeded throughout Europe they began to suggest new ways of looking at the world. At table Luther was reported to have said that 'we are at the dawn of a new era for we are beginning to recover the knowledge of the external world that was lost through the fall of Adam ... we now observe creatures properly'.[8] However, despite their rapid political impact the new ways of thinking about the Bible did not always and immediately translate into other fields.

[7] S. Lotzer and C. Schappeler, 'The Twelve Articles of the Upper Swabian Peasants (March 1525)', in *The German Reformation and the Peasants' War*, ed. Baylor, M. G. (2012), 77.

[8] L. W. Spitz, *The Renaissance and Reformation Movements* (1971), 582.

An Absurd Matter

When a 24-year-old professor of mathematics at Wittenberg University was granted study leave, the motive was only tangentially scientific. Georg Joachim Rheticus was given leave of absence by Philipp Melanchthon—the head of the university and Martin Luther's right-hand man—partly in order to get him out of Luther's way (he was the friend of a poet who had written critical verses about the great man). Although he had intended Rheticus to further his studies of the heavens, Melanchthon's interest was not in furthering astronomy as such but rather in putting astrology on a more scientific footing. In a letter of recommendation he described Rheticus as 'above all an astrologer'.

Rheticus had gone first to study with scholars in Nuremberg. From there he had travelled on into the Catholic region of Warmia, where the famous Copernicus was a canon at the cathedral of Frombork. This journey was not without risk. Bishop Dantiscus had recently expelled all Lutherans from the region and was soon to follow this with draconian anti-Lutheran legislation.

Copernicus' own views seem to have been more moderate. His friend Tiedemann Giese had, with Copernicus' encouragement, published his own 110 theses acknowledging the justice of some of Luther's complaints but urging conciliation. Rheticus had, at any rate, won the canon's trust and stayed with Copernicus for almost two years. During that time he wrote and published *Narratio Prima*, a short summary of Copernicus' theory, and helped to prepare the great work *De Revolutionibus* for publication [Fig. 27.4].

Rheticus returned to Wittenberg in 1541. News of the new theory was met with scorn. Anthony Lauterbach, who dined with the Luthers, reported that Luther had referred to 'a certain new astrologer' who had tried to prove that the Earth moves and not the sky, Sun, and the Moon. His response had been resolutely conservative: 'whoever wants to be clever must agree with nothing that others esteem.... I believe the holy scriptures for Joshua commanded the sun to stand still and not the earth'.[9] Melanchthon meanwhile told a correspondent that 'Many hold it for an excellent idea to praise such an absurd matter, like that sarmatic [Polish] astronomer, who moves the earth and lets the sun stand still'.[10]

[9] J. Repcheck, *Copernicus' Secret* (2009), 159.
[10] J. Repcheck, *Copernicus' Secret*, 160.

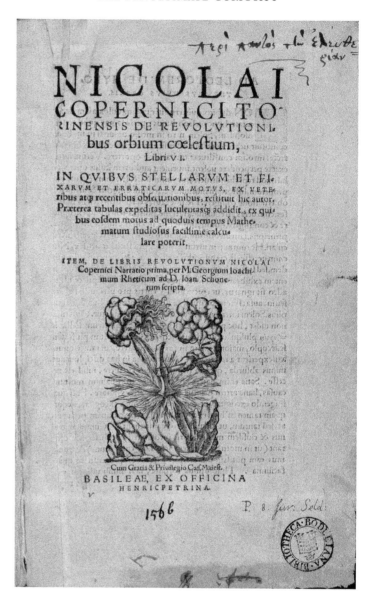

27.4 Title page of *De Revolutionibus* by Nicholas Copernicus.

Scorn, however, was not the same thing as official condemnation. Luther's democratization of Bible reading meant that one man's interpretation of Scripture did not automatically trump that of another. This new liberty for ultimate curiosity would have powerful consequences. The same principle could reasonably be applied to philosophizing. It soon was.

CHAPTER TWENTY-EIGHT

A Lutheran Astrologer

On 19 July 1595 a young Lutheran maths teacher at the Stifts-schule at Graz received what seemed to him a divine revelation of the deep geometrical beauty of the universe. So overwhelming was the experience (which was to shape the rest of his life) that he immediately burst into tears and vowed to God to 'make public in print this wonderful example of His wisdom'.[1]

The young teacher, Johannes Kepler, had been introduced to Copernicus' theory by his seminary professor Michael Maestlin while studying to be a Lutheran pastor (this was some 50 years after Rheticus' adventure described in Chapter Twenty-Seven). The theory had not only seemed to be true but had filled him with an 'unbelievable rapture'. At the same time it had raised in his mind a whole host of questions. There was, he thought, an 'unexhausted treasure' of divine insights to be explored into 'the magnificent order of the . . . world' [Fig. 28.1].[2]

When he published his first book, *Mysterium Cosmographicum* (The Mystery of the Universe), Kepler had included Rheticus' *Narratio Prima* as an appendix. It was in response to being sent this book that Galileo had confessed to being himself a secret Copernican, but when asked by Kepler 'would it not be better to pull the rolling wagon to its destination by a united effort?' had not replied.

In their theological attitude to the new theory the two men, though belonging to different confessions, were in effect singing from the same hymn sheet. In his *New Astronomy* (1605) Kepler argued, as Galileo was to do a few years later, that we should 'regard the Holy Spirit as a divine messenger and refrain from dragging him into the physics class'. At the same time, in an open letter to Galileo, Kepler agreed with the argument in the *Letters on Sunspots* that mathematical ability was a precious gift of God: 'Geometry is one and eternal shining in the mind of God. That share in it accorded to man is one of the reasons that man is in the image of God'.[3]

[1] J. Repcheck, *Copernicus' Secret* (2009), 63.
[2] J. Repcheck, *Copernicus' Secret*, 61.
[3] J. Kepler, *Gesammelte Werke* (1937).

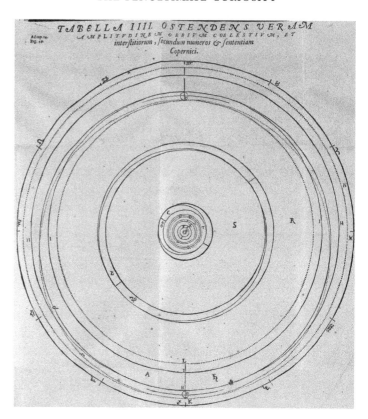

28.1 Kepler's diagram of the Copernican system.

Hence some 20 years later when Kepler published *Epitome of Copernican Astronomy* (in which he announced that 'I build my whole astronomy on Copernicus' hypothesis concerning the world'), he argued that 'it is a right, yes a duty to search in a cautious manner for the numbers, sizes and weights, the norms for everything [God] has created'.[4]

A Priest of God

Moving to Graz to teach mathematics had at first seemed to frustrate his ambition to be a priest, but two years later he had begun to see things differently. In a letter to Maestlin, following his geometrical revelation, he describes how he had wanted to be a theologian and for a long time was restless 'but now see how by my pains God is being celebrated in astronomy also'.[5] In the dedication of his *Epitome of Copernican Astronomy* he expressed the same idea even more forcefully. The book, Kepler tells his readers, should be understood as a hymn which he as

[4] M. Caspar, *Kepler* (1993), 381.
[5] M. Caspar, *Kepler*, 375.

'a priest of God at the book of nature' had written in honour of the creator.

This idea of 'the book of nature' was not a new one. Vincent of Beauvais, writing in the thirteenth century, describes God as the author of all creation whose 'book of creatures' was 'given us for our reading'.[6] Nevertheless, in describing his own calling as an astronomer as that of 'a priest of God', Kepler was developing this tradition in a distinctively Lutheran direction.

'I am a Lutheran astrologer', he wrote in a letter to Maestlin in 1598, 'I throw away the nonsense and keep the hard kernel'.[7] In a letter to a Catholic friend written in 1607 Kepler had criticized what seemed to him a tendency in the Roman Church to chain the human intellect which God addresses through his servants. In the introduction to his *Astronomia Nova* he argued 'with due respect to the doctors of the church' that Lactantius had been wrong in denying that the Earth was round, that Augustine had been wrong in denying that the Antipodes were inhabited, and that today 'the Holy Office' was wrong in denying 'the earth ... moves through the stars'.[8] Authority might have some place in theology but in philosophy 'it is the influence of reason that should be present'.[9] In making this claim he was speaking from experience.

The Renovation of Astronomy

The vision of geometrical beauty that he had received as a young maths teacher had arisen from the attempt to understand why it was that in Copernicus' theory, the six planets revolved around the Sun in different orbits. Convinced that God does nothing by chance, that 'Geometry is ... a reflection out of the mind of God', and that God created humanity 'so that we should share his own thoughts',[10] he had hit on the idea that the spacing of the orbits of the six planets corresponds to the five regular geometrical solids described by Euclid, one nested within another. Kepler was so entranced by the beauty of this hypothesis (which has no basis in reality) that he continued to promote it throughout his life [Fig. 28.2].

Nevertheless, far from imprisoning him in fruitless speculation it had stimulated him to further research. Believing that 'each philosophical speculation must take its point of departure from the experience of the senses',[11] Kepler was aware that it was impossible to get a better understanding of the planetary

6 P. Harrison, *The Bible, Protestantism, and the Rise of Natural Science* (1998), 44.
7 J. Kepler, *Gesammelte Werke*, 184.
8 J. Kepler, *Gesammelte Werke*, 18–36; P. Harrison, *The Bible, Protestantism, and the Rise of Natural Science*, 112–13.
9 J. Kepler, *Gesammelte Werke*, 18–36.
10 M. Caspar, *Kepler*, 93.
11 M. Caspar, *Kepler*, 67.

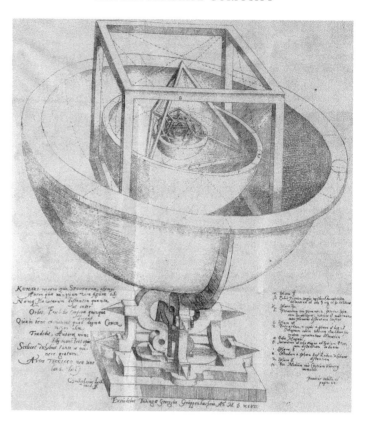

28.2 Kepler's diagram of nested geometrical solids from *Mysterium Cosmographicum* (1596).

orbits without more accurate astronomical observations. Fortunately these were to hand.

The Mystery of the Universe, the book that true to his vow Kepler published about his theory, had attracted the notice of the great Danish astronomer Tycho Brahe. Brahe, who had spent the past 30 years making and compiling accurate observations of the movements of the planets, invited Kepler to join him in Prague. Although at first reluctant, when religious persecution forced him to leave Graz, Kepler had done so. A year after the young mathematician's arrival in Prague, Tycho suddenly died, and Emperor Rudolf II offered to buy all his astronomical observations so that Kepler would have access to them.

The first task that Kepler undertook with these records was to establish the orbit of the planet Mars so that its position could be accurately predicted [Fig. 28.3]. He made a bet that he would be able to accomplish this in eight days, but soon found that 'Tycho's astronomy so shackled me that I nearly

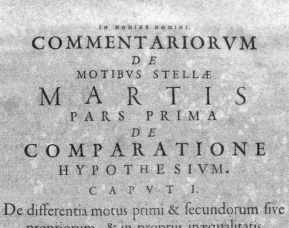

De differentia motus primi & secundorum sive
propriorum, & in propriis inæqualitatis
primæ & secundæ.

PLANETARVM motus orbiculares esse perennitas testatur.
Id ab experientia mutuata ratio statim præsumit gyros ipsorum perfectos esse circulos. nam ex figuris circulus, ex cor-
poribus cœlum, censentur perfectissima. Vbi vero diligen-
ter attendentes experientia diversum docere videtur; quod
Planetę a circuli simplici semita exorbitent; plurima existit admiratio,
quæ tandem in caussas inquirendas homines impulit.

Hinc adeo nata est inter homines Astronomia, cujus scopus esse pu-
tatur docere caussas, cur stellarum motus irregulares in terris appareant
cum sint ordinatissimi in cœlo, & investigare, quibusnam circulis stellæ
cieantur, ut horum beneficio loca & apparitiones illarum ad quævis
tempora prædici possint.

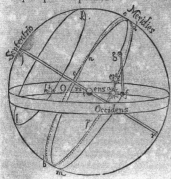

CVM nondum constaret de di-
scrimine inter motum primum & secundos, homines
intuiti Solem, Lunam & stel-
las, notarunt itinera ipsorum
diurna, æquiparari quampro-
xime circulis ad sensum, sic ta-
men ut alter ex altero nectere-
tur in fili glomerati modum,
circulosq; ut plurimum mino-
res in sphæra, rarissime maxi-
mos esse (*ut jam* A B C E, F M N G,
secantes A B *aquatorem in* C N) partem eorum in Austro, partem
A   in Bo-

28.3 Discussion of the orbit of
Mars from Kepler's *Astronomia
Nova* (1609).

went out of my mind'.[12] After repeating more than 70 times a
whole series of laborious and difficult calculations [Fig. 28.4],
he seemed to have solved the problem, but was still left with a
troubling inaccuracy. At one point the planet was as much as
eight minutes of arc out from the position it should be. If these
observations were right some basic assumption must be wrong.

[12] M. Caspar, *Kepler*, 127.

28.4 Kepler's calculations from *Astronomia Nova* (1609).

For Kepler it was an integral part of his Lutheran theological outlook to acknowledge this possibility: 'After the divine goodness has given us in Tycho Brahe so careful an observer that from his observations the error ... of eight minutes betrayed itself', it was, he argued, only seemly to recognize and use 'in thankful manner this good deed of God's ... to search out at last the true form of the heavenly motions'. It was indeed, he later recognized, 'these eight minutes', which 'showed the way to the renovation of the whole of astronomy'.[13]

[13] M. Caspar, *Kepler*, 128.

Ever since Eudoxos had solved the problem set by Plato described in Chapter Ten to explain the apparent motions of the planets, it had been accepted as a fundamental axiom that the planets moved of their own accord in uniform circular orbits. To 'save the appearances' of astronomers' actual observations, this had been complicated over the centuries by a whole series of subsidiary 'epicycles'. In suggesting that the planets moved round the Sun rather than the Earth, Copernicus had unified this picture, essentially inventing 'the solar system', but had not challenged the basic assumption. Now, for the first time in 2000 years, Kepler did so.

Discarding the axiom of uniform circular motion he was driven (after years of laborious calculation) to the conclusion that both Mars and the Earth moved in an ellipse around the Sun, speeding up as they got closer and slowing down as they moved away [Fig. 28.5]. The reason for this, he argued, was that they were not moving of their own accord but were subject to a physical force emanating from the Sun (which he conjectured might be magnetic) [Fig. 28.6]. What was true of Mars must be true of all the planets, so that once he had established the shape of the Martian orbit he was able to state as a planetary law that 'planets move in an ellipse with the sun as one focus'.[14] His goal, he announced, was 'to show that the heavenly machine is not a kind of divine living being but similar to clockwork in so far as almost all the manifold motions are taken care of by one single absolutely simple magnetic bodily force'.[15]

World Harmony

In 1619 Kepler published his most complete attempt to realize this goal. This was a year of utmost strain in his personal life, when he was being attacked by fellow Lutherans and his mother was on trial for witchcraft. Starting from the metaphysical conviction that geometry was 'part of the divine mind . . . from before the origin of things'[16] the *Harmonices Mundi* (The Harmony of the World) begins in book I with the geometrical harmony of regular polygons, and concludes in book V with the harmonies of planetary motion [Fig. 28.7]. In between he sets out to show 'geometrical things have provided the Creator with the model for decorating the whole world'[17] so that book III contains a discussion of harmonic proportions in music and book IV talks of harmonic configurations in astrology.

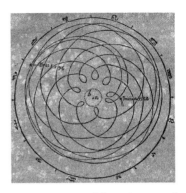

28.5 Diagram of the trajectory of Mars through several periods of apparent retrograde motion. From Kepler's *Astronomia Nova* (1609).

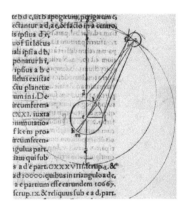

28.6 Diagram of planetary orbit from Kepler's *Astronomia Nova* (1609).

[14] M. Caspar, *Kepler*, 134.
[15] M. Caspar, *Kepler*, 136.
[16] M. Caspar, *Kepler*, 271.
[17] M. Caspar, *Kepler*, 265–6.

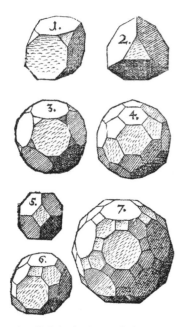

28.7 Polyhedra in Kepler's *Harmonices Mundi* (1619).

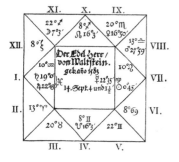

28.8 Kepler's horoscope for General Wallenstein.

[18] M. Caspar, *Kepler*, 279.
[19] M. Caspar, *Kepler*, 286.
[20] D. C. Lindberg and R. L. Numbers, *God and Nature* (1986), 220.

Kepler was required throughout his life to produce astrological tables, but here as elsewhere his approach was distinctive [Fig. 28.8]. As 'a Lutheran astrologer' he stressed that, even if the planets might have an effect in forming our disposition, we are not ruled by their influence. He writes in book IV that 'My heavenly bodies were not the rising of Mercury in the seventh house in quadrature to Mars, but Copernicus and Tycho Brahe, without whose journal of observations everything which I have now pulled up into light would be buried in darkness'.[18] This finally leads him to book V, which he describes as 'a sacred sermon, a veritable hymn to God the creator', and in which he announces his most recent discovery.

Ever since his revelation at Graz, Kepler had been convinced that the different revolutions of the planets around the Sun were governed by some deep principle of harmony. Now at last he had discovered what it was. On the 8th of March 1618 the idea came into his head that the relationship between the average distance of each planet from the Sun and the speed of its orbit around the Sun corresponded to an exact mathematical relationship. At first he thought he must have made a mistake, but 'finally, on May 15, it came again and with a new onset conquered the darkness of my mind'.

There was 'such an excellent agreement between my seventeen years of work at the Tychonic observations and my present deliberations' that despite fearing it was a dream he knew that 'it is entirely certain and exact that the proportion between the periodic times of any two planets is precisely one and a half times the proportion of the mean distances'.[19] This discovery, which is now known as Kepler's third planetary law (and was to lead directly to Newton's law of universal gravitation), was the fulfilment of all his hopes. He accompanied it with a prayer that 'the body of your church on earth be built from harmonies as you have constructed the heavens themselves'.[20]

As it turned out it was not this vision of world harmony but the simple accuracy of his *Rudolphine Tables* that ultimately won acceptance for Kepler's theory. These charted and predicted the movements of the planets without commentary and without explaining the theory on which the calculations were based. They were immediately useful 'snatched at by astronomers who wanted to test their theories, horoscope casters who needed constellations at some one time, calendar makers who wanted to state the position of the wandering stars and,

last but not least, seafarers who were compelled to make geographical place determinations'.[21]

Real utility, as Shadwell pointed out, was a rarity in the
seventeenth-century study of nature. The idea that one should
'study for use' was, however, an ideal that, it was already beginning to be argued, seemed to be in tune with the new Reformation theology.

[21] M. Caspar, *Kepler*, 326–7.

CHAPTER TWENTY–NINE

The Experimental Philosophy

In early January 1522, while Luther was in hiding in Wartburg Castle, Wittenberg city council, at the behest of Andreas Karlstadt, the chancellor of the university, ordered the physical removal of all imagery in the city's churches. On his return Luther (who was a friend of the painter Lucas Cranach) had stopped the destruction in Wittenberg, but it soon began again elsewhere.

Two years later Ulrich Zwingli, the leader of the Reformation in Geneva, actively encouraged the smashing of sculptures and the whitewashing of frescoes in Swiss churches. Zwingli remarked that 'in Zurich we have churches that are positively luminous', but the new flood of light that was admitted into many old church buildings was the only positive aesthetic side effect of the wholesale destruction that followed.

This iconoclasm spread throughout Europe, and was the most dramatically visible result of the new Reformation. The images that for centuries had been placed in Christian churches to provide what Roger Bacon had described as 'a flood of grace' now began to be seen as expressions of idolatry: the sin of worshipping images of false gods. Despite Luther's personal opposition, the clearing out of visual imagery from churches was not merely justified as obedience to the Second Commandment; it was also seen as mirroring the radical new way of reading Scripture that he had signalled when he first issued his students with their blank-margined psalters.

There had always been more to this gesture than a simple readiness to set aside authorities from the past and a declaration of freedom from the commentaries and glosses of the church fathers. Luther's growing conviction that the Scriptures should be understood 'in their simplest meaning as far as possible' and his insistence the literal meaning was 'the highest, best, strongest'[1] had also involved a rejection of the long tradition of more imaginative allegorical approaches.

[1] P. Harrison, *The Bible, Protestantism, and the Rise of Natural Science* (1998), 108.

The Allegory of Nature

In the third century AD Origen, the great theologian and head of the Didascalium in Alexandria, had argued that there were three different senses in which the Scriptures must be understood. The first was the literal sense—the obvious historical sense. The second was the moral sense—which showed us how to live. The third was the allegorical sense—which pointed towards timeless theological truths. In later centuries a fourth distinction was made between ordinary allegory—which could refer to anything—and 'anagogy'—'the heavenly sense'.

These four senses formed the so-called 'quadriga', which became the standard interpretative scheme for reading the Bible in the Middle Ages. Because 'the heavenly sense' was judged to be the most important, allegorical and anagogical readings tended to proliferate; and because the doctors of the Church were judged to have a special authority, such readings began over time to accumulate, in glosses and commentaries which threatened to overwhelm the original text.

Such allegorical readings were not, however, limited to Holy Writ. Because God was the author of nature as well as of Scripture, the same method of interpretation could (it was argued) be applied to both. This was first expounded in a book called *Physiologus* written by an anonymous contemporary (and probably disciple) of Origen. 'A small but comprehensive work on animals plants and stones',[2] the *Physiologus* draws on classical sources like Pliny and Aelian to discover an allegorical significance in every feature of the natural world which it describes. Thus, the pelican feeding its young with its own blood is an allegory of Christ's atoning sacrifice, while as the serpent sheds its skin so 'we, too, throw off for Christ the old man and his clothing' [Fig. 29.1].[3]

29.1 Pelican feeding its young with its blood (*c.*1450).

The *Physiologus* had an astonishing influence. E. P. Evans claimed that 'perhaps no book except the Bible has ever been so widely diffused among so many people and for so many centuries'.[4] It formed the basis of innumerable medieval bestiaries, lapidaries, and herbals. Its allegorizing approach fed into the Renaissance idea of the microcosm and macrocosm (which held that every aspect of the world was represented in the human body) and the doctrine of 'signatures' (which held that every natural object contained some sign indicating its use). While Roger Bacon and his successors might talk of trying to

[2] P. Harrison, *The Bible, Protestantism, and the Rise of Natural Science*, 23.
[3] M. J. Curley, *Physiologus* (1979), 16.
[4] E. P. Evans, *Animal Symbolism in Ecclesiastical Architecture* (1896), 41.

understand the fundamental laws of the natural world and the fruit that would come from doing so, the attempt of the *Physiologus* to read the meaning of the natural world was a far more pervasive aspect of the medieval mindset.

Luther's desire to limit the use of allegory and to simplify biblical interpretation was an axe laid to the root of this tree. It was echoed by other Reformation leaders, who not only extended his preference for simplicity from the reading of Scripture to visible symbolic forms of worship (including the decoration of churches), but helped it to extend into new ways of looking at the natural world.

One aspect of this extension, as Peter Harrison has argued, was the idea that natural objects should be 'regarded as having been designed for their utility rather than their meaning: creatures were not symbols to be read but objects to be used or investigated for potential applications'.[5] No one was to expound this more forcefully than a philosophically inclined young lawyer called Francis Bacon whose working life was spent at the English court.

In England the iconoclasm that had begun with Henry VIII's dissolution of the monasteries was continued by Edward VI, who in 1548 ordered the removal of all 'idols' from churches. Some were restored under Mary I, but removals continued under Elizabeth I and later under James I. Francis Bacon was born in the early years of Elizabeth's reign, with this physical iconoclasm going on all around, and throughout his life it remained for him a model of the intellectual process that would be required if God's creation was to be truly understood.

Entering the Kingdom

Sir Francis Bacon (as he became) was the son of Queen Elizabeth's Lord Keeper of the Great Seal, and was to achieve office in the court of King James. Shortly after the death of Elizabeth he published his most influential book: *The Advancement of Learning* [Fig. 29.2].

In this he argued that Luther had not been 'aided by the opinions of his own time'.[6] In recovering the true meaning of Scripture he had gone back to the original languages. This had been necessary because 'in the inquiry of the divine truth' men's pride inclined them 'to leave the oracle of God's word and to vanish in the mixture of their own inventions'. Similarly,

[5] P. Harrison, *The Bible, Protestantism, and the Rise of Natural Science*, 203.
[6] F. Bacon, *The Advancement of Learning and New Atlantis* (1974), 25.

29.2 Portrait of Francis Bacon in
The Advancement of Learning (1605).

he argued, in the 'inquisition of nature' men have 'left the oracle of God's works and adored the deceiving and deformed
images which . . . their own minds, or a few received authors or
principles did represent unto them'.[7]

In one of his last books Lord Verulam, as he had by then
become (having first risen to the rank of Lord Chancellor and
then been disgraced and dismissed from the post), argued that
a proper knowledge of the world could only begin when we
set aside all our 'idols'. These he categorizes as 'idols of the tribe
and the market place'—the distortions of human perception
and language—and 'idols of the theatre'—the distortions produced by systems like that of Aristotle.

In reading the book of God's Word the watchword of the
Reformation had been *sola scriptura*—Scripture alone. At the

[7] F. Bacon, *The Advancement of Learning
and New Atlantis*, 29.

heart of Bacon's approach to reading the book of God's works was the idea that any true philosophy must similarly be built on the actual 'Phenomena of the Universe'.[8] Access to these came first through sensuous perceptions, and then through the instruments we invent to extend our observations, like telescopes, microscopes, and all kinds of measuring devices. Beyond these, however, it came through any kind of experiment we can devise to test our observations and ideas. It is experiments, he argued, which enable us to pass beyond raw data and mere conjectures and allow us to discover the fundamental causes of things: 'the axioms' and 'the universal laws'.

For this process to succeed it was critical not to rush into metaphysics or to be captivated by man-made systems. Final causes had a place in theology, but to introduce them into physics was 'impertinent'. It was 'to impress the stamp of our own image on . . . the works of God, instead of carefully examining and recognising in them the stamp of the creator'. It was critical, he argued, 'to approach with humility . . . the volume of creation', and for the understanding to be 'thoroughly freed and cleansed; the entrance to the kingdom of man, founded on the sciences, being not much other than the entrance into the kingdom of heaven where into none may enter except as a little child'.[9] Consequently, he argued, we should 'give to faith only that which is faith's'.[10] Rather than blindly clinging to the figurative language that the Scriptures sometimes uses to describe natural phenomena, in reading the book of nature we must attend to the language in which God has actually written it and follow where this leads.

Plus Ultra

For Bacon and for many of his contemporaries the discovery of the new world by Columbus and his successors seemed like a kind of divine providence confirming the validity of this approach. Throughout his works the image recurs of a voyage of discovery to which we are summoned by God.

The title page of *The Advancement of Learning* depicts a ship on the sea with a quotation from the Book of Daniel: 'Many shall go to and fro and knowledge shall be increased' [Fig. 29.3]. The title page of *Sylva Sylvarum*, published after his death, shows the globe of the intellectual world resting between the pillars of Hercules (representing for classical writers the edge of the

8 F. Bacon, *The Instauratio Magna* (2000), vol. IV, 28.
9 F. Bacon, *Novum Organum* (1902), vol. IV, 69.
10 M. Purver, *The Royal Society* (1967), 145.

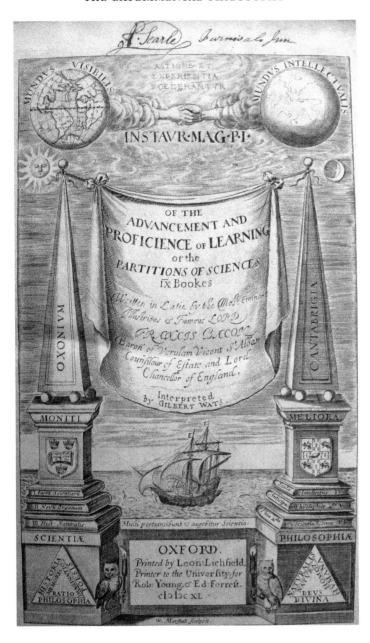

29.3 Title page of *The Advancement of Learning* (1605), Francis Bacon.

known world—*ne plus ultra*: beyond which there is no more) [Fig. 29.4].

On the title page of the *Novum Organum* these two images are combined in the picture of a ship plunging through the pillars of Hercules, voyaging to discover worlds unknown to classical antiquity [Fig. 29.5]. In this way Bacon believed that

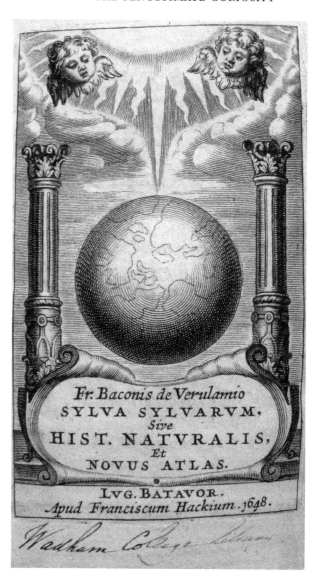

29.4 Title page of *Sylva Sylvarum* (1627), Francis Bacon.

'many objects in nature fit to throw light upon philosophy have been exposed to our view, and discovered by means of long voyages and travels, in which our times have abounded'.[11]

The New Instrument

Aristotle had described the process of arriving at truth through logic in a collection of works known as the *Organon*—The Instrument. Bacon's *Novum Organum*—The New Instrument—

[11] F. Bacon, *Novum Organum*, vol. I, 62.

29.5 Title page of *Novum Organum* (1620), Francis Bacon.

was his attempt to put the discovery of truth on a new footing. As new data were first gathered and then subjected to experimental testing, he thought that it would be possible 'to establish progressive degrees of certainty'.[12] By a process which he described as 'true and perfect induction' it should be possible to gradually advance from raw data towards ever more general axioms and universal laws.

Bacon's laborious attempt to specify the steps of this process have often been ridiculed. William Harvey (the discoverer of the circulation of the blood) famously remarked that Lord Verulam wrote natural philosophy 'like a Lord Chancellor', and the detailed rules of procedure laid down by Bacon have little in common with the actual practices of Harvey or any other

[12] F. Bacon, *The Instauratio Magna*, vol. IV, 40.

natural philosopher. Bacon, however, did not claim (as is sometimes suggested) that induction from data was a straightforward mechanical process that if operated correctly would inevitably produce results. He was aware that scientific discovery would not be a smooth procession in one direction. 'Our road', he wrote, 'does not lie on a level but ascends and descends, first ascending to axioms then descending to works'.[13] It would often be fraught with difficulty. Wrong and incomplete hypotheses could seem to fit the same data, while 'mistakes in experimenting' could lead in false directions. Nevertheless ultimately, he argued, the process would be self-correcting. Crucial experiments, what he called 'instances of the fingerpost', could sometimes point decisively towards one hypothesis rather than another.

The possibility of mistakes meant that it was important to avoid being impatient for utilitarian results and leaping to conclusions: 'reaping the green corn'. Although in the long term new knowledge would produce much fruit, what he called 'experiments of light' had an advantage over 'experiments of fruit' in that they succeed whichever way they turn out 'for they settle the question'.[14] The possibility of mistakes also meant that experimenters must carefully record their methods so that false steps could be detected and corrected by others.

In much the same way that Luther had democratized the reading of God's Word, Bacon in this way sought to democratize the reading of God's works. The advancement of learning was, he insisted, a collective enterprise 'done by many together in the succession of ages'.[15] Claude Bernard captured this spirit in a quote from a contemporary poet, 'L'art, c'est moi; la science, c'est nous.'[16] The procedures which Francis Bacon set out in the *Novum Organum* implied that this was what has been described as a kind of algorithmic process, which required artisans and intellectuals to join together in an organized way. The frontispiece of *The Advancement of Learning* shows the visible and the intellectual world reaching out to one another.

Although we cannot 'command the nature of things' we can, Bacon insisted, 'command our questions'. To elicit intelligible answers these needed to be separated into distinct (though overlapping) research topics, which could be systematically pursued. Bacon himself produced 'A Catalogue of Particular Histories', outlining experimental programmes to tackle a whole variety of distinct topics. In *The New Atlantis*, a utopian

[13] F. Bacon, *The Instauratio Magna*, vol. IV, 96.
[14] F. Bacon, *The Instauratio Magna*, vol. IV, 95.
[15] F. Bacon, *The Instauratio Magna*, vol. IV, 291.
[16] C. Bernard, 'Introduction à l'étude de la médecine expérimentale', http://www.cosmovisions.com/textes/Bernardo10204.htm#mLcxy51q7sM7yTxF.99http://www.cosmovisions.com/textes/Bernardo10204.htm. Accessed 13 August 2015.

fantasy published after his death, he described a state-funded research institute—'Solomon's House'—in which teams of researchers investigate and experiment in different fields, while others record, collate, and interpret their results.

The study of the natural world in Solomon's House (also known as 'The College of the Six Days') was a religious activity, but in a changed way from that of the medieval bestiary. Rather than seeing nature as 'a vast array of symbols which pointed to a transcendent realm beyond',[17] its founding principle was that 'by finding out the true nature of things ... God might have the more glory in the workmanship of them and men more fruit in the use of them'.[18]

Glorifying God's works provided a motivation for the closest examination of them: 'if we should rest only in the contemplation of the exterior ... as they first offer themselves to our sense, we should do a like injury to the majesty of God'.[19] It was in this spirit that 50 years later the Dutch naturalist Jan Swammerdam wrote to a colleague 'Herewith I offer you the Omnipotent Finger of God in the anatomy of a louse: wherein you will find miracles upon miracles and will see the wisdom of God manifested in minute point.'[20] Nevertheless Bacon reckoned that it was the promise of 'fruit in the use of them', 'the relief of man's estate', that provided the strongest motivation for state funding of research.

State funding was not forthcoming from King James I. When Bacon died in 1626 (according to John Aubrey from a chill caught while experimenting with freezing some meat in snow), he had, as he put it, 'sounded a trumpet call to action' in some 30 published books. Among his contemporaries this call had mostly fallen on deaf ears. In the succeeding generation, however, its note was to be heard by a small but brilliant group at the University of Oxford.

---

[17] P. Harrison, *The Bible, Protestantism, and the Rise of Natural Science*, 168.

[18] F. Bacon, *The Advancement of Learning and New Atlantis*, 230.

[19] F. Bacon, *The Advancement of Learning and New Atlantis*, 42.

[20] J. Swammerdam, *The Letters of Jan Swammerdam to Melchisedec Thévenot* (1975), Letter 19a, April 1678.

CHAPTER THIRTY

The Oxonian Sparkles

The keys of the city of Oxford were formally handed over to Sir Thomas Fairfax, Lord General of the parliamentary army on the 25th of June 1646. The previous day after the signing of a treaty, 900 men of the royalist garrison, with flags flying and drums beating, had marched out of the city across Magdalen Bridge and on up Headington Hill between the watching lines of the victorious New Model Army.

Their departure brought to an end a brief interlude in which, after abandoning London, King Charles I had in effect made Oxford the new capital of England. Although most of the town supported the parliamentary cause, the university was resolutely royalist. In 1644 Christ Church had become the King's new residence (while Queen Henrietta Maria lived at Merton) and the hall of Christchurch the seat of a new parliament. While the colleges melted down their silver for a new currency, 'Oxford crowns', All Souls became the garrison's arsenal and New College their magazine.

Following Oxford's surrender, Oliver Cromwell was installed as the chancellor of the university in 1650, after four years during which the royalist heads of colleges had been systematically replaced with parliamentary supporters. Among these new appointments, which it was hoped would establish a new ethos in the university, was a young Puritan clergyman named John Wilkins, who in 1648 had been made warden of Wadham College.

In the *Novum Organum* Francis Bacon had proposed that rather than simply recycling old knowledge, universities should resemble mines 'where the noise of new works and further advances is heard on every side'.[1] This suggestion had not been warmly received by most academics, but the new warden of Wadham now set himself to try to realize Bacon's dream.

[1] F. Bacon, *Novum Organum* (1902), vol. IV, 90.

The Experimental Club

Wilkins was a convinced Copernican who at the age of 24 had published *The Discovery of a World in the Moon* and followed it up two years later with *A Discourse Concerning a New World and Another Planet* [Fig. 30.1]. In the preface to this he described his desire to raise up an active spirit to search after 'hidden and unknown truths' in the manner advocated by 'the judicious *Verulam*'.[2] A year after his appointment as warden, he founded 'an experimentall philosophicall clubbe'[3] which met weekly in his rooms.

Among the group who assembled were dons, like Seth Ward, the new professor of astronomy, and William Petty, who the

30.1 Title page of John Wilkins' *A Discourse Concerning a New World and Another Planet* (1640).

[2] J. Wilkins, *A Discourse Concerning a New World and Another Planet* (1640), 17–18.

[3] J. Aubrey, *Brief Lives, John Wilkins*, vol. III (2015), 583.

following year became professor of anatomy, and at least one undergraduate: 'that prodigious young scholar Mr Chr: Wren'.[4] From the outset 'the noise of new works' had begun to be heard. Between them Wilkins and Wren laboured to construct an 80-foot telescope 'to see at once the whole moon'. Seth Ward built himself a small observatory. Anatomical dissections were carried out by Petty, and within a few years Christopher Wren had initiated daring experiments with blood transfusions.

At first the club had 'no rules or method fix'd', but by 1651 a list of rules had been drawn up. Meetings were to take place 'every Thursday before two of the clock', and Petty wrote to a friend that 'the Club-men have cantonized or are cantonizing their whole Academy to task men to several imploiments'.[5] One group was tasked to make a catalogue of all the scientific literature in the Bodleian, while Seth Ward wrote that 'Besides this Great Clubb we have a combination of lesser number viz: of 8 persons who have joyned together for the furnishing an elaboratory and for making chymicall experiments'.[6]

This systematic Baconian organization distinguished the Oxford club from the various London discussion groups (like Theodore Haak's club or Samuel Hartlib's 'invisible college') which preceded it, and was to continue a decade later when the Oxford club metamorphosed into the Royal Society.

Wilkins meanwhile had begun to recruit new members to come and join the dazzling array of talent that Henry Oldenburg, the future secretary of the Royal Society, described as 'the Oxonian sparkles'. The most illustrious such recruit was a younger son of the Earl of Cork, the Honourable Robert Boyle (known as Robin to his family). In 1653 Wilkins wrote to Boyle, saying that his presence in Oxford would be 'a means to quicken and direct us in our enquiries'.[7] Two years later, after some further encouragement, Boyle moved into a house on the High in Oxford where he remained for the next twelve years. A plaque now marks the place.

In the 1640s Boyle's writings had been mainly focussed on moral and spiritual themes, but towards the end of the decade he had started to conduct chemical experiments. In a letter to his sister he describes how he spent his Sundays 'Studying the Booke of the Creatures' in reading and experiments which he sought to make 'contributory to . . . the Glory of the Author of them'.[8] In the summer of 1649 he began to write *Of the Study of the Booke of Nature*, where in language reminiscent of Kepler

[4] *The Diary of John Evelyn*, ed. Guy de la Bédoyère (1995), 89.
[5] M. Purver, *The Royal Society* (1967), 121.
[6] M. Purver, *The Royal Society*, 112.
[7] M. C. W. Hunter, *Boyle* (2009), 89.
[8] M. C. W. Hunter, *Boyle*, 73.

he argues that man as the great high priest of nature is 'bound to returne Thankes & Prayses, not only for himselfe but for the whole Creation'.[9]

In the years after moving to Oxford, Boyle devoted himself both to a programme of experimental research and to a parallel programme of writing. In this latter he set out to champion the experimental method by describing experiments he had conducted, explaining what conclusions could be drawn from them, and pointing out their practical and religious value. The first of what was to be a long line of such books to appear in print was his account, published in 1660, of *New Experiments Physico-Mechanical, Touching the Spring of the Air, and its Effects.*

The Spring of Air

The acceptance or rejection of Aristotle's principle that 'nature abhors a vacuum' was already a dividing line between Aristotelian scientists and adherents of the new 'mechanical philosophy'. During the course of the seventeenth century it became something more: a means of showing how a well-designed experiment could solve a philosophical dispute.

Back in the 1640s Galileo and others had observed that water could not be pumped up from a depth of more than 34 feet. Two of Galileo's followers, Evangelista Torricelli and Vincenzo Viviani, who were living with him in the villa at Arcetri, set themselves to investigate this phenomenon. The common assumption (shared by Galileo) was that air was weightless. Torricelli and Viviani, however, argued that water was pushed into pumps by the weight of the outside air, and that the weight of 34 feet of water balanced the weight of the surrounding atmosphere.

To test this hypothesis in 1644 Viviani, on Torricelli's suggestion, carried out an experiment which has become famous [Fig. 30.2]. He filled a long glass tube with mercury and turned it upside-down with his finger over the open end. He then placed the tube in a basin filled with mercury and removed his finger. Torricelli had predicted that because mercury was 14 times denser than water, the surrounding air pressure would only support a column of 29 inches, and so it proved. As soon as Viviani removed his finger the column of mercury dropped to 29 inches, leaving an apparently vacant space at the top of the tube. This demonstrated, so Torricelli argued, both that air had a weight and that a vacuum could exist.

[9] M. C. W. Hunter, *Boyle*, 73–4.

30.2 Torricelli's barometer experiment (1644).

Not everyone was convinced. Committed Aristotelians argued that the apparently empty space at the top of the tube in fact contained an invisible vapour that was holding up the mercury column. In order to try to prove that the effect of air pressure was real, a young French mathematical prodigy, Blaise Pascal, persuaded his brother-in-law, Florin Périer, to make an ascent of the Puy de Dôme carrying with him Torricelli's apparatus. Pascal argued that because the pressure of air should be less at the top of the mountain, the column of mercury would be lower there. Once again the prediction was correct. When Périer measured the mercury at the foot of the mountain it was 29 inches but at the summit he found it only reached 23 inches.

Robert Boyle's contribution to this debate was to propose a way in which the apparent vacuum at the top of Torricelli's tube could be more directly investigated. He had heard of a pump made in Germany by Otto von Guericke, which sucked air from between two copper hemispheres (making them stick together). Boyle suggested that if instead of copper hemispheres the air could be sucked out of a glass sphere into which objects could be placed, then all sorts of experiments would be possible. The London instrument maker Ralph Greatorex tried and failed to make a pump to this design, but Robert Hooke, an undergraduate from Christchurch College who had come to work with Boyle as a laboratory assistant, triumphantly succeeded [Fig. 30.3].

In some of the experiments which he carried out with his new apparatus, Boyle was able to show that sound did not travel through the evacuated glass sphere and that flames could not burn in it. In others he demonstrated that when Torricelli's 'barometer' was placed in the sphere, the mercury fell as the air was pumped out. The majority of the experiments demonstrated something else that was experienced by everyone operating the pump, namely that air had a 'spring' which could be compressed or expanded. When a glass vial half-filled with water was placed in the sphere and the air pumped out, the vial exploded and cracked the sphere.

Boyle published his experiments in 1660. The following year an English Jesuit, Franciscus Linus, published a treatise attacking the idea that air had a 'spring' which could produce significant effects, and arguing that the apparently empty space in Torricelli's barometer was in fact filled by a 'funiculus'—an invisible thread which held up the mercury and was responsible for its movements. Boyle's response was to devise a further experiment [Fig. 30.4].

This time he used a J-shaped glass tube open at the top of the long arm and sealed at the top of the short arm. He then poured mercury into the long arm trapping a bubble of air in the short arm and was able to show that the 'spring' in this small bubble was even more powerful than had been realized: 'so that here our adversary may plainly see that the spring of air . . . may not only be able to resist the weight of 29 inches but . . . above a hundred inches of quicksilver and that without the assistance of his Funiculus'.[10]

This wasn't all. Two other natural philosophers, Henry Power and Richard Townley, had suggested that there might be

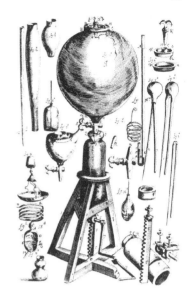

30.3 Boyle's air pump from *New Experiments touching the Spring of Air* (1660).

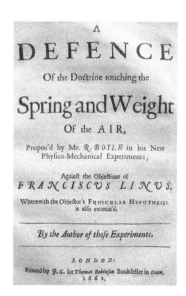

30.4 Title page of Robert Boyle's *A Defence against Linus* (1662).

[10] R. Boyle, *The Works of the Honourable Robert Boyle* (1772), 159.

a simple mathematical relationship between the pressure and the volume of gases. By carefully comparing the height of the column of mercury as it was poured into the long arm of the J tube with the size of the bubble in the short arm, Boyle was able to prove that this was so. In his *A Defence against Linus* he published four columns of figures tabulating the increasing height of the mercury and the decreasing volume of air [Fig. 30.5]. The far-right column showed 'what the pressure ought to be according to the Hypothesis that supposes pressures and expansions to be in reciprocal proportions'. The second from right showed what it actually was. In no case were they exactly the same—there was no cheating—but the agreement with the hypothesis was so striking that Boyle was able to argue that the variations could be put down to 'such want of exactness as in such nice experiments is scarce avoidable'.[11] He thereby laid the foundation for the analysis of experimental error that now forms part of the education of every experimental scientist.

30.5 Table from Robert Boyle's *A Defence against Linus* (1662).

A Table of the Condensation of the Air.

A	A	B	C	D	E
48	12	00		$29\frac{2}{16}$	$29\frac{2}{16}$
46	$11\frac{1}{2}$	$01\frac{7}{16}$		$30\frac{9}{16}$	$30\frac{6}{16}$
44	11	$02\frac{13}{16}$		$31\frac{1}{16}$	$31\frac{12}{16}$
42	$10\frac{1}{2}$	$04\frac{6}{16}$		$33\frac{8}{16}$	$33\frac{7}{}$
40	10	$06\frac{3}{16}$		$35\frac{5}{16}$	$35--$
38	$9\frac{1}{2}$	$07\frac{14}{16}$		$37--$	$36\frac{11}{19}$
36	9	$10\frac{2}{16}$		$39\frac{1}{16}$	$38\frac{7}{8}$
34	$8\frac{1}{2}$	$12\frac{8}{16}$	makes	$41\frac{10}{16}$	$41\frac{1}{7}$
32	8	$15\frac{1}{16}$		$44\frac{2}{16}$	$43\frac{11}{16}$
30	$7\frac{1}{2}$	$17\frac{7}{16}$		$47\frac{1}{16}$	$46\frac{1}{2}$
28	7	$21\frac{3}{16}$		$50\frac{2}{16}$	$50--$
26	$6\frac{1}{2}$	$25\frac{3}{16}$	Added to $29\frac{2}{8}$	$54\frac{5}{16}$	$53\frac{12}{13}$
24	6	$29\frac{11}{16}$		$58\frac{13}{16}$	$58\frac{2}{8}$
23	$5\frac{3}{4}$	$32\frac{1}{16}$		$61\frac{5}{16}$	$60\frac{2}{3}$
22	$5\frac{1}{2}$	$34\frac{15}{16}$		$64\frac{1}{16}$	$63\frac{6}{11}$
21	$5\frac{1}{4}$	$37\frac{15}{16}$		$67\frac{1}{16}$	$66\frac{4}{7}$
20	5	$41\frac{2}{16}$		$70\frac{11}{16}$	$70--$
19	$4\frac{3}{4}$	$45--$		$74\frac{2}{16}$	$73\frac{11}{19}$
18	$4\frac{1}{2}$	$48\frac{12}{16}$		$77\frac{14}{16}$	$77\frac{2}{3}$
17	$4\frac{1}{4}$	$53\frac{11}{16}$		$82\frac{12}{16}$	$82\frac{7}{17}$
16	4	$58\frac{2}{16}$		$87\frac{14}{16}$	$87\frac{7}{8}$
15	$3\frac{3}{4}$	$63\frac{15}{16}$		$93\frac{1}{16}$	$93\frac{3}{5}$
14	$3\frac{1}{2}$	$71\frac{5}{16}$		$100\frac{7}{16}$	$99\frac{6}{7}$
13	$3\frac{1}{4}$	$78\frac{11}{16}$		$107\frac{13}{16}$	$107\frac{1}{13}$
12	3	$88\frac{7}{16}$		$117\frac{9}{16}$	$116\frac{8}{}$

AA. The number of equal spaces in the shorter leg, that contained the same parcel of Air diversly extended.

B. The height of the Mercurial Cylinder in the longer leg, that compress'd the Air into those dimensions.

C. The height of a Mercurial Cylinder that counterbalanc'd the pressure of the Atmosphere.

D. The Aggregate of the two last Columns *B* and *C*, exhibiting the pressure sustained by the included Air.

E. What that pressure should be according to the *Hypothesis*, that supposes the pressures and expansions to be in reciprocal proportion.

11 R. Boyle, *The Works of the Honourable Robert Boyle*, 159.

Boyle's *Defence* was a near perfect demonstration of the power of Bacon's experimental methodology. Whereas Galileo had merely stated that 'there was no appreciable discrepancy in the result' of his repeated experiments, Boyle, by minutely recording his methods and tabulating his results, made it possible, as Bacon had prescribed, for anyone who so wished (and had the necessary equipment) to repeat the experiment and compare their own results. The experiment itself was a Baconian 'instance of the fingerpost', pointing decisively towards one hypothesis rather than another. Furthermore the demonstration of the reciprocal relation of the pressure and volume of gases (which has gone down in history as 'Boyle's law') confirmed Bacon's belief that it might, through experiments, be possible to advance from raw data towards 'axioms' and 'universal laws'. In fact some measure of social acceptance had preceded this intellectual demonstration. By the time Boyle had published his results the Oxford experimental club had already been transformed by charter into the Royal Society for the Advancement of Experimental Philosophy.

The Christian Virtuosi

At the end of the 1650s the Oxford club had dispersed and reconvened in London, to where in the following decade many of its members gravitated. In 1659 Wilkins had been appointed by Cromwell as Master of Trinity College, Cambridge, but at the Restoration in 1660 he lost this position and by 1662 he was back in London, reduced to a preacher at Gray's Inn and lodging with Seth Ward (who had similarly lost his post as Master of Trinity College, Oxford).

According to John Aubrey the sparkles now 'mett at ye Bull-head taverne in Cheapside ... till it grew to big for a clubbe and so they came to Gresham Colledge Parlour'.[12] It was at Gresham College that the idea was conceived of obtaining a royal charter, which among other things allowed the club to publish its own books without censorship. In 1668 Boyle himself moved permanently to London and took up residence in his sister's house in Pall Mall. It was in the larder of this house that his servant discovered the luminous leg of veal that so amused Thomas Shadwell.[13]

The Bible reading that Shadwell alludes to in his joke about the meat had been a central part of Boyle's life since a moment

[12] M. Purver, *The Royal Society*, 109.

[13] The site of this house, which occupied part of the same space as the present-day RAC Club, later became the home and hat shop of Melchior Wagner, the forbear of one of the authors.

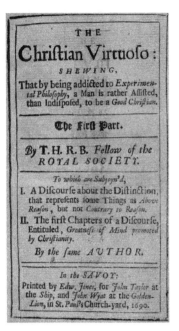

THE

Chriſtian Virtuoſo :

SHEWING,

That by being addicted to *Experimental Philoſophy*, a Man is rather Aſſiſted, than Indiſpoſed, to be a *Good Chriſtian*.

The Firſt Part.

By **T. H. R. B.** *Fellow of the* ROYAL SOCIETY.

To which are Subjoyn'd,

I. A Diſcourſe about the Diſtinction, that repreſents ſome Things as *Above Reaſon*, but not *Contrary to Reaſon*.

II. The firſt Chapters of a Diſcourſe, Entituled, *Greatneſs of Mind promoted by Chriſtianity*.

By the ſame *AUTHOR*.

In the *SAVOY*:

Printed by *Edw. Jones*, for *John Taylor* at the Ship, and *John Wyat* at the *Golden-Lion*, in St. *Paul's* Church-yard, 1690.

30.6 Title page of Boyle's *The Christian Virtuoso* (1690).

when (like Luther) he had been caught in a violent thunderstorm. Reading 'the stenography of God's omniscient hand'[14] through the experimental study of nature was, Boyle argued, a seamless extension of Bible reading: as much a part of the devotional life as singing choral music in divine service. Boyle was not alone in his feeling. A friend of Robert Hooke described how 'he never made any considerable discovery of Nature, invented any useful Contrivance, or found out any difficult Problem, without setting down his Acknowledgement to the Omnipotent Providence, as many places in his Diary testify'.[15]

Such 'acknowledgments' had a value beyond personal piety. They provided a rationale for apparently useless experimentation in the teeth of the mockery and scorn it sometimes produced. Five years after the first performance of *The Virtuoso*, with its depiction of a man addicted to experiments, Boyle began work on a book entitled *The Christian Virtuoso* [Fig. 30.6] with a subtitle which announced 'That by being addicted to *Experimental Philosophy*, a Man is rather Assisted, than Indisposed to be a *Good Christian*' [Fig. 30.7]. Similarly, shortly after its charter of incorporation had been finalized, the Royal Society had commissioned a young clergyman, Thomas Sprat, to write a history of its origin and purpose in which this religious motivation is prominently stressed [Fig. 30.8].

Light and Evidence

In considering the relationship of religion to the new philosophy, Sprat took as his starting point the idea that the Christian virtue of humility compels a man 'to misdoubt the best of his own thoughts' and 'be sensible of his own ignorance'.[16] From there he argues that 'the present *Inquiring Temper* of this *Age* was first produced by the liberty of judgement and searching and reasoning that was used in the first *Reformation*'. Both 'follow the great Præcept of the *Apostle* of *Trying all things*', and where one examines the original text of God's Word, the other looks to 'the large Volume of the *Creatures*' [Fig. 30.9]. In fact, he concludes, the Church of England 'may justly be styled the *Mother* of this sort of *Knowledge*'.[17]

Sprat nevertheless stressed that the society was resolutely non-sectarian. The experimental philosophy it promotes is 'not *English, Scotch, Irish, Popish* or *Protestant* Philosophy: but a Philosophy of *Mankind*'.[18] When Dr Wilkins had first convened his

[14] R. Hooykaas, *Robert Boyle* (1997), 67.

[15] L. Jardine, *The Curious Life of Robert Hooke* (2003), 96.

[16] T. Sprat, *History of the Royal Society* (1959), 367.

[17] T. Sprat, *History of the Royal Society*, 371–2.

[18] T. Sprat, *History of the Royal Society*, 63.

experimental club in the aftermath of the Civil War, he says their first purpose had been 'onely the satisfaction of breathing a freer air, and of conversing in quiet with one another, without being ingag'd in the passions and madness of that dismal Age'.[19]

The question of how to live with the competing ways of thinking that had sprung out of the Reformation insistence on liberty of thought, and which had caused so much civil strife, was a continuing preoccupation of the club members. Boyle once complained that one of his furnaces had crumbled 'into as many pieces as we into sects'.[20] Like several presidents of the Royal Society after him, Wilkins went on to become a bishop. As Bishop of Chester he courageously campaigned in the House of Lords for a policy of toleration towards dissenters. The most cogent theological and philosophical reasons for such toleration, however, were spelt out by another of the original members of Wilkins' Oxford club: the doctor and philosopher John Locke.

In a letter written to Boyle from Cleves, Locke (who was on his first trip abroad) describes how the Protestant and Catholic inhabitants of the town 'quietly permit one another to choose their way to heaven' and manage to 'entertain different opinions without any secret hatred or rancour'.[21] The following year he wrote his *Letter Concerning Toleration*. Rather as Sprat had found a basis for the experimental method in the New Testament injunction of humility, Locke takes as his starting point Christ's instruction to his disciples that while 'The kings of the Gentiles exercise leadership over them . . . ye shall not be so.'

In a later work he argues that rather than 'striving for power and empire over one another', toleration is the 'chief characteristical mark of the true Church'. While 'everyone is orthodox to himself', anyone, however orthodox, 'if he be destitute of charity, meekness, and good-will in general towards all mankind, even to those that are not Christians . . . is certainly yet short of being a true Christian himself'.[22] Trying to compel belief by force is not only wicked but futile, Locke argues, because we are in fact unable to believe things by compulsion: 'it is only light and evidence that can work a change in men's opinions'.[23]

'Light and evidence' were exactly the goals of the Royal Society. In his *History* Sprat expresses the hope that in joining together in a common experimental enterprise 'our several *interests* and *sects* may come to suffer one another with the same

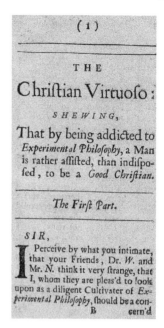

30.7 First page of Boyle's *The Christian Virtuoso* (1690).

30.8 Title page of Thomas Sprat's *History of the Royal Society* (1667).

[19] T. Sprat, *History of the Royal Society*, 53.
[20] M. C. W. Hunter, *Boyle*, 71.
[21] R. S. Woolhouse, *Locke* (2007), 63.
[22] J. Locke, *A Letter Concerning Toleration* (1983), 23.
[23] J. Locke, *A Letter Concerning Toleration*, 27.

30.9 Frontispiece of Thomas Sprat's *History of the Royal Society* (1667).

peaceableness as men of different *Trades* live one by another in the same *Street*'.[24] The ultimate goal was, as Bacon had said, 'the glory of God and the relief of man's estate' and in this task their was work to be done by '*Minds* of all sizes ... from the most ordinary capacities, to the highest and most searching *Wits*'.[25] In this pursuit 'the *Soldier*, the *Tradesman*, the *Merchant*, the *Scholar*, the *Gentleman*, the *Courtier*, the *Divine*, the *Presbyterian*, the *Papist*, the *Independent*, and those of *Orthodox Judgement*, have laid aside their names of distinction, and calmly conspir'd in a mutual agreement of *labors* and *desires*'.[26]

Realizing this godly vision was, as it turned out, not quite as easy as Bishop Sprat (he too became a bishop) had imagined. The problem lay not in the apparent uselessness of the experiments mocked by Shadwell, but in the intractable nature of the people who carried them out. As the experimental philosophy developed, it rapidly became clear that the competitive instinct could intrude into even the most collaborative endeavour.

It also began to become evident that the very success of the experimental philosophy was (from the religious perspective) a double-edged sword.

[24] T. Sprat, *History of the Royal Society*, 427.
[25] T. Sprat, *History of the Royal Society*, 435.
[26] T. Sprat, *History of the Royal Society*, 427.

Boundless Sea

At the beginning of the seventeenth century Galileo, when describing fundamental physical explanations, had spoken of his fear of inadvertently engulfing himself 'in a boundless sea from which I might never get back to port'.[27] By the end of the eighteenth century it seemed to some that these fears had been realized. As the horizon of scientific study expanded to include the whole physical universe, so it suddenly began to seem possible to some that science could literally explain everything. The answers to penultimate questions it seemed might provide answers to ultimate ones. For the first time since antiquity the study of nature started once again to be deployed as a means of marginalizing religious concerns.

These moves first began to be made in a concerted way, in a country where the Reformation had been largely crushed, and where 'freedom of philosophising' was often seen as politically suspect. The figure whose achievements were the inspiration for such ideas came, however, from a rather different background.

This was the man who in an age of 'sparkles' and 'large minds' came to be seen, both at home and abroad, as the largest mind, and (though a Cambridge professor) the brightest sparkle of them all.

[27] T. Sprat, *History of the Royal Society*, 279.

PART VII

The Ocean of Truth

CHAPTER THIRTY-ONE

Le Grand Newton

A year after the death of Sir Isaac Newton, the Revd Joseph Spence, the sometime Oxford Professor of Poetry, began to compile a book of anecdotes which included a handful of memories of the great man. Spence's book (largely derived from conversations with the poet Alexander Pope) remained unpublished for a further century. Most of the recollections were relatively trivial. Pope recalled that Sir Isaac 'though so deep in Algebra and Fluxions could not readily make up a common account'. One story was more resonant. It recounted how 'a little before he died', Sir Isaac had said, '"I don't know what I may seem to the world, but as to myself I seem to have been only like a little boy playing on the sea-shore, and diverting myself in now and then finding a smoother pebble or a prettier shell than ordinary, whilst the great ocean of truth lay all undiscovered before me".'[1]

Whatever he had seemed to himself, Newton had not, in fact, been unconcerned with how he seemed to the world. During his lifetime he had sat for more than 20 busts and portraits. One was donated to an influential foreigner, another to the Royal Society. Engravings and miniature copies were circulated to admirers and disciples [Figs. 31.1a–g]. In the course of his career he engaged in bitter disputes with other men of science to establish his priority in different fields. So successful was the promotion of his reputation that by the end of the eighteenth century Newton had begun to be regarded as guaranteeing two entirely opposite views of the world. He was hailed at home as a vehicle of divine revelation ('God said let Newton be and all was light'[2]). He was revered on the Continent as an icon of human rationality: the supreme representative of a way of thinking that could replace the old superstitions of religion.

Although a few individuals had argued in this way previously, these Continental admirers of Newton represented the first concerted philosophical attempt to use science against religion since the Epicurean philosophers of Greece and Rome

[1] J. Spence, *Anecdotes, Observations, and Characters, of Books and Men* (1820), 54.
[2] A. Pope, *The Poems of Alexander Pope* (1966).

31.1 (a) *Isaac Newton* by Godfrey Kneller (1689); (b) *Isaac Newton* by John Vanderbank (1725); (c) *Isaac Newton* by Enoch Seeman (1726); (d) *Sir Isaac Newton* (ivory and stained wood); (e) *Isaac Newton* by Sir Godfrey Kneller (1702); (f) *Isaac Newton* by Enoch Seeman; (g) *Isaac Newton* by James Thornhill.

we described in Chapter Ten. Their approach was a reverse image of Newton's British followers, who found a confirmation of divine order in the theories of Sir Isaac in much the same way that Plato had found a confirmation of the rationality of the universe in the theories of Eudoxos. This latter perspective was closer to Newton's own view. All the same the growth of his star status had not been accidental. The competitive impulse which fostered it did not sit easily with Thomas Sprat's vision of experimental philosophy as a model of godly cooperation.

The High Silence

While Newton's journal records that he bought a copy of Sprat's *History* for seven shillings in 1667, he showed little immediate inclination to follow its recommendation of 'mutual communication' or of the need in science 'to join many men together'.[3]

According to his own account Newton had made all his major mathematical discoveries during the plague years of 1665 and 1666, while at the family home in Woolthorpe and isolated from Cambridge contacts. It was not until 1668, when his erstwhile Cambridge tutor Isaac Barrow showed him a recent publication by the Danish mathematician Nicholas Mercator, that Newton had been prompted to write a paper *On Analysis by Infinite Series* to establish his own priority and superiority in the field. Even then while he allowed Barrow to show the paper to the publisher John Collins and later to the president of the Royal Society, he adamantly refused to allow Collins to publish it.

Around the same time he had begun to develop what was to be a lifelong interest in the secretive tradition of alchemy, with its quasi-mystical belief in the possibility of transmutation—turning lead into gold. Robert Boyle shared this interest but disapproved of the alchemists' arcane terminology: 'their obscure ambiguous and almost Ænigmatical way of expressing what they pretend to teach'.[4] To counter that, Boyle had boldly published 'An Experimental Discourse of *Quicksilver* growing hot with *Gold*' in the *Philosophical Transactions of the Royal Society*.

Newton was outraged by this publicity and wrote to the Society's secretary, urging that Boyle should maintain 'the High Silence' of the 'true hermetic philosopher'.[5] Although deeply

[3] T. Sprat, *History of the Royal Society* (1959), 85.
[4] R. Boyle, *The Sceptical Chymist* (1661), 3.
[5] I. Newton, *The Correspondence of Isaac Newton* (1959), vol. II, 2.

31.2 A replica of Newton's reflecting telescope (1671).

influenced by Boyle, after Boyle's death Newton told Locke that he had refused to correspond with the older man about these matters because of his 'conversing with all sorts of people & being in my opinion too open & too desirous of fame'.[6]

His own first taste of fame had been traumatic. In 1671 he had allowed a tiny reflecting telescope he had made to be shown to the Royal Society. This exquisite device, the first working example of its kind, was more powerful than a refracting telescope six times its size [Fig. 31.2]. Having been examined 'by some of the most eminent in optical science and practice and applauded by them' it was taken to the King, who was given a personal demonstration. The acclamation Newton received, and the rapid election to the Royal Society that followed, had encouraged him to tell the fellows about 'a Philosophicall discovery which induced mee to the making of the said Telescope'.[7] In 1672 he sent a long letter that was printed in the Society's *Philosophical Transactions*, which gave an account of his theory of light and of the experiments which he believed demonstrated it. The criticism this letter received, particularly from Robert Hooke (who demonstrated no eagerness to try repeating the experiments), caused Newton to stop all correspondence, to threaten to resign from the Society, and to withhold the publication of his *Optics* until after Hooke's death some 30 years later.

More acrimony was to follow.

The Impetuous Man

In January 1684 the young astronomer Edmund Halley met with Robert Hooke and Christopher Wren at a coffee house in London. Follwing this meeting he travelled to Cambridge to consult with Newton, who was by now Lucasian Professor of Mathematics. Over coffee Halley mentioned his idea that the action which kept the planets in motion round the Sun decreased as an inverse square of their distance from it. The two older men laughed and explained 'that upon that principle all the laws of celestial motion were to be demonstrated'.[8]

Hooke had been unable to produce any such demonstration. Back in 1680 he had written to Newton putting his 'supposition . . . that the attraction is always in a duplicate proportion to the Distance from the Center Reciprocal',[9] but had received no reply. When Halley now went to Cambridge to ask directly

[6] I. Newton, *The Correspondence of Isaac Newton*, vol. II, 315.
[7] I. Newton, Letter from Newton to Henry Oldenberg, 18 January 1671/2, Cambridge University Library, MS Add. 9597/2/18/13.
[8] I. Newton, *The Correspondence of Isaac Newton*, vol. II, 433–5.
[9] I. Newton, *The Correspondence of Isaac Newton*, vol. II, 239.

whether it actually could be demonstrated, he prompted Newton to begin writing his *Philosophiæ Naturalis Principia Mathematica*. Halley subsequently paid for its publication.

On the eve of publication Halley wrote to tell Newton that 'Mr Hooke has some pretensions upon the invention of the rule of the decrease of Gravity ... He says you had the notion from him [and] seems to expect you should make some mention of him in the preface'.[10] Newton's response was to go through the draft of Book II of the *Principia* striking out all references to 'the most illustrious Hooke'—Cl(arissimus) Hookius—and threatening to withhold Book III.

This was to become a repeating pattern.

In the year following the letter from Hooke, Newton wrote to John Flamsteed, the astronomer royal. On the 12th of December 1680 a comet had appeared in the night sky with a tail 'above a moon broad' and stretching the full length of King's College chapel. Throughout January and February Newton made nightly observations of the comet while corresponding with Flamsteed. Flamsteed, who already suspected that comets behaved much like planets, had no hesitation in supplying the Cambridge professor with data on the relative position of planets and stars. These enabled Newton to calculate the trajectory of the comet, showing that it, like the planets, obeyed the inverse square law, and confirmed the theory that lay at the heart of the *Principia* of universal gravitation.

Thirteen years later, in 1694, Newton took a boat up the Thames to visit Flamsteed at the observatory in Greenwich. His object was to obtain from the astronomer his observations of the movements of the Moon in order to prove (in a projected second edition of the *Principia*) that these too conformed to his theory of gravitation. He found Flamsteed less cooperative than previously. The astronomer had objected to the 'very slight acknowledgements' of what Newton had received from the observatory that he had included in the *Principia*, and was reluctant simply to hand over the fruit of 30 years of observations.

In 1704 Newton, by now President of the Royal Society, devised a method of forcing him to do so. Using his influence he established, under royal command, a commission to publish a complete Moon and star catalogue—*Historia Coelestis Britannica*—to be edited by Edmund Halley. In 1711 Flamsteed, still dragging his feet, was called before the Council of the

[10] I. Newton, *The Correspondence of Isaac Newton*, vol. II, 285.

Royal Society and complained 'he was robbed of his labours'. 'At this', according to Flamsteed's own account, 'the impetuous man grew outrageous and said "are we then the robbers of your labours?" I answered I was very sorry they owned themselves to be. After which he was all in a rage. He called me many hard names; puppy was the most innocent of them'.[11] The catalogue was published in 1712. When the second edition of the *Principia* appeared the following year, almost every reference to Flamsteed had been struck out.

A third such erasure came about as direct a consequence of Newton's earlier reluctance to publish.

Between 1673 and 1675 a German mathematician and philosopher, Gottfried Leibniz, had developed a version of the infinite series and calculus similar to that which Newton had worked out at Woolsthorpe ten years earlier. The main difference was that Leibniz's method and notation were to prove much easier to use. When news of this began to circulate, Newton was persuaded to write to Leibniz, asserting (in a coded form) his priority, but still insisted that none of his mathematical papers should be published 'without my special licence', which he did not give.

Leibniz was at first relaxed about the issue of priority, writing that 'one man makes one contribution, another man another'[12] and in 1684 published his first paper on calculus. Newton in response added a scholium to Book II of the *Principia*, referring to the letters 'which went between me and that most excellent geometer G. W. Leibniz', in which he had 'signified that I was in the knowledge of a method of obtaining maxima and minima, of drawing tangents and the like', while 'that most distinguished man wrote back that he had fallen on a method of the same kind, and communicated his method, which hardly differed from mine except in his forms of words and symbols'.[13]

Soon, however, accusations of plagiarism were flying back and forth (made by supporters and in anonymous reviews). Newton established a Royal Society committee to look into the question, which produced a report that he secretly wrote himself. He later produced an account of the committee's work. Even after Leibniz's death he continued to make new drafts of this, adding the comment that 'second inventors count for nothing'.[14] When the third edition of the *Principia* appeared a year before Newton's death, all reference to Leibniz had been removed.

[11] J. Flamsteed, *Self-Inspections of J.F.*, 1667–1671, Royal Greenwich Observatory, 1/32/A.

[12] G. W. Leibniz, *Sämtliche Schriften und Briefe* (1950), vol. IV, 475–6.

[13] I. Newton, *Sir Isaac Newton's Mathematical Principles of Natural Philosophy* (1934), 655–6.

[14] I. Newton, *Papers Relating to the Dispute Respecting the Inventions of Fluxions 1665–1727*, Cambridge University Library, Department of Manuscripts and University Archives, MS-ADD-03968.37.

It was not only his colleagues who appeared to have been written out of Newton's account of the universe. To some it would seem that the system he described left no room for divine activity. In the first edition of the *Principia* there was indeed a striking absence of any reference to God.

Feigning Hypotheses

However normal this might seem today, it was less so in the seventeenth century. The style in which Newton wrote the *Principia* seems to have been deeply influenced by the negative response he had received to his first paper on optics. He tells his readers that 'to prevent disputes', he had laid out the book in the form of mathematical propositions, 'which should be read only by those who had first made themselves masters of the principles in the preceding books'[15] and confided to a friend that he had made it as difficult as possible 'to avoid being bated by little smatterers in mathematics'.[16]

In his optics paper Newton had incautiously remarked that 'it could no longer be disputed' that light was corpuscular. Hooke had pointed out that this was an unproven hypothesis. In the *Principia* there would be no such hostages to fortune. He could give 'no reasons for these properties of gravity' and would not 'feign hypotheses'. To explain this he argues that 'whatever is not deduced from the phenomena must be called a hypothesis, and hypotheses whether metaphysical or physical or based on occult qualities or mechanical have no place in experimental philosophy'.[17]

This famous sentiment was echoed nearly a century later in a quotation attributed to Pierre-Simon Laplace, 'the French Newton', dismissing unneeded invocations of God.

Newton had been unable to establish that the interaction of the planets in the solar system would be permanently stable, and suggested that periodic divine intervention would be needed to preserve it. Laplace in his *Exposition du système du monde* (1796) was able to show that this remedy was unnecessary. When Napoleon queried him about not mentioning the Creator in his account Laplace is reputed to have replied, 'Sire, je n'avais pas besoin de cette hypothèse là'—'Sire, I have no need of that hypothesis'.

According to his colleague François Arago, shortly before his death Laplace (who, though he veered towards deism, was

[15] M. White, *Isaac Newton* (1997), 217.
[16] M. White, *Isaac Newton*, 217.
[17] I. Newton, *The Principia* (1999), 943.

never an atheist and received the sacraments on his death bed from the curé of Arcueil) requested that this anecdote be either explained or removed from any biography 'and the second way was the easiest'.[18] His remark appears to have been directed solely at Newton's particular hypothesis. But while it seems unlikely Laplace himself thought that Newtonian physics as a whole left no room for theology, there were both religious and anti-religious readers of Newton in France who certainly did.

The Great Newton

Jesuit writers like Cardinal de Polignac (who in 1747, some 20 years after Newton's death, published a poem called *Anti-Lucretius*) argued that Newton's philosophy led directly to materialism and atheism. Others argued that Newtonian science supported what became known as 'deism'—a perspective that rejected miracles, revelation, and all the paraphernalia of organized religion, arguing instead that God was the supreme architect of the universe who did not intervene in the world but could be known through the observation of nature and exercise of reason.

This was the view of François-Marie Arouet, who under the nom de plume 'Voltaire' became the most well-known writer in Europe. In 1738 Voltaire published *Elémens de la Philosophie de Neuton* (with a frontispiece depicting Newton as a semi-divine figure) [Fig. 31.3] and in 1749 he and his lover Émilie du Châtelet produced a French translation of the *Principia* complete with scholarly commentary.

Those more radical in their views than Voltaire, who saw deism as a halfway house to atheism (which in practice it sometimes was), tended, however, to agree with Cardinal Polignac's assessment.

When the atheist Baron d'Holbach (sometimes known as 'le premier maître d'hôtel de la philosophie') anonymously published his *Le Système de la nature* (The System of Nature), he argued that 'the universe that vast assemblage of every thing that exists presents only matter and motion . . . an immense uninterrupted succession of causes and effects'.[19] To complete this argument it seemed necessary to extend this materialist account to cover human beings, and a number of the new atheists of the eighteenth century set themselves to become 'Newtons of the mind'. Thus, d'Holbach in his *Le Système de la nature*

[18] H. Faye, *Sur L'origine Du Monde* (1884), 109–11.
[19] P.-H. T. d'Holbach, *The System of Nature* (1821), vol. I, Chapter 1.

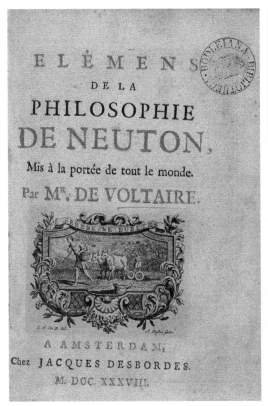

31.3 Title page and frontispiece of Voltaire's *Elements of the Philosophy of Newton*.

included five chapters arguing that there was no immaterial 'soul' and that free will was an illusion. In a similar vein Claude Adrien Helvétius wrote *De l'esprit* (On Mind), and Julien Offray de La Mettrie authored *Histoire naturelle de l'âme* (Natural History of the Soul) and *L'homme machine* (Machine Man).

Denis Diderot and Jean d'Alembert, who edited the first volumes of the great *Encyclopédie*, which began to appear at the start of the 1750s and subsequent volumes (despite attempts to suppress them) continued to appear over the following decades, saw themselves as promoting a Baconian and Newtonian programme which relegated theology to the status of superstition.

By the turn of the century the idea of 'le grand Newton' had in some quarters become a kind of rationalist cult. In 1803 Henri de Saint-Simon published a tract called *Lettres d'un habitant de Genève à ses contemporains* in which he argued that a 'Great council of Newton' should be established as the supreme

31.4 Design for a cenotaph for Sir Isaac Newton by Étienne-Louis Boullée.

authority on Earth, deposing all bishops, popes, and priests, and that 'temples of Newton' should be set up to carry out their instructions and to act as centres of rational worship.

A decade earlier the architect Étienne-Louis Boullée had, in fact, already designed such a temple—a cenotaph for Newton in the form of a gigantic sphere lit either by a smaller central armillary sphere or by small holes in the vault imitating stars. In the centre was an altar from which (in Boullée's illustration) the magus Zoroaster prays with arms outstretched and head turned up in adoration [Fig. 31.4].

This late-eighteenth-century image of Newton as rationalist icon, was, however, increasingly remote from any viewpoint that would have been shared by the man himself.

CHAPTER THIRTY-TWO

The Beautiful System

In a journal begun when he was 19, the young Newton made a list which provides an insight into his early spiritual life. The journal, which ends with mathematical accounts of how to describe parabolas, ellipses, and hyperbolae, begins by listing 49 sins committed before Whitsunday 1662. Some were very specific—'making pies on Sunday night', 'punching my sister'. Others were more general—'not living according to my belief'. This question was to be a lifelong preoccupation.

A Skill in Divinity

At the King's School in Grantham where his education began, Newton had been introduced to Reformation theology by a Puritan divine named John Angell. From him he would have learnt the method, which was to remain with him all his life, of rooting theology in an individual study of Scripture. He amassed, according to those who knew him, a collection of some 30 different Bibles. Locke (himself a considerable biblical scholar) described Newton as 'a very valuable man not only for his wonderful skill in mathematicks but in divinity too and his great knowledg in the Scriptures where in I know few his equals'.[1]

In a work on the interpretation of Scripture Newton declared, 'I have a fundamental belief in the Bible ... I study the Bible daily'. He wrote that men are to 'study the scriptures <to the end of their lives> ... & live accoding [sic] to what they learn'.[2] While some like Hooke, Flamsteed, and Leibniz (not to mention the forgers whose execution he secured when in charge of the mint) might have reason to question it, there were plenty who testified to the sincerity of his attempt to do this.

Some of these remarked on an evident personal piety. Voltaire described how Newton's disciple, Samuel Clarke, always

1 S. D. Snobelen, 'Isaac Newton, Heretic: The Strategies of a Nicodemite', *British Journal for the History of Science* 32 (1999), 381–419.
2 I. Newton, *Irenicum, or Ecclesiastical Polyty tending to Peace* (post-1710), King's College Library, Cambridge, Keynes MS 3, p. 41.

mentioned the name of God with great reverence, and had told him that he had learned the habit from Newton. According to his niece, Newton 'could not bear to hear anyone talk ludicrously of religion' and another pupil recalled his saying that 'the wicked Behaviour of most modern courtiers had been caused by their having laughed themselves out of religion'.[3] In his own writings Newton speaks of Jesus as 'our Lord' and of having 'made an attonement for us & to have <satisfied God's wrath &> merited our pardon & to have washed away our sins in his blood'.[4] Others related how this faith translated into action.

The son of the man with whom Newton shared a room in his early days at Trinity described a 'Charitable Benefaction' that Newton established for distributing 'many Dozens of Bibles sent by him for poor people'.[5] Several people described being helped by him through spiritual crises, and throughout his life there are records of Newton's charitable giving to both family and strangers.

His writings make it clear that he saw his physical studies as having the same motivation as his theological ones. In discussing his alchemical researches he declares that 'this philosophy is not that of kind which tends to vanity & deceit but rather to edification inducing first the knowledge of God & secondly the way to find out true medicines in his creatures ... the scope is to glorify God in his wonderful works to teach man to live well'.[6] At Trinity Newton is known to have prepared and probably delivered at least one and possibly two lay sermons. If he was reticent in speaking about religion in later life it was for a reason.

The Polish Brethren

At some point in the 1670s he had begun to read books published by a radical Reformation sect known as the 'Polish Brethren' or 'Socinians' (after their leader Faustus Socinus). Newton's intense exploration of theological issues had probably been motivated by his impending ordination (as Lucasian Professor of Mathematics he was committed to entering holy orders), but the consequence of his studies was to make ordination into the Anglican church impossible for him. He had become convinced that the Polish Brethren were right: that the doctrine of the Trinity as propounded by the church councils

[3] S. D. Snobelen, 'Isaac Newton, Heretic: The Strategies of a Nicodemite', 411.

[4] I. Newton, 'Irenicum, or Ecclesiastical Polyty tending to Peace', pp. 3, 36.

[5] S. D. Snobelen, 'Isaac Newton, Heretic: The Strategies of a Nicodemite', 410.

[6] I. Newton, 'Manna': transcript of an anonymous alchemical treatise in another hand with additions and notes by Newton (1675), King's College Library, Cambridge, Keynes MS. 33.

of Nicaea and Chalcedon (in the fourth and fifth century) was an aberration from the Christianity of the primitive church and that the creeds and articles of the Church of England were consequently heretical.

This was a dangerous position to hold. When William Whiston, Newton's pupil and successor as Lucasian Professor, publicly advocated the theological ideas he had learned from his mentor, he was forced to give up his chair and leave Cambridge. Newton himself was more cautious. Although he once asked Locke to publish in France a theological treatise he had written, he later withdrew it and seems to have kept to a lifelong strategy of communicating these dangerous ideas in underground networks and through private conversations. This inhibition did not, however, prevent him from seeking, when the opportunity arose, to demonstrate the unity between his scientific and his theological thinking.

The Frame of the World

When Robert Boyle died, on the last day of 1691, he was found to have left an endowment for a series of lectures on the defence of the Christian religion against atheists and others. The trustees of the endowment acted quickly and in February 1692 they appointed a formidable young classical scholar called Richard Bentley to give the first lectures. Bentley started lecturing in March and wrote to Newton asking for advice on how the arguments of the *Principia* might serve him in the final three lectures dealing with God's wisdom in creating the universe. Newton's response was enthusiastic. 'When I first wrote my treatise about our systeme', he replied, 'I had my eye upon such Principles as might work with considering men for the beleife of a Deity & nothing can rejoice me more than to find it useful for that purpose'.[7]

These principles had been a longstanding interest. In one of his notebooks Newton argues that 'there is no way without revelation to come to the knowledge of a deity but by the frame of nature' and suggests that 'the frame of the ancient temples', like Stonehenge, were designed 'to propose to mankind ... the study of the frame of the world as the true temple of the great God they worshipped' [Fig. 32.1].[8] The same kind of principles, he argues, apply in both interpreting Scripture and understanding the world.

[7] I. Newton, original letter from Isaac Newton to Richard Bentley (10 December 1692), Trinity College Library, 189.R.4.47.

[8] I. Newton, draft chapters of a treatise on the origin of religion and its corruption (~1690s), National Library of Israel, Jerusalem, Yahuda MS 41.

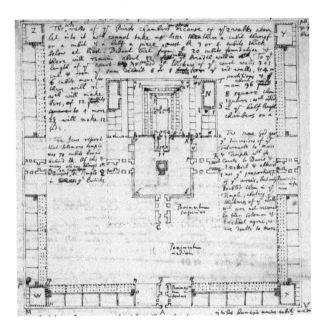

32.1 A drawing of Solomon's Temple from Newton's *Prolegomena ad lexici prophetici*.

As the world, which to the naked eye exhibits the greatest variety of objects, appears very simple in its internall constitution when surveyed by a philosophic understanding ... so it is in these visions ... He is the God of order & not of confusion. And therefore as they that would understand the frame of the world must indeavour to reduce their knowledg to all possible simplicity, so it must be in seeking to understand these visions.[9]

Newton described the means by which he had been led to this simplicity in the famous account he gave to the antiquarian William Stukeley one day when 'after dinner the weather being warm we went into the garden and drank tea under the shade of some apple trees'.

Starting from the question 'why should that apple always descend perpendicularly to the ground ... why should it not go sideways or upwards? But constantly to the earth's centre?', he had reasoned that there must be 'a drawing power in matter', that 'the same drawing power must be in the earth's centre', and that 'if matter draws matter it must be in proportion of its quantity ... the apple draws the earth as the earth draws the apple'.[10] From this, he told Stukeley, he concluded '"that there is a power like that we here call gravity which extends

[9] I. Newton, untitled treatise on revelation (Section 1.1), National Library of Israel, Jerusalem, Yahuda MS 1.1.

[10] W. Stukeley, 'Revised Memoir of Newton' (1752), Royal Society Library, MS 142.

itself thro" the universe', and so 'by degrees he began to apply this property of gravitation to the motion of the earth & the heavenly bodys: to consider their distances, their magnitudes, their periodical revolutions'.[11]

What distinguished this idea from the 'kinetic force' that Philoponus had imagined was the critical insight that 'the drawing power' of matter could be mathematically described. In 1664 Newton had read Descartes' *Geometry* and proceeded in astonishingly short space of time to absorb almost the entire canon of mathematics as it existed at the time. From there he went on to develop his own techniques. 'I was', he later wrote, 'in the prime age of my invention', and developed first a method of finding the exact gradient of a curve (known as differentiation) and then a method of finding the area under a curve (now known as integration). Applying these methods (which Newton called 'fluxions' and Leibniz 'calculus') to the planetary motions, he was able to obtain a mathematical value for the force exerted by an object completing a circular motion.

Writing to the Hugenot exile Pierre des Maizeaux shortly before he died, Newton described how he had 'begun to think of gravity extending to the orb of the moon' and 'having found out how to estimate the force with which [a] globe revolving within a sphere presses the surface of the sphere, from Kepler's rule of the periodical times of the planets being in a sesqui-laterate (3:2) proportion of their distances from the centre of their orbs', he had been able to deduce that 'the forces which keep the planets in their orbs must [be] reciprocally as the squares of their distances from the centres about which they revolve'.[12]

When he first calculated the receding force of the Moon using Galileo's estimate for the radius of the Earth it did not clearly show the inverse square relationship that he had expected. It was not until he was writing the *Principia* nearly 20 years later, when he repeated the calculation using a new more accurate figure for the Earth's radius that had been obtained by the French astronomer Jean Picard, that he was able to demonstrate the inverse square relationship was correct.

Newton set out his demonstration in a manner that echoed the style that the ancient Greek mathematician Euclid had used in his *Geometry*. Euclid begins with five postulates ('that all right angles are equal to one another . . . ') and five 'common

11 W. Stukeley, 'Revised Memoir of Newton'.

12 I. Newton, draft letter to Pierre des Maizeaux (1718), Cambridge University Digital Library, MS Add. 3968.41.

notions' ('that all things that are equal to the same thing are equal to one another ... '). Newton begins with three laws of motion ('Every body continues in its state of rest or of uniform motion in a right line unless it is compelled to change by forces impressed upon it ... ') and two general principles of space and time ('absolute true and mathematical time of itself and from its own nature flows equally without relation to anything external ... '). But here the similarity ends.

Where Euclid's conclusions flow directly from his axioms, Newton had to proceed through formidably complex mathematics until the point in Book III where he is finally able to show that Galileo's terrestrial mechanics and Kepler's planetary laws could be united in the single concept of universal gravitation. 'In the preceding books', he writes, 'I have laid down the principles of philosophy, principles not philosophical but mathematical ... It remains', he goes on, 'that from the same principles I now demonstrate the frame of ... the world' [Fig. 32.2].[13]

32.2 Title page of Newton's *Philosophiæ Naturalis Principia Mathematica.*

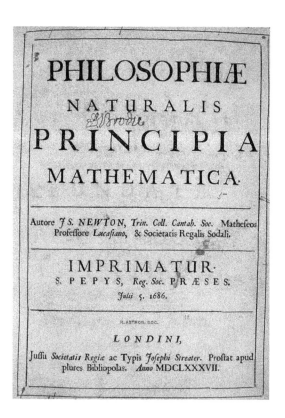

[13] I. Newton, *The Principia* (1999), Book III, 397.

Freedom and Necessity

A quarter-century after the publication of the *Principia*—*The Mathematical Principles of Natural Philosophy* to give it its full title—Richard Bentley (who had by then become Master of Trinity) persuaded Newton to publish a second edition (which Bentley paid for). This enabled Newton not only to extend the description of universal gravitation with the benefit of new data, but also to locate his account within a wider theological context. In a 'General Scholium', printed at the end of the third book, he nails his colours decisively to the mast. 'This most beautiful system of the Sun and Planets', he asserts, 'could only proceed from the counsel and dominion of an intelligent and powerful being'.[14]

Newton (by now both President of the Royal Society and Master of the Mint) felt confident enough at this stage of his life to couch his argument in the terms of his own theology (for those who had ears to hear). In language which directly echoes that of John Crell (the chief theologian of the Polish Brethren) he emphasizes that 'this being governs all things, not as the soul of the world, but as Lord over all. And on account of his dominion he is wont to be called Lord God παντοκρατωρ (pantokrator) or Universal Ruler.'[15]

The point Newton wants to make here though is not about the Trinity. His argument is rather that 'a God without Dominion . . . is nothing else but Fate and Nature. Blind Metaphysical necessity, which is certainly the same always and everywhere could produce no variety of things'. It is in the 'diversity of natural things', he argues, that we witness God's power to choose. Consequently although the nature of God is utterly mysterious there is 'much concerning God' which we can discover 'from the appearance of things' and which 'does certainly belong to Natural Philosophy'.[16]

While Newton argues for the relevance of science to theology, the relevance of theology for science is spelt out in a preface written by Roger Cotes—a young Cambridge mathematician whom Newton, on Bentley's recommendation, employed to edit the new edition.

Cotes echoes Newton argument against those 'who dream that . . . matter exists by the necessity of its nature always and everywhere'. The 'variety of forms', he suggests, 'is entirely inconsistent with necessity', while 'this World so diversified with

[14] I. Newton, *The Mathematical Principles of Natural Philosophy* (1729), 387–93.
[15] I. Newton, *The Mathematical Principles of Natural Philosophy*, 387–93.
[16] I. Newton, *The Mathematical Principles of Natural Philosophy*, 387–93.

that variety of forms and motions we find in it, could arise from nothing but the perfectly free will of God directing and presiding over all'. This is also true of the laws of nature, in which 'there appear many traces of the most wise contrivance but not the least shadow of necessity'.[17] But this, Cotes goes on to suggest, has important consequences for the way science is done.

Someone who supposes the world exists by necessity might hope to find the principles of physics by 'the internal light of his reason', but those who acknowledge God's freedom will recognize 'that we must seek from uncertain conjectures but learn them from observations and experiments'[18] even where it leads us to such apparently mysterious phenomena as universal gravitation.

Natural Theology

Despite Cotes' and Newton's insistence on its fruitfulness, this kind of two-way traffic between science and theology had historically always contained risks as well as benefits.

The banquet that Plato gave at the Academy for Eudoxos and his pupils had marked a moment of synthesis between science and religion, which motivated astronomical studies for two millennia. It had also ensured that a single (ultimately misleading) hypothesis became set in stone.

The kind of 'beautiful system' described by Newton might hold out the prospect that a theological view of the world could, in a similar way, be underwritten by the unarguable certainties of mathematical science. Would it ultimately lead to a similar cul-de-sac? There were those who (for very different reasons) thought that it might.

[17] R. Cotes, Preface, in I. Newton, *The Mathematical Principles of Natural Philosophy*, second edition, xxxvii.
[18] R. Cotes, Preface, in I. Newton, *The Mathematical Principles of Natural Philosophy*, xxxvii.

CHAPTER THIRTY-THREE

Mathematical Theologies

Following the death of Queen Anne in August 1714, her nearest Protestant relative, George Ludwig, the Prince Elector of Hanover, travelled to England and was crowned in Westminster Abbey that October. Among the German entourage attending the coronation were the Elector's son George Augustus with his wife Caroline of Ansbach, who shortly afterwards were themselves invested as Prince and Princess of Wales.

Caroline was perhaps the most intellectually engaged holder of this title to date. As a young woman growing up at the court of Frederick I she had been a friend of Handel and been tutored by Gottfried Leibniz. Once in England she created a considerable library at St James' Palace, championed an early form of vaccination, and kept up a correspondence with her former tutor.

A year after the coronation she received a letter from Leibniz (who was not asked to accompany the Elector's court), remarking on what he considered to be the decay of natural religion in England. Caroline relayed the contents of the letter to a new acquaintance: Dr Samuel Clarke, the Rector of St James, Westminster. An exchange of papers was then conducted through the Princess, which continued until Leibniz's death in 1716 [Figs. 33.1, 33.2, 33.3].

CAROLINE OF ANSBACH

33.1 Caroline of Ansbach.

An Odd Opinion

In his letter to Caroline, Leibniz had objected to Newton's description of space as 'an organ of God', and had remarked that

> Sir Isaac Newton and his Followers have . . . a very odd opinion concerning the Work of God. According to their doctrine, God Almighty wants to *wind up* his Watch from Time to Time: otherwise it would cease to move. He had not, it seems, sufficient Foresight to make it a perpetual motion. Nay, the machine of God's making is so imperfect,

33.2 Samuel Clarke.

33.3 Gottfried Leibniz.

[1] S. Clarke, *A Collection of Papers, Which Passed between the Late Learned Mr Leibnitz, and Dr Clarke, in the Years 1715 and 1716*, http://www.newtonproject.sussex.ac.uk/catalogue/viewcat.php?id=THEM00224, accessed 15 August 2015. Mr. Leibnitz's First Paper.

[2] S. Clarke, *A Collection of Papers, Which Passed between the Late Learned Mr Leibnitz, and Dr Clarke, in the Years 1715 and 1716*, Dr Clarke's First Reply, paragraph 3.

[3] S. Clarke, *A Collection of Papers, Which Passed between the Late Learned Mr Leibnitz, and Dr Clarke, in the Years 1715 and 1716*, Dr Clarke's First Reply, paragraph 4.

[4] S. Clarke, *A Collection of Papers, Which Passed between the Late Learned Mr Leibnitz, and Dr Clarke, in the Years 1715 and 1716*, Dr Clarke's First Reply, paragraph 4.

[5] S. Clarke, *A Collection of Papers, Which Passed between the Late Learned Mr Leibnitz, and Dr Clarke, in the Years 1715 and 1716*, Mr Leibnitz's Second Paper, paragraph 1.

[6] S. Clarke, *A Collection of Papers, Which Passed between the Late Learned Mr Leibnitz, and Dr Clarke, in the Years 1715 and 1716*, Mr Leibnitz's Second Paper, paragraph 1.

[7] S. Clarke, *A Collection of Papers, Which Passed between the Late Learned Mr Leibnitz, and Dr Clarke, in the Years 1715 and 1716*, Mr Leibnitz's Second Paper, paragraph 8.

[8] S. Clarke, *A Collection of Papers, Which Passed between the Late Learned Mr Leibnitz, and Dr Clarke, in the Years 1715 and 1716*, Mr Leibnitz's Second Paper, paragraph 8.

according to these Gentlemen, that he is obliged to *clean* it now and then by an extraordinary concourse, and even to *mend* it, as a clockmaker mends his work.[1]

The Princess showed the letter to Clarke, knowing him to be a friend and disciple of Newton. The two men were neighbours in St Martin's Street, and Caroline later informed Leibniz that all of Dr Clarke's replies had been looked over by Sir Isaac before they were sent.

In the first of these replies Clarke denied that Newton had ever thought of space as 'an organ of God'. He had used a 'similitude' to describe infinite space being '(*as it were*) the *Sensorium* of the Omnipresent being',[2] who perceives all things by his immediate presence. The idea that God was the upholder and preserver of his creation was 'not a *diminution* but the true *Glory* of his workmanship'.[3] The contrary view that the world was 'a great *Machine*, going on *without the interposition of God*, as a Clock continues to go without the Assistance of a Clockmaker' betokened a notion 'of *Materialism* and *Fate*' whose tendency was 'to Exclude God out of the World'.[4]

When this was sent on to Leibniz by the Princess of Wales, Leibniz replied by referring to the metaphysical principles he had set out in his book *Théodicée* published five years earlier. The most important of these was 'the principle of sufficient reason', which held that 'nothing happens without a *Reason* why it should be so'.[5] By that single principle he believed 'one may demonstrate the being of a *God*',[6] and it ruled out the idea of absolute contingency. He did not say that the material world was 'a Machine or Watch that goes *without* God's *Interposition*', but he did say that it went 'without needing to be *Mended* by him',[7] as if God needed to have second thoughts. 'There is in his Works', Leibniz insists, 'a Harmony, a beauty already *pre-established*'.[8]

Neither of these positions is significantly altered in the exchanges that followed. Leibniz continually asserts that Newton's term 'sensorium' seems to make God a part of creation—'the soul of the world'—whose work needs mending. Clarke continually insists that Newton was speaking metaphorically and that Leibniz's 'pre-established' harmony turns the world and human beings within it into 'a machine like a clock' and reduces God to something like 'Necessity and Fate'.

In his fifth paper Leibniz argues that while God '*can* do everything possible, he will only do what is best', which in his *Théodicée* he had stated as the principle that 'if there were not the best among all possible worlds, God would not have produced any'.[9] All of this follows, he argues, from the great principle of sufficient reason. Clarke, however, concludes his fifth reply by arguing that Leibniz's interpretation of sufficient reason amounted to '*Petitio Principii*, or *Begging of the Question*: Than which, nothing can be more unphilosophical',[10] adding in the published version, 'NB *Mr* Leibnitz *was prevented by Death, from returning any Answer to this last Paper.*'

In the introduction to this published version Clarke argued that, despite Mr Leibniz's suspicions, the foundations of natural religion had never been so deeply or so firmly laid as in Sir Isaac Newton's experimental philosophy. Futhermore the protestant succession secured by the house of Hanover (and Princess Caroline's new baby) would now ensure that this learning and knowledge would be promoted 'both against Atheism and Infidelity on the one hand . . . and against Superstition and Bigotry on the other hand'.[11]

By the latter part of the century it seemed to some of Clarke's compatriots that this hope had been achieved, as a succession of Boyle lecturers and others continued to pour out works of natural theology confuting the writings of the Continental 'philosophes'. While there were deists and atheists in Britain, just as there were proponents of natural theology on the Continent, by and large these opposing camps seemed to face each other on either side of the Channel, and it was in conscious opposition to the secularizing tendency of the French *Encyclopédie* that the *Encyclopaedia Britannica* began to be published in 1768.

The Dangers of Defence

As intellectual Martello towers, the defences that natural theology offered against atheism and infidelity were not as impregnable as they seemed. Both the English and the German versions contained serious weaknesses.

The weakness of the English version was that it relied in part on scientific lacunae—'a God of the gaps'. When Pierre-Simon Laplace, 'the French Newton', was able to explain the stability of the solar system without recourse to Newton's hypothesis of God's intervention, Newton's argument that God was the upholder and preserver of his creation could then seem to do the very thing which Clarke had objected to and 'exclude God out of the world'.

9 G. W. Leibniz, *Theodicy: Essays on the Goodness of God, the Freedom of Man, and the Origin of Evil* (1951), 128.
10 S. Clarke, *A Collection of Papers, Which Passed between the Late Learned Mr Leibniz, and Dr Clarke, in the Years 1715 and 1716,* Dr Clarke's fifth reply, final paragraph.
11 S. Clarke, *A Collection of Papers, Which Passed between the Late Learned Mr Leibniz, and Dr Clarke, in the Years 1715 and 1716,* Front Matter.

The weakness of the German version was that Leibniz's grand metaphysical system could seem to have no contact with the actual puzzles of human existence. Thus, Voltaire in his novel *Candide* was later to use the fictional misfortunes of the philosopher Pangloss, and the real disaster of the Lisbon earthquake to ridicule the idea that 'all was for the best in the best of all possible worlds'.

The potential dangers of natural theology had already been noticed in the bundles of writings left by a third great mathematician who had died at the age of 39, when Leibniz was just 16 and while Newton was still an undergraduate at Trinity. This was Blaise Pascal.

The Pascaline

Pascal had been a child prodigy who, according to family legend, had worked out Euclid's theorems for himself when he had not been allowed to learn mathematics. At the age of 16 he had written an essay on conics, which the philosopher Descartes thought must have been composed by his father, Étienne. In 1639 Étienne was awarded a post as a tax collector, and it was in part to assist his parent in the laborious calculations that his job involved that the young Pascal had developed the first prototype of his calculating machine, 'the Pascaline' [Fig. 33.4].

In 1648 while the family was living in Rouen, both father and son had heard about Torricelli's experiments with mercury barometers from the court engineer Pierre Petit, who was visiting the town, and it was this which prompted Blaise to urge his brother-in-law to carry out the experiment on the Puy de Dôme described in Chapter Thirty (Pascal suffered from continual ill health and at the age of 24 could only take liquid food fed to him drop by drop).

33.4 The Pascaline.

Six years later in 1654 he had begun a correspondence with the mathematician Pierre de Fermat in which they set out to solve a gaming problem that had been posed by an amateur mathematician Antoine Gombaud (who became known under his nom de plume 'Chevalier de Méré'). If two players agree to play a certain number of games and are interrupted before they can finish, how should the stakes be divided? Their solution to this problem laid the foundations of probability theory which was later studied by Leibniz (Leibniz in later life admitted that studying Pascal's manuscripts on geometry provided the 'light beam' that led him to infinitesimal calculus—he also adapted and improved Pascal's calculating machine).

Pensées

In the same year as his correspondence with Fermat, Pascal had an experience that was to divert him for a time from his mathematical and scientific pursuits:

The year of grace 1654
Monday 23 November, feast of Saint Clement . . .
From about half past ten in the evening until about
half past midnight
Fire
God of Abraham, God of Isaac, God of Jacob
not of philosophers and scholars
Certainty, joy, certainty, emotion, sight, joy
God of Jesus Christ[12]

For the rest of his life Pascal kept this description on a piece of paper sewn into the lining of his jacket (it was found after his death) [Fig. 33.5], and it was in the wake of this experience that he began to write down the thoughts that were published after his death as the 'Pensées'.

Some eight years before his Saint Clement experience Pascal had been much influenced by two doctors who had helped his father recover from a broken hip after slipping on an icy street in Rouen. Both men were followers of the Dutch theologian Cornelius Jansen, whose emphasis on the grace of God and insistence that mankind can do no good without it, brought him (according to his opponents in the Catholic Church) perilously close to the Reformation doctrines of Jean Calvin.

[12] B. Pascal, *Pensées and Other Writings* (1995), 178.

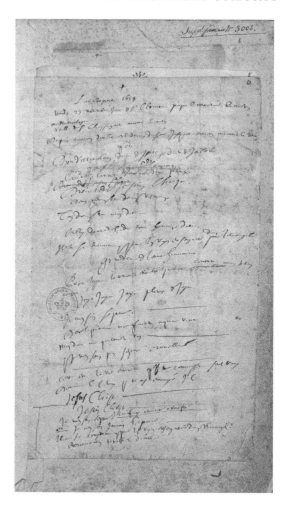

33.5 Pascal's memorial.

Among other things Jansen argued that the desire for knowledge was an illness 'from which arises the investigations of nature's secrets (which are irrelevant to us), knowledge of which
is useless, and which men do not wish to know except for
the sake of knowing'.[13] Pascal agreed with many Jansenist
positions and strongly defended them in a satire called *Lettres
provinciales*—The Provincial Letters. His attitude to 'the investigations of nature's secrets' was though more ambivalent.

In the course of *Lettres provinciales*, Pascal argues that it was
in vain that the Jesuits 'obtained against Galileo a decree from
Rome . . . if it can be demonstrated by sure observation that it
is the earth and not the sun that revolves, the efforts and arguments of all mankind put together will not hinder our planet
from revolving'.[14] In other writings he insists that researchers
working together 'make continual progress in proportion as the

---

[13] C. Jansenius, *Discours De La Reforma-
    tion De L'homme Interieur* (2004), 24.
[14] B. Pascal, *The Provincial Letters*
    (1875), Letter 18.

world gets older';[15] 'experimental facts' and telescopes, for instance, revealed to us realities that 'did not exist for our philosophers of old'.[16]

A person who has difficulty grasping the idea of infinite divisibility might realize by looking at a tiny point in the night sky through a telescope that it is a vast area of space which, through a still bigger telescope, might appear 'so as to equal that firmament . . . they admire'.[17] This revealed a paradox: in extending our knowledge and enabling us to overcome the limitations of our nature, instruments like telescopes enable us to see just how little we know and how limited we are.

In the most famous of Pascal's bundles of jottings which have the heading *Discours de la machine*—Dialogue Concerning the Machine—he begins his dialogue by developing this point.

Concerning the Machine

Human beings, he argues, have an innate capacity to know that infinity exists even while we cannot grasp it. Hence, just as we know 'there is an infinite number but . . . do not know what it is . . . so we can clearly understand that there is a God without knowing what he is'.[18] Speaking 'according to natural lights', however, it follows that 'if there is a God he is infinitely beyond our comprehension' and that 'reason cannot decide anything'.[19]

This leads to an existential dilemma. There is, as he puts it, 'an infinite chaos separating us. At the end of this infinite distance a game is being played and the coin will come down head or tails. How will you wager?'.[20]

Drawing on his study of probability, Pascal argues that it is rational (given the stakes, the impossibility of certainty, and the fact that you have to wager) to call heads 'that God exists'. Nevertheless, human psychology is only rational to a limited extent. 'We are', he insists, 'as much automaton as mind',[21] as much driven by custom and passion as the Pascaline was driven by its cogs. 'The machine' discussed in the dialogue therefore is the machine of our own psychology:

—'Is there no way of seeing underneath the cards?'
—'Yes, Scripture and the rest etc'
—'Yes but my hands are tied . . . I am forced to wager . . . And I am made in such a way that I cannot believe. So what do you want me to do?'[22]

[15] B. Pascal, *Les Provinciales* (2004), Traite du Vide 456.
[16] B. Pascal, *Pascal's Pensées* (1931), 266.
[17] B. Pascal, 'Minor Works, of the Geometrical Spirit', Harvard Classics, vol. XLVIII, Part II, 1909–14, 68, http://www.bartleby.com/48/3/9.html, accessed 15 August 2015.
[18] B. Pascal, *Pensées and Other Writings*, 152.
[19] B. Pascal, *Pensées and Other Writings*, 153.
[20] B. Pascal, *Pensées and Other Writings*, 153.
[21] B. Pascal, *Pensées and Other Writings*, 148.
[22] B. Pascal, *Pensées and Other Writings*, 155.

Pascal's solution is to employ the psychological mechanisms of custom and habituation to bring our passions in line with our reason by behaving *as if* you believed, 'taking holy water, have masses said, etc'.[23] In doing so we may be liberated to respond to what are, he suggests, fundamental intuitions.

In a textbook called *De l'Esprit géométrique*—Of the Geometric Spirit—Pascal suggests that 'knowledge of first principles, like space, time, motion' are fundamental intuitions 'coming from the heart and instinct' which cannot be demonstrated and on which 'reason has to depend and base all its arguments'. 'The heart', he argues, 'feels there are three spatial dimensions and that there is an infinite series of numbers and reason goes on to demonstrate that there are no two square numbers of which one is double the other'. 'Principles', he concludes, 'are felt, propositions proved, and both with equal certainty though by different means'.[24]

In the *Pensées* he comes to the same conclusion with regard to God: 'It is the heart that feels God, not reason . . . The heart has its reasons which reason itself does not know'.[25]

If, however, knowledge of God was a fundamental intuition, his existence could not be proved and it followed that 'All those who seek God outside Jesus Christ and whose search stops with nature' are likely, according to Pascal, 'to sink into either atheism or deism'.[26] Whereas Newton and Leibniz both believed that the beauty and simplicity of the universe might lead people to believe in God, Pascal's perspective was somewhat different.

In another bundle of notes headed 'The disproportion of nature' he points to what he calls 'the double infinity of the universe'. The immense scale of the universe revealed by telescopes is paralleled by the immensity that appears in microscopes. 'Within the confines of this miniature atom', he writes, 'we can see an infinity of universes of which each has its own firmament, planets, earth'.[27] It follows that 'every science is infinite in the scope of its researches'.

Even mathematics has 'an infinity of infinities of propositions to expound'. Mathematics might appear the most certain part of science but 'who cannot see that the principles we claim to be ultimate ones, cannot stand on their own, but depend on others, which are themselves supported by others so there can be no possibility of an ultimate principle?' 'Hence', he argues, 'the vanity, absurdity and ignorance of the titles of some

23 B. Pascal, *Pensées and Other Writings*, 156.
24 N. Hammond, *The Cambridge Companion to Pascal* (2003), 220.
25 B. Pascal, *Pensées and Other Writings*, 157–8.
26 B. Pascal, *Pensées and Other Writings*, 172.
27 B. Pascal, *Pensées and Other Writings*, 67.

books, *De omni scibili* [Of everything knowable]'.[28] 'How', he asks, 'could a part possibly know the whole?'[29]

The elaborate arguments of both natural theology and scientific atheism were consequently highly precarious—built on sand. Although 'we burn with desire to find a firm foundation, an unchanging solid base on which to build a tower rising to infinity', in the event 'the foundation splits and the earth opens up to its depths'.[30] What we are left with is an immense awe: 'the eternal silence of the infinite spaces'.[31]

A Thinking Reed

If Joseph Spence's anecdote related in Chapter Thirty-One is accurate, Sir Isaac Newton, 'a little before he died', seems to have shared this sense of humility before the unfathomed depths of nature: the sense that as the landmass of knowledge grew, the coastline of wonder increased rather than diminished.

The vertiginous perspectives that natural philosophy was opening up did not lead either Newton or Pascal to conclude that rational thinking was a waste of time. For Pascal it may be that 'a human being is only a reed, the weakest in nature', but he is nevertheless 'a thinking reed' and consequently must 'work on thinking well'.[32]

Similarly in Newton's description of the great ocean of truth lying *all undiscovered* before him, there is a reminiscence of the note of divine summons to a task, the 'trumpet call to action' that had first been sounded by Francis Bacon and which Newton had imbibed in reading Sprat's *History of the Royal Society* some 60 years earlier.

In fact whatever the limitations of natural theology, it did not in practice lead to a scientific cul-de-sac.

The Tales of Fairies

Of Princess Caroline's two correspondents it is only Leibniz who criticizes any actual science. To his mind Newton's idea of action at a distance could only be described as 'miraculous'. When Newton describes gravity as 'invisible, intangible, not mechanical, he might as well have added', Leibniz suggests, 'unexplicable, unintelligible, groundless and unexampled'.[33] Where Boyle 'had made it his business to inculcate that everything was

[28] B. Pascal, *Pensées and Other Writings*, 68.
[29] B. Pascal, *Pensées and Other Writings*, 70.
[30] B. Pascal, *Pensées and Other Writings*, 70.
[31] B. Pascal, *Pensées and Other Writings*, 73.
[32] B. Pascal, *Pensées and Other Writings*, 72–3.
[33] S. Clarke, *A Collection of Papers, Which Passed between the Late Learned Mr Leibnitz, and Dr Clarke, in the Years 1715 and 1716*, Mr Leibnitz's Fifth Paper, 120.

done *mechanically* in natural philosophy', nowadays, he remarks, people 'are become fond again of the *Tales of Fairies*'.[34]

Clarke's response to this (or possibly Newton's in whose voice he seems to speak) was that 'it is very unreasonable to call attraction a miracle ... after it has been so often distinctly declared that by that term we do not mean to express the *Cause of bodies tending towards* each other but merely the *Effect* and the laws of proportions of that tendency discovered by experience whatever be the cause of it'.[35] Gravitation, he insists, is 'an *actual phenomenon of Nature*', whose reality was 'now sufficiently known by observations and experiments'. If the cause was not known 'is therefore the effect less True?'[36] According to this notion, he protests, 'all Arguments in Philosophy taken from Phenomenon and Experiments are at an end'.[37]

In fact neither the attempt to seize the banner of the new science on behalf of religion or irreligion turned out to be an unqualified success. The Continental *philosophes* were no more successful than Epicurus in using science to eradicate religion, and the temples of Newton remained unbuilt. Likewise the hostages to fortune offered by natural theology (such as Newton's emphasis on an unstable universe or, later in the century, William Payley's idea of a divine watchmaker) were often counterproductive, and the jibe that 'no one doubted the existence of God until the Boyle lecturers set out to prove it' was not without some truth. Closing the gap between religion and science could, it seemed, short circuit both.

Leibniz's metaphysical sense of divine order and Newton's empirical insistence on God's freedom to do what he chose were both critical elements in driving discovery forward. And while the project of using science to provide an impregnable basis for belief may have been impossible to realize, what it did succeed in doing was to entrench the idea that the discoveries of science were of interest not merely to a small group of virtuosos, but to the whole of society.

The publication of a correspondence on this topic sponsored by a future Queen of England was yet further confirmation that these were central human concerns in which 'minds of all sizes' had an interest.

[34] S. Clarke, *A Collection of Papers, Which Passed between the Late Learned Mr Leibnitz, and Dr Clarke, in the Years 1715 and 1716*, Mr Leibnitz's Fifth Paper, 114.

[35] S. Clarke, *A Collection of Papers, Which Passed between the Late Learned Mr Leibnitz, and Dr Clarke, in the Years 1715 and 1716*, Dr Clarke's Fifth Reply, 104.

[36] S. Clarke, *A Collection of Papers, Which Passed between the Late Learned Mr Leibnitz, and Dr Clarke, in the Years 1715 and 1716*, Dr Clarke's Fifth Reply, 118.

[37] S. Clarke, *A Collection of Papers, Which Passed between the Late Learned Mr Leibnitz, and Dr Clarke, in the Years 1715 and 1716*, Dr Clarke's Fifth Reply, 104.

CHAPTER THIRTY-FOUR

The Coast of Infinity

In the early nineteenth century, the poet William Wordsworth famously described a statue of Newton 'voyaging through strange seas of thought alone' [Fig. 34.1]. Although this romantic picture was not without some truth, it hardly did justice to the reality. For all the difficulty of his relationship with his peers Newton always knew that experimental philosophy, in its open-ended requirement for data, and its insistence on the physical testing of theoretical ideas, was (almost by definition) a collective endeavour.

In a letter written to Henry Oldenburg following Newton's election to the Royal Society he had made a point of asking 'for what time the society continue their weekly meetings'. This was because he wanted 'to be considered & examined, an account of a philosophical discovery'[1] which had inspired the construction of his miniature telescope.

Newton's discovery (made with the aid of a glass prism he had bought at Stourbridge Fair in 1664) was that white light was composed of the different coloured lights of the spectrum. When he darkened his room, 'made a small hole in my window-shuts', and placed a prism in front of it, the light split up into rainbow colours with red at one end and blue at the other, making 'a very pleasing divertissement'[2] on the opposite wall. During 1669 and 1670 he attempted with a series of experiments to prove that his interpretation of this phenomenon was correct.

In one of these (later known as the 'experimentum crucis' or as Bacon would have said 'an instance of the fingerpost') the spectrum produced when the light was broken up by the prism was then passed through a lens focussing it on to a wall, where it once again appeared as a spot of pure white light. It had been to avoid the problem of 'chromatic aberration'—the rainbow fringes that appeared at high levels of magnification because lenses have slightly different focal lengths for different colours—that Newton had made his reflecting telescope.

[1] I. Newton, *The Correspondence of Isaac Newton* (1959), vol. I, 82–3.
[2] I. Newton, *The Correspondence of Isaac Newton*, vol. I, 92.

34.1 Statue of Newton in Trinity College Chapel.

Despite his criticisms and disagreements with Newton, Robert Hooke did eventually take part in a committee appointed by the Royal Society 'to try Mr Newton's experiment'. When it was carried out in front of the fellows in 1676 'according to Mr Newton's directions' it was seen to succeed 'as he all along asserted it would do'.[3] Although Hooke himself still refused to concede that Newton was right, he was in a rapidly diminishing minority. In 1703, eight months after Hooke's death, Newton was elected as president of the Royal Society. Almost his first act was to present the fellows with a copy of his *Optics* (paying for the publication this time himself) [Fig. 34.2].

[3] M. Ben-Chaim, *Experimental Philosophy and the Birth of Empirical Science* (2004), 94.

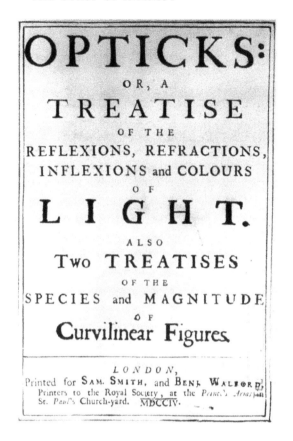

34.2 Title page of Newton's *Optics* (1704).

Universal Extent

While the *Optics* was largely based on 30-year old manuscripts, it also incorporated some of his most recent thinking. Newton originally seems to have intended the first three books to be followed by a fourth demonstrating how nature 'observes the same method in regulating the motions of smaller bodies (including the corpuscles of light) which she does in regulating those of the greater'.[4] As it turned out he was never able to achieve this unification and contented himself with adding, as an addendum to the first three books, a list of 'queries' which grew longer with each new edition (16 in the first edition, 23 in the Latin edition, 31 in the second English edition).

The first query, in a direct anticipation of Einstein, asks, 'Do not bodies act upon light at a distance and by their action bend its rays . . . ?'[5] An unpublished draft of the final query reveals just how directly Newton's scientific intuition of unity in nature was fed by his religious imagination, and just how

[4] I. Newton, misc. drafts of Newton's *Opticks* (1700–4), Cambridge University Digital Library, MS Add. 3970.3.

[5] I. Newton, *Opticks, or a Treatise of the Reflections, Refractions, Inflections and Colours of Light* (1730), 339.

powerful could be the pull exerted by ultimate curiosity on the direction of scientific enquiry: 'If there be a universal life and all space be the sensorium of a thinking being who by immediate presence perceives all things in it', this suggests that 'the laws of motion arising from life or will may be of universal extent'.[6]

Newton was aware that he lacked the data to substantiate this universal hypothesis and without it he could not proceed further. It was again the requirement for data that ensured that the experimental philosophy would remain a collective enterprise, one in which, as Bishop Sprat had said, there was work to be done 'by minds of all sizes'.[7] As if to prove his point, the observations which eventually did confirm Newton's hypothesis were not made by an academic or Royal Society insider, but by a poorly educated German immigrant who (at the time he started making them) was earning his living as a professional flute player.

The Ocean of Stars

34.3 William Herschel's seven-foot telescope.

Wilhelm (or William) Herschel first came to the attention of the Royal Society when William Watson, the son of the then secretary, found him on a street in Bath looking at the Moon through a telescope. When Watson 'very politely asked if he might be permitted to look in', he discovered that the clarity and resolution of what he saw were superior to any telescope he had ever used [Fig. 34.3].

Herschel had begun to construct his telescope in 1773 (nearly half a century after Newton's death) when he realized that to achieve what he called 'space-penetrating power', instead of the conventional refracting telescope, he would need a larger version of the tiny reflector telescope that Newton had presented to the Royal Society. With this he might explore the depths of space as one might explore an uncharted ocean. Casting, grinding, and polishing the enormous metal mirrors he required, while his sister Caroline fed him 'by putting the vital bits into his mouth'[8] (if the polishing paused even for seconds in the final stages the metal hardened and misted over) Herschel produced first a 5-, then a 7-, and then a 20-foot telescope with which to pursue his explorations [Fig. 34.4].

Herschel was convinced that to succeed in charting this ocean some kind of compass was necessary. Soon after their meeting Watson had arranged for him to be elected to the Bath

[6] J. H. Brooke, *Science and Religion* (1991), 139.
[7] T. Sprat, *History of the Royal Society*, 435.
[8] R. Holmes, *The Age of Wonder* (2008), 86.

Philosophical Society, and on 14 April 1780 Herschel read a paper to their meeting in which he argued that 'half a dozen experiments made with judgement by a person who reasons well are worth a thousand random observations of insignificant matters of fact'.[9]

Ultimately, he told the Bath philosophers, those who love wisdom were enabled by 'metaphysics . . . to prove the existence of a first cause, the infinite author of all dependent beings'.[10] Such reflections on 'the unknowable must-exist Being' could, as he explained in letters to his family, decide one's whole outlook: 'If one observes the whole Natural World as one, one finds everything in the most Beautiful Order; it is my favourite maxim: *Tout est dans l'ordre*'.[11] Shortly before midnight on the 13th of March the following year, what turned out to be the first new planet to be identified in a thousand years swam into his ken. Herschel insisted that this had not been a matter of chance: 'it was that night *its turn* to be discovered. I had that night perused the great volume of the Author of Nature and now came to the seventh planet. Had business prevented me that evening I must have found it the next'.[12]

The discovery of the new planet (which Herschel patriotically dubbed *Georgium Sidus*, the French called *Herschel*, and was eventually named *Uranus*) not only established Herschel's reputation but also gained him the royal patronage that enabled him to set up his own observatory near Windsor. With the 20-foot reflector, Caroline and he were able to identify some 466 new nebulae, which, he argued, were star clusters or galaxies outside the Milky Way.

While it was true, he wrote, 'that it would not be consistent confidently to affirm that we were an *Island Universe* unless we had actually found ourselves bounded by an ocean . . . A telescope with a much larger aperture . . . will be the surest means of completing and establishing the argument'.[13] This he proceeded to construct. With a grant from the King of £2,000 he set about building a telescope 'of the Newtonian form with an octagon tube 40 feet long and five feet in diameter'[14] [Fig. 34.5].

The Herschels' success was due to more than the power of their enormous telescopes. Brother and sister had together developed a system of what they called 'sweeping' the heavens. Keeping the telescope facing in one direction and moving it slowly up and down the meridian like a broom, while night

34.4 William Herschel's 20-foot telescope.

[9] W. Herschel, *The Scientific Papers of Sir William Herschel* (1912), 81.
[10] W. Herschel, *The Scientific Papers of Sir William Herschel*, 81–2.
[11] R. Holmes, *The Age of Wonder*, 73.
[12] R. Holmes, *The Age of Wonder*, 104.
[13] R. Holmes, *The Age of Wonder*, 124.
[14] R. Holmes, *The Age of Wonder*, 163.

34.5 William Herschel's 40-foot telescope.

after night, and through the seasons and years, the constellations revolved in front of them, they were able to systematically record and investigate all the phenomena that interested them.

One of the first fruits of this was a catalogue of so-called 'double stars' (a category that included both stars that are visually aligned and stars that are bound to each other in binary system). In updating this catalogue some 40 years later, Herschel's son John was able to establish that the Newtonian laws did indeed hold in the furthest reaches of the universe.

On Brighton Beach

In his book *The Age of Wonder* Richard Holmes describes the poet Thomas Campbell's meeting with the 74-year-old Herschel and his son John on holiday in Brighton in 1813. In the course of their conversation Herschel told the poet that "'I have looked further into space than ever human being did before me. I have observed stars of which the light, it can be proved, must have taken millions of years to reach the earth.' '[15] Walking on Brighton Beach after this encounter Campbell was irresistibly reminded of Joseph Spence's anecdote.

Herschel had almost come to embody the vision that Bacon had outlined in *The New Atlantis*. He was an exemplar of the visible and intellectual worlds coming together in a project, funded by the monarch himself, to extend the reach of human knowledge. Although sustained state funding still lay in the future, the idea that science could progressively extend our understanding of the universe had in Herschel become both socially accepted and intellectually entrenched.

This progressive extension of human knowledge stood in apparent contrast to the static unchanging claims of religious revelation. The book of God's Word and the book of his works might both be read, but what would happen if the new passages that were being read in the latter began to contradict the old passages in the former? Fears that they might do so had already begun to creep in, and would grow stronger as the century advanced. Nearly 50 years later, Matthew Arnold's poem 'Dover Beach' would powerfully express his sense that the tide of faith was retreating 'to the breath | of the night wind', leaving only 'the vast edges drear | and naked shingles of the world'.[16]

Such anxieties were not felt by Herschel as a reason for drawing back. While some might worry about where such discoveries would lead, for others an openness to new evidence was an expression of faith rather than a threat to it. The natural philosopher swimming in the slipstream of metaphysical interest need not fear that 'the gulfs will wash us down'.[17] Just as the 'ocean of truth', as Newton had conceived it, was one in which the order and beauty of the universe revealed 'much concerning God', so, according to John Herschel, his father had believed that 'by investigating the magnificent structure of the Universe', he was forwarding the glory of the 'benevolent, intelligent and superintending Deity' of whom he was a worshipper.[18]

Such investigations were not limited to the macrostructure of the cosmos. The law-like order which obtained on that grand scale might also be expected at every other level of the physical world: in biology as well as astronomy.

'Men shall go to and fro and knowledge shall increase'—the epigraph to Bacon's *The Advancement of Learning*—had an immediate reference for his contemporaries to the discovery of

[15] R. Holmes, *The Age of Wonder*, 210.
[16] Matthew Arnold, *Selected Poems*, ed. I. Hamilton (1993).
[17] Alfred Lord Tennyson, 'Ulysseus', in *The Poems of Tennyson*, ed. C. Ricks (1969).
[18] C. A. Lubbock, *The Herschel Chronicle* (1933), 197.

the new world. By the beginning of the nineteenth century European travellers were spreading throughout the globe, and where travellers went, men of science went with them. Both astronomers and naturalists needed to travel, and though their attention was directed at very different things, the kinds of law-like order they were looking for was closely related.

PART VIII
Voyages of Discovery

CHAPTER THIRTY-FIVE

Two Journeys

In the early years of the 1830s two men, one in Plymouth and another in Portsmouth, each found themselves assembling and packing the equipment that they would need for long scientific expeditions. For both men the journey on which they were about to embark would turn out to be the central adventure of their lives, yielding a harvest of observations that would occupy them for years to come after their return. Although each had rejected the idea of taking holy orders, both men shared a scientific outlook which had at its heart an essentially religious idea. This was the conviction that, as William Herschel had put it, 'toute est dans l'ordre': that behind all the apparently random and haphazard facts that we observe are divinely ordained laws and principles.

The younger of the two men and the first to depart was Charles Darwin. With a new microscope, a geological hammer, and a brace of pistols all stowed into his tiny cabin on HMS *Beagle*, he set sail from Plymouth in late December 1831 as an unpaid naturalist accompanying a five-year hydrographical survey voyage. The elder was John Herschel (the son of William), who in November 1833, with his 20-foot reflecting telescope packed into the hold of the *Mountstuart Elphinstone* and his young wife, three young children, and household staff stowed in three stern cabins, embarked from Portsmouth en route to Cape Town to begin what was to be a five-year astronomical survey of the southern hemisphere.

The two travellers had never met. For Darwin, however, Herschel was a scientific hero whose understanding of science was to profoundly effect the formation and conclusions of his own most famous book.

Here and Everywhere

William Herschel hadn't married until his late forties. John, his only child, was born when he was 54. Brought up in a household where studying the stars was a way of life, John had soon re-vealed himself as a scientific prodigy in his own right. At Cam-bridge his three closest friends were mathematicians: William Whewell, Charles Babbage, and George Peacock. With the last two he formed the 'Analytical Society', which successfully campaigned to introduce Leibniz's methods of calculus into Britain to replace Newton's system of fluxions. John ended his undergraduate career by becoming senior wrangler and win-ning the Smith's prize. He then proceeded to publish a series of brilliant mathematical papers, as a result of which he was elected to the Royal Society at the age of 21 and went on a few years later to win the Copley Medal: the Society's highest prize.

By 1816 his father, William, now 78, was finding himself un-able to continue his 40 years of astronomical observation, but much of his work was incomplete. John, after some struggle with himself, left the Cambridge post which he had taken up, to carry on the work that William and Caroline had begun.

His first observations as an astronomer had been concerned with the so-called 'double stars' observed by his father—pairs of stars that seemed to orbit around a common centre of grav-ity. These he believed could show whether Newton's law of gravitation applied to even the most distant parts of the uni-verse. William Herschel had listed over 800 of these stars and 40 years had now passed since he had made some of these obser-vations. This might be enough time to observe the pairs move in relation to each other. John and a friend, James South, now set themselves the task of re-measuring each of these pairs to see whether changes had occurred.

Between 1821 and 1823 Herschel and South produced a catalogue of 380 double stars together with every available ob-servation to determine their orbit. In the following ten years Herschel went on to produce a catalogue of double stars that listed 5075 objects, of which he had discovered 3347 himself. It was demanding work. 'Two stars last night and up till two wait-ing for them', he once wrote, 'Ditto the night before. Sick of star-gazing—mean to break the telescope and melt the mirrors.'

These vast array of measurements inspired him to develop a method involving plotting the position, angle, and time inter-vals in a coordinate system to compute the elliptical orbits of these stars in accordance with Kepler's first law of planetary motion. This was one of the first applications of computational astronomy to the stars and Herschel was awarded the Royal medal for his work by the Royal Society.

In the address given by the President of the Society, the Duke of Sussex, to mark the awarding of the medal, the Duke said that

> Herschel has ... demonstrated that the laws of gravitation, which are exhibited as it were in miniature in our own planetary system, prevail also in the most distant regions of space: a memorable conclusion, justly entitled by the generality of its character, to be considered as forming an epoch in the history of astronomy, and presenting one of the most magnificent examples of the simplicity and universality of those fundamental laws by which the Great Author has shown that He is the same today and forever, here and everywhere.[1]

In 1829 Herschel had married a girl 20 years younger than himself, 'A union of unclouded happiness', according to his first biographer, and almost immediately began planning a great scientific expedition. His father, William, had devoted the greater part of his life to systematically mapping the northern sky, but the southern sky remained largely uncharted. No one at the time in the southern hemisphere had a telescope anything like as powerful as Herschel's. His plan was to complete the systematic map of the heavens that his father had begun.

Herschel's astronomical investigations were only one aspect of his scientific interests. 'My first love was light', he once wrote. During the 1820s he made important contributions to the field of optics, investigating the phenomenon of polarization and demonstrating the possibility of making chemical analyses through observing and interpreting light spectra. In the same year that Darwin joined the *Beagle*, Herschel had used these broad interests to write *A Preliminary Discourse on the Study of Natural Philosophy*, which was in effect the first general book on scientific method since Francis Bacon. When the young Darwin began to read it the impact on him was immediate.

Two Books

In his last year as a student at Cambridge, Charles Darwin read two books which changed the course of his life. 'No one or a

[1] G. Buttmann, *The Shadow of the Telescope* (1974), 52.

dozen other books', he later wrote, 'influenced me nearly as much as these two.'[2] Up to that point a love of natural history and an obsession with collecting, 'which leads a man to be a systematic naturalist, a virtuoso, or a miser', had had no particular focus. Darwin describes in his autobiography how once finding two rare beetles he seized one in each hand, and then, seeing a third, popped it in his mouth. But this, he comments, 'was the mere passion for collecting, for I did not dissect them and rarely compared their external characters with published descriptions'.[3]

Darwin had yet to find his path in life. He had left Edinburgh University after two sessions, having found that the thought of being a physician did not appeal to him. He had then gone up to Cambridge with the idea of taking holy orders. Looking back he found this somewhat ludicrous, but he insisted that it wasn't at the time insincere. Indeed it had been the influence of a clergyman that led him to the two books whose joint effect was to stir up 'a burning zeal to add even the most humble contribution to the noble structure of Natural Science'.

The Reverend John Stevens Henslow was the newly appointed Regius Professor of Botany. In addition to his public lectures Henslow kept open house every Friday for undergraduates and older members of the university who were interested in science, and used to take undergraduates on botanical field trips. Darwin took part in these with such enthusiasm that he soon became known as 'the man who walks with Henslow'. Darwin was impressed by Henslow's breadth of scientific knowledge and by the sweetness of his character—'I never saw a man who thought so little about himself or his own concerns'. When Henslow confided that he had longed 'to explore regions but little known and enrich science with new species',[4] it inspired Darwin to read Alexander von Humboldt's massive seven-volume *Personal Narrative*, the account of a scientific expedition through the Brazilian rain forest and beyond.

Darwin copied long passages about 'the glories of Teneriffe' from Humboldt and read them out on one of Henslow's botanical excursions. His proposal was that the group should mount their own expedition. Nothing came of this (though Darwin got as far as trying to learn Spanish and enquiring about ships), but when Henslow wrote to him that the Admiralty were looking for a companion to sail with Captain Fitzroy on the five-year hydrographical expedition of HMS *Beagle*,

[2] C. R. Darwin, *The Autobiography of Charles Darwin* (1958), 24.
[3] C. R. Darwin, *The Autobiography of Charles Darwin*, 21.
[4] A. Desmond and J. Moore, *Darwin's Sacred Cause* (2009), 91.

Darwin leapt at the chance. The example of Humboldt's book which he 'read and re-read' was his inspiration, and the *Personal Narrative* was one of the three books he took with him on the voyage. It was, though, Herschel's book that was, in the long run, to be an even greater influence.

Among those who occasionally came to Henslow's Friday night soirées was the professor of mineralogy William Whewell. He sometimes walked back afterwards with Darwin from Henslow's house on Parker's Piece. Darwin described him as 'the best discourser on grave subjects to whom I have ever listened'. Whewell's many areas of expertise included a deep interest in the history and philosophy of science. He praised in print the way that Henslow's approach to science 'encouraged students in framing generalisations and drawing up new "laws" from observations',[5] and it was this approach that was the central theme of the book by his friend John Herschel, which appeared as the first volume of a new series: *Lardner's Cabinet Cyclopaedia*.

As soon as he had finished it Darwin wrote to his cousin Fox: 'If you have not read Herschel in Lardner's Cyclo—read it directly'. He himself had read the book 'with care and profound interest', and the fourteen or so neat marginal pencil marks in his copy not only testify to this, but chart with remarkable accuracy the course he would pursue in his future career.

Walking in Wonders

Herschel's book contained a brief introduction to the various branches of science but the bulk of it is concerned with the method of scientific thinking—'the delicate interplay between observation and theory' by which 'researchers sought to attain a single unifying explanation for . . . phenomena'.[6] 'It is principles not phenomena—laws not insulated independent facts', Herschel argues, 'which are the true object of the natural philosopher'.[7] He 'contemplates the world not as a set of phenomena . . . but as a system disposed with order and design' which may lead him 'to the conception of a power and intelligence superior to his own'.[8] While it was 'no uncommon thing' to find even scientifically literate people fastening on some apparently inexplicable fact as evidence of God's hand, this for Herschel argued a profoundly inadequate notion of the Creator.

Ten years earlier in his poem 'Lamia', John Keats had suggested that philosophy, conquering 'all mysteries by rule and

[5] E. J. Browne, *Charles Darwin: Voyaging* (1995), 128.
[6] E. J. Browne, *Charles Darwin: Voyaging*, 128.
[7] J. F. W. Herschel, *A Preliminary Discourse on the Study of Natural Philosophy* (1830), 13.
[8] J. F. W. Herschel, *A Preliminary Discourse on the Study of Natural Philosophy*, 4.

line', would 'clip an Angel's wings', and relegate even the 'awful rainbow' to 'the dull catalogue of common things'. Herschel took the opposite view. 'The Natural Philosopher', he argues, 'accustomed to trace the operation of general causes and the exemplification of general laws ... walks in the midst of wonders; every object which falls in his way elucidates some principle, affords some instruction and impresses him with a sense of harmony and order.'[9]

The Sublime Science

Where Herschel turns to consider particular sciences, Darwin in his copy marked two passages that deal with geology. The first of these describes geology as ranking in the scale of sciences next to astronomy 'in the magnitude and sublimity of the objects of which it treats', the difference being that geological studies 'can hardly be said to be more than commenced'.[10] This meant that the field was wide open for discovery, and in a second passage marked by Darwin, Herschel asks, 'what may we not expect from powerful minds called into action under circumstances totally different from any which have yet existed and over an extent of territory far surpassing that which has hitherto produced the whole harvest of human intellect?'[11]

In his final two terms Henslow had persuaded Darwin to begin studying geology and had introduced him to Professor Sedgwick, who agreed to take him on a field trip to North Wales. It was the trip that summer which brought home to Darwin more than anything else the importance of Herschel's emphasis on looking for systematic correlations between observations.

They had started from Darwin's father's house in Shrewsbury where Darwin had told Sedgwick about a tropical shell that had been found in a gravel pit near the town. The professor's reaction had not been what he had expected:

he at once said (no doubt truly) that it must have been thrown away by someone into the pit; but then added, if really embedded there it would be the greatest misfortune to geology, as it would overthrow all that we know about the superficial deposits of the Midland counties ... I was utterly astonished at Sedgwick not being delighted at so wonderful a fact as tropical shell being found near the

[9] J. F. W. Herschel, *A Preliminary Discourse on the Study of Natural Philosophy*, 15.
[10] Darwin's copy of J. F. W. Herschel, *A Preliminary Discourse on the Study of Natural Philosophy* (1830), Cambridge University Library, CCD 118, 287.
[11] Darwin's copy of J. F. W. Herschel, *A Preliminary Discourse on the Study of Natural Philosophy*, 350.

surface in Middle England. Nothing before had ever made me thoroughly realise, though I had read various scientific books, that science consists in grouping facts so that general laws or conclusions may be drawn from them.[12]

By the time the Herschels arrived at Cape Town on 16 January 1834, and began the task of unpacking and building an observatory at the foot of Table Mountain, Darwin had reached Tierra del Fuego and was roughly at the mid-point of the *Beagle*'s five-year voyage. His first encounter with the Brazilian rain forest had been everything that Humboldt had lead him to expect: 'Twiners entwining twiners—tresses like hair—beautiful Lepidoptera—silence—hosannah!'[13] 'I never experienced such intense delight', he wrote to Henslow, 'I formerly admired Humboldt, I now almost adore him; he alone gives any notion of the feelings that are raised in the mind on entering the Tropics'. These feelings were in part religious. 'It is not possible to give an adequate idea of the higher feelings of wonder, astonishment and devotion which fill and elevate the mind', he wrote in his journal, and later commented, 'I well remember my conviction that there is more in man than the mere breath in his body'.[14] These were feelings which directly inspired his science, and following the example of Humboldt, he measured, recorded, and collected samples of everything he could. Yet while conscious of walking 'in the midst of wonders', Darwin's first success in trying to work out a general principle from observed facts had come from his geological studies.

Before he had left, Henslow had encouraged him to get and study the first volume of Charles Lyell's *Principles of Geology*. Captain Fitzroy had presented him with a copy, and this was another of the three books he had taken with him on the voyage (the third was Milton's *Paradise Lost*). Lyell's second volume was sent out and reached him at Montevideo in 1832, while the third volume was waiting for him in Valparaiso in 1834 [Fig. 35.1].

At the core of the geological theory that Lyell outlines in these volumes is the idea that sections of the Earth's crust are gradually pushed up or depressed by the pressure of internal molten rock. Lyell was convinced that all the geological events of the past were caused by processes like volcanic activity and the weathering of sea and rain, which we see today, acting at the same rate over immense periods of time (Whewell dubbed

12 C. R. Darwin, *The Autobiography of Charles Darwin*, 25.
13 A. Moorehead, *Darwin and the Beagle* (1969), 55.
14 C. R. Darwin, *The Autobiography of Charles Darwin*, 65.

35.1 Title page of Lyell's *Principles of Geology*.

this idea 'uniformitarianism'). Right at the beginning of the voyage at St Jago in the Cape de Verde islands, Darwin was thrilled to discover that, using Lyell's principles, it was possible to reconstruct the geological history of the island. In South America while wandering over the Andes he began to think about a more specific geological problem: the question of coral reefs.

Lyell had suggested that corals must build their atolls on the rim of submarine craters. The problem with this idea was the difficulty of reconciling it with the size, form, and distribution of coral reefs. 'The idea of a lagoon island 30 miles in diameter being based on a submarine crater of equal dimensions has', Darwin wrote to his sister Caroline, 'always seemed to me a monstrous hypothesis'. If one could imagine the coral polyps slowly building up towards the surface at the same time the land was gradually sinking, it would in time produce the effect of islands surrounded by rings of coral or rings of coral surrounding the place where islands had once been. This could be tested. If the coral reefs went down below the depth at which they were known to be able to survive and all the reef below that point were dead, it would confirm that the sea floor had gradually been sinking while the coral built towards the surface [Fig. 35.2].

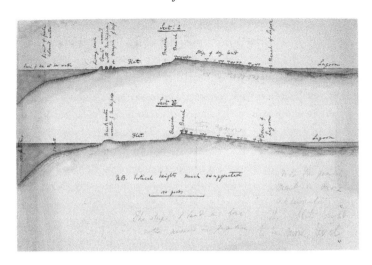

35.2 Darwin's diagram of coral formation.

By the spring of 1836 the *Beagle* had reached the Cocos islands. When soundings were taken around the outside of the Keeling atoll they found, to Darwin's great delight, that 'within ten fathoms the prepared tallow at the bottom of the lead invariably came up marked with impressions of living corals', but the corals gathered from depths below 20 or 30 fathoms were dead.

Darwin managed to arrange the facts 'in a more simple and connected point than that in which they have hitherto been considered'[15] and had succeeded in providing some experimental evidence for his theory. This was the approach that Herschel had recommended, and Darwin had not been alone in being inspired by Lyell's book. The third volume of *The Principles of Geology* had also travelled to Cape Town, and by the time the *Beagle* had left the Cocos and was heading towards the Cape, Herschel had thoroughly digested Lyell's book and had written an extensive letter of comments to the author.

Stars, Stars, Stars

Since his arrival at the Cape, Herschel had not been idle. With the help of a mechanic who had travelled out with them, he had erected the great 20-foot reflecting telescope in the grounds of a Dutch farmhouse he had rented [Fig. 35.3]. The observing conditions soon turned out to be superb, especially after heavy rainfalls: 'on these occasions the tranquillity of the images and the sharpness of vision is such, that hardly any limit is set to

[15] E. J. Browne, *Charles Darwin: Voyaging*, 318.

35.3 John Herschel's telescope at Feldhausen.

magnifying power but what the aberrations of the specula necessitates'.[16] On the other hand, the corrosion of the salt in the air meant that Herschel had to work for hours at the polishing machine to restore the reflectivity of whichever of the three interchangeable mirrors were not in use. 'Stars, stars, stars, fires! fires! fires!', he wrote in one of his letters to Lyell (he quickly became proficient at extinguishing the continual bush fires).

He estimated that a total of 5.3 million stars could be seen with the 20-foot telescope in the two celestial hemispheres. In surveys of 3,000 sections of the sky he counted 68,948 stars in a distribution that caused him to revise his father's conception of the structure of the galaxy. Double stars, nebulae, and star clusters were meticulously recorded. In a catalogue which he later produced, there are 1,707 nebulae recorded, of which 1,268 had never been seen before, and 2,102 pairs of double stars.

During his time at the Cape he invented what he called an 'astrometer'—the first stellar photometer—which enabled him to make accurate estimations of the relative brightness of stars. As well as studying stars he also undertook a detailed study of sunspots and proposed the idea of using photography to make a continuous record of solar activity. An observation that gave him great pleasure was seeing the satellites of Saturn—Mimas and Enceladus—which had not been seen since they had first been detected by his father in 1789.

Herschel's scientific investigations were not restricted to astronomy. Together with Thomas Maclear, the newly appointed astronomer at the Cape Observatory, he set about making systematic observations of the tides to send back to William Whewell, who was working on the mathematical foundations of tidal theory, and instituted regular meteorological observations which laid the foundations of weather research at the colony. He became a keen botanist, collecting rare bulbs wherever he went and planting these out in a garden he created at the farmhouse. 'I doubt if he ever enjoyed existence so much as now', his wife wrote to Aunt Caroline Herschel, 'for there are not the numerous distractions which tore him apart in England, & here he has time to saunter about with his gun on his shoulder & basket and trowel in his hand—I sometimes think we are all too happy'.[17] 'Whatever the future may be and whatever the past has been', Herschel wrote in a letter to Lyell, 'the days of our sojourn in that sunny land will stand marked with many a white stone as the happy part of my earthly pilgrimage.'[18]

[16] G. Buttmann, *The Shadow of the Telescope*, 90.

[17] J. F. W. Herschel, *Herschel at the Cape* (1969), 98.

[18] G. Buttmann, *The Shadow of the Telescope*, 117.

The astonishingly wide range of Herschel's scientific interests led him to take a keen interest in Lyell's geological theories.
By the time he wrote his letter he had read Lyell's 1,200-page
book three times 'and every time with increased interest'. In
the course of the letter he discusses the destruction and distribution of species he had encountered in his botanical studies,
comments on nomenclature, elaborates a large-scale theory of
volcanic activity, discusses the analogy between geology and
historical linguistics, describes some experiments with the ger-
mination of seeds, gives an extensive description of the geology
of the Cape and his discovery of fossilized trees, makes detailed
comments on Lyell's account of rock cleavages (which Lyell
subsequently included in the next edition), and describes some
specimens he has collected for various colleagues, while merely
regretting in an aside that 'my astronomical pursuits' (charting,
that is, the whole of the southern sky) have not yet allowed
time for further geological expeditions.

The letter to Lyell is dated 20 February 1836. Four months later
he noted in his diary: 'Captain Fitzroy, Mr Darwin, Capt Alexander, Mr C.Bell & Mr &Mrs Hamilton dined here at 6. Captain
F and Mr D came at 4 & we walked together up to Newlands.'[19]

---

[19] J. F. W. Herschel, *Herschel at the
Cape*, 242.

CHAPTER THIRTY-SIX

The Mystery of Mysteries

The *Beagle* had arrived at the Cape three days before, on 31 May 1836. Darwin had already mentioned his anxiety to meet Herschel—'I have heard so much about his eccentric but very amiable manners that I have a high curiosity to meet the great man'[1]—and he was not disappointed.

'It was the most memorable event which, for a long period I have had the good fortune to enjoy', Darwin wrote to his sister Catherine. In a letter to Henslow dated July 9th he described the meeting in detail:

> We dined at his house and saw him a few times besides. He was exceedingly good natured, but his manners at first appeared to me rather awful. He is living in a very comfortable house, surrounded by fir and oak trees, which alone in so open a country give a charming air of seclusion and comfort. He appears to find time for everything. He showed us a pretty garden full of Cape bulbs of his own collecting, and I understood afterwards that everything was the work of his own hands.[2]

'The great man' was not at all as Darwin had expected: 'He was very shy and he often had a distressed expression. Lady Caroline Bell, at whose house I dined . . . admired Herschel much, but said he always came into the room as if he knew his hands were dirty, and that his wife knew they were dirty'.[3]

Darwin's 'high reverence' for Sir John, however, was in no way diminished. Although 'he never talked much . . . every word he uttered was worth listening to'.[4] What did they talk about? There are no records of their conversations but on one topic at least their thoughts at the time had been moving in similar directions.

A Natural Process

Both men in reading Lyell's book had been impressed by the way he envisaged the physical landscape as being sculpted over

[1] A. Desmond and J. Moore, *Darwin* (1991), 184.
[2] J. F. W. Herschel, *Herschel at the Cape* (1969), 242.
[3] E. J. Browne, *Charles Darwin: Voyaging* (1995), 329.
[4] C. R. Darwin, *The Autobiography of Charles Darwin* (1958), 36.

vast tracts of time by natural processes that were still at work. Partly for fear of raising 'a host of prejudices against me' and partly because he could think of no mechanism by which it might operate, Lyell had refrained from any suggestion that the same principle might operate in biology—that living creatures might likewise be slowly altered over vast periods of time. Herschel, however, saw this implication, and his religious faith was quite robust enough to deal with it.

On the voyage out from England he had skimmed one evening through a book by Sharon Turner called *The Sacred History of the World, as Displayed in the Creation and subsequent events to the Deluge.* He described it as 'a vile trash-book on the principle of "bringing science to support religion" as it is now called—ie "proving" everything it is considered desirable to prove by making a roll-call of quotations misapplied and misunderstood out of books called scientific (all being held of equal authority)'. He noted the claim that 'the 6 days of Creation were really & truly 6 times 24 hours of the same length as at present in which the Geolog work was done (vide Lyell's 3rd volume!!)', and that 'the Atmospheric water if precipitated . . . would redrown the world whereas it would not raise the Ocean a foot &c&c'.[5]

In his letter to Lyell, Herschel said that his book, by contrast, had opened for him 'a region of speculation connected with it' which demanded courage to explore:

He that on such quest would go must know nor fear nor failing

To coward soul or faithless heart the search were unavailing

Of course I refer to that mystery of mysteries the replacement of extinct species by others. Many will doubtless think your speculations too bold—but it is as well to face the difficulty at once. For my own part—I cannot but think it is an inadequate conception of the creator to assume it as granted that his combinations are exhausted upon any one of the theatres of his former exercise—though in this as in all his works we are lead by analogy to suppose that he operates through a series of intermediate causes & that in consequence, the origination of fresh species could it ever come under our cognizance would be found to be a natural in contradistinction to a miraculous process.[6]

[5] J. F. W. Herschel, *Herschel at the Cape*, 32.

[6] W. F. Cannon, 'The Impact of Uniformitarianism: Two Letters from John Herschel to Charles Lyell, 1836–1837', *Proceedings of the American Philosophical Society* 105 (1961), 305.

In South America Darwin had made a dramatic scientific discovery along the coast of Patagonia that had turned his mind in a similar direction. On the beach of Punta Alta he had found in the gravel at the foot of a cliff an extraordinary number of fossilized bones: 'a perfect catacomb of extinct races'. Among these were parts of a giant sloth, a lama as big as a camel, and 'one of the strangest animals ever discovered': a large animal 'with an osseous coat in compartments very like an armadillo'. What was so striking about these monstrous beasts was their close resemblance to the smaller versions of the sloth, lama, and armadillo that still lived in South America. 'This wonderful relationship in the same continent between the dead and the living' must, he thought, 'hereafter throw more light on the appearance of organic beings on earth and their disappearance from it'.[7]

As they moved down the coast he was struck by the way that closely related animals seemed to replace one another as they proceeded south. When in 1835 they reached the Galapagos Islands, although almost all the animals and plants showed 'a marked relationship' with those of South America, most of them could be found nowhere else. There was even a difference between the inhabitants of the different islands. Mr Lawson, the Vice Governor, told him that 'the tortoises differed from the different islands and that he could with certainty tell from which island any one was brought'. Those from Hood Island had shells turned up in front 'like a Spanish saddle', while those from James Island were 'rounder, blacker, and have a better taste when cooked'.[8] The beaks of one group of finches—*Cactornis*—looked like the beak of a starling, while those of another—*Camarhynchus*—were 'slightly parrot shaped'. It seemed almost, as he later wrote, as if 'from an original paucity of birds in this archipelago one species had been taken and modified for different ends' [Fig. 36.1].[9]

In his copy of the *Preliminary Discourse* Darwin had marked a passage where Herschel had talked about 'the especial force of some two or three strongly impressive facts', phenomena which 'on account of some peculiarly forcible way in which they strike the reason ... impress us with a ... sense of causation or a particular aptitude for generalisation'.[10] Yet if there was some underlying law of adaptation involved here, without being able to understand how it worked he could, for the moment, get no further.

[7] A. Moorehead, *Darwin and the Beagle* (1969), 83.

[8] C. R. Darwin, *The Voyage of the Beagle* (1959), 257.

[9] C. R. Darwin, *The Voyage of the Beagle*, 245.

[10] Darwin's copy of J. F. W. Herschel, *A Preliminary Discourse on the Study of Natural Philosophy* (1830), Cambridge University Library, CCD 118, 287.

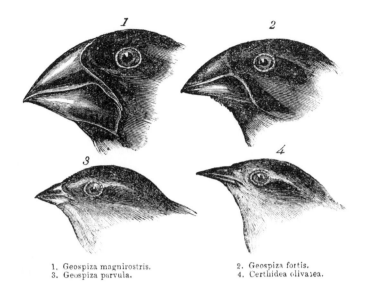

1. Geospiza magnirostris.
2. Geospiza fortis.
3. Geospiza parvula.
4. Certhidea olivacea.

36.1 Galapagos finches.

A Theory by Which to Work

Whatever they discussed in their conversations at the Cape, it was not long before Darwin had become acquainted with Herschel's letter to Lyell. In his reply to Herschel, Lyell mentions that he had allowed Charles Babbage to publish sections of the letter in his 1837 *Ninth Bridgewater Treatise*. In a letter to his sister Caroline, Darwin, who on his return to England got to know Lyell, and saw more of him 'than any other man both before and after my marriage', quotes from unpublished sections of the letter.[11]

The first edition of Darwin's journal of his travels on the *Beagle* appeared in 1839 as part of Fitzroy's work. Writing about the similarity of species on the Galapagos Islands and in South America he says only that 'the circumstance would be explained according to the views of some authors by saying that the creative power had acted according to the same law over a wide area ... but there is not space in this work to enter into this curious subject'. In his own 1845 edition of the *Zoology of the Voyage of H.M.S. Beagle*, however, he quotes directly (though without acknowledging the quotation) from Herschel's letter, siting the Galapagos as a place where 'both in space and time, we seem to be brought somewhat near to that great fact, that mystery of mysteries—the first appearance of new beings on this earth'.[12]

[11] A. Desmond and J. Moore, *Darwin*, 215.
[12] C. R. Darwin, *The Voyage of the Beagle*, 244.

Darwin had opened his first notebook 'for facts in relation to the origin of species' in July 1837, almost a year after his return to England. He had been 'much struck' by the beautiful adaptations of organisms to their environments he had seen in his travels, but until he could find some way of explaining these, it 'seemed almost useless' to try and show by indirect evidence that species had been modified. In the notebook therefore, working 'on true Baconian principles ... without any theory', he tried to understand the way in which variations occurred in domesticated species of animals and plants and 'collected facts on a wholesale scale ... by printed enquiries, by conversations with skilful breeders and gardeners and by extensive reading'.[13]

The difficulty was to see how anything like the careful selection practised by breeders and gardeners could operate in nature. Then in October 1838, 15 months after he had begun his investigation, he 'happened to read for amusement' the Revd Thomas Malthus' book on *Population*. Malthus pointed out that in every known species of animal or plant, far more were reproduced than survived to reproduce in their turn. Darwin was 'at once struck ... that under these circumstances favourable variations would tend to be preserved and unfavourable ones to be destroyed. The result of this would be the formation of new species'. Here then at last he had got 'a theory by which to work' [Fig. 36.2].[14]

By 1845 Darwin felt he had got much closer to understanding 'that great fact, that mystery of mysteries'. The idea that species might change over time was not a new one. Darwin's own grandfather Erasmus had speculated about it, but rereading his grandfather's works Darwin had been dismayed by the percentage of speculation to fact. As a student the very first passage he had marked in Herschel's *Preliminary Discourse* had stated that 'it is not possible to satisfy ourselves completely that we *have* arrived at a true statement of any law of nature, until setting out from such a statement and making it a foundation of reasoning, we can show by strict argument that the fact must follow from it as a necessary logical consequence'.[15]

Following this advice in 1842 Darwin wrote out a sketch of his theory in a series of steps, 'first giving the established facts about nature, followed by an analogy or extrapolation into theory, and then ending with the reasons why he thought that his view was correct or at least a worthwhile hypothesis'.[16]

36.2 Darwin's first known drawing of an evolutionary tree.

[13] C. R. Darwin, *The Autobiography of Charles Darwin*, 42.
[14] C. R. Darwin, *The Autobiography of Charles Darwin*, 43.
[15] Darwin's copy of J. F. W. Herschel, *A Preliminary Discourse on the Study of Natural Philosophy*, 25.
[16] E. J. Browne, *Charles Darwin: Voyaging*, 437.

As it stood though this version of the theory still provided no account of how species were so finely adapted to their environments. Darwin in old age remembered 'the very spot on the road whilst in my carriage when to my joy the solution occurred to me', namely 'that the dominant and increasing forms tend to become adapted to many and highly diversified places in the economy of nature'.[17]

This idea, like the transmutation of species, was also not entirely original. An ornithologist named Edward Blyth, with whom Darwin conducted an extensive correspondence, had already written about natural selection and the progressive adaptation of organisms to their environment, back in 1835. Blyth thought of these only as selections and adaptations of organisms within species. Darwin, however, saw it as a great process operating in nature that shaped the development and transmutation of species in the same way that geological processes had shaped the physical landscape.

How could this idea be established? 'The next step in the verification of an induction', according to another passage Darwin had marked in Herschel's *Discourse*, 'must ... consist in extending its application to cases not originally contemplated; in studiously varying the circumstances under which our causes act, with a view to ascertain whether their effect is general; and in pushing the application of our laws to extreme cases.'[18] This was material for a lifetime's work and all of Darwin's researches over the years that followed, from his minute studies of barnacles to his botanical experiments, were conducted against the background of his great idea and became the means for trying to establish it.

A later passage in Herschel's book that Darwin's student pencil marks had singled out drew attention to the importance of the interchange of knowledge. Not only could this aid research, it could also 'prevent a number of individuals from making the same discoveries at the same moment, which (besides the waste of valuable time) has always been a fertile source of jealousies and misunderstandings'.[19]

A letter from Blyth drew Darwin's attention to the work of a young naturalist called Alfred Russel Wallace, who had gone out to Malaya and seemed to be working on very similar lines to Darwin. In 1856 Lyell, who had also become aware of Wallace's work, urged Darwin to write up his theory to establish priority. He had just begun to do so when he received a

[17] C. R. Darwin, *The Autobiography of Charles Darwin*, 43.
[18] Darwin's copy of J. F. W. Herschel, *A Preliminary Discourse on the Study of Natural Philosophy*, 167.
[19] Darwin's copy of J. F. W. Herschel, *A Preliminary Discourse on the Study of Natural Philosophy*, 351.

20-page letter from Wallace, 'On the Tendency of Varieties to Depart Indefinitely from the Original Type', which contained 'exactly the same theory as mine' and seemed to forestall everything he had been working on for the previous 20 years. In the end it was arranged that papers by Wallace and Darwin were jointly published by the Linnean Society of London in 1858. Darwin immediately began work on a book.

When *On the Origin of Species* was published 13 months later, Darwin states in his opening sentence that his observations in South America 'seemed to throw some light on the origin of species—that mystery of mysteries, as it has been called by one of our greatest philosophers', quoting again (and now acknowledging the quotation) from the letter that Herschel had written more than 20 years earlier.

CHAPTER THIRTY-SEVEN

The Creed of Science

Since his return from the Cape, Herschel had not been idle. In 1839 he had written to a friend that 'with the publication of my South African observations . . . I have made up my mind to consider my astronomical career as terminated'. This was partly a recognition of the immense mathematical task that faced him in ordering his observations at the Cape, but also the discovery that his health could no longer sustain the long nights in the observatory. After eight years of labour, *Results of Astronomical Observations made during the years 1834, 5,6, 7, 8 at the Cape of Good Hope; Being a completion of a telescopic Survey of the whole surface of the Visible Heavens, commenced in 1825* was eventually published to immense acclaim in 1847. Herschel's astronomical labours, however, had not exhausted his scientific curiosity.

On 22 January 1839, six months after returning from Africa, he received a letter reporting Louis Daguerre's process for producing images by exposing photosensitive silver salts in a camera obscura. Twenty years earlier Herschel had discovered that sodium thiosulphate (known to photographers as 'hypo') dissolves silver salts. He saw that by using it to wash out the silver salts and preventing them from continuing to darken, it would make an ideal 'fixer' for Daguerre's process. With his long experience of lenses he also recognized the need to avoid spherical or chromatic aberration and the need for the focal plain to be evenly illuminated.

On January 30th, using this process, he produced an image of his father's 40-foot telescope that still stood in the garden at Slough [Fig. 37.1]. It had taken Herschel only a week to match and improve a process that others had been labouring on for years, but he made no effort to profit by his discovery. He was quite happy for both Daguerre and Henry Fox Talbot (who had developed another method of producing 'photogenic images') to include his process in their patents (only suggesting to Talbot that 'photography' would be a better term to describe the process). His interest in the technique was scientific rather

37.1 Herschel's 1839 photograph of the 40-foot telescope.

than commercial and over the next five years he conducted a series of pioneering experiments that established the basis of photochemistry.

The Law of Higgledy Piggledy

Following the publication of *The Origin of Species*, Darwin directed his publishers to send Sir John a copy of his book 'with the hope that you may still retain some interest on the question', and by way 'of showing in this feeble manner my respect, & the deep obligation, which I owe to your Introduction to Natural Philosophy. Scarcely anything in my life made so deep an impression on me: it made me wish to add my mite to the accumulated store of natural knowledge'.[1] He wrote to Lyell, 'Sir J Herschel, to whom I sent copy, is going to read my Book. He says he leans to side opposed to me. If you shd meet him, after he has read me, pray find out what he thinks. For of course he will not write; & I shd excessively like to hear whether I produce any effects on such a mind.'[2]

The initial news did not seem good. On December 10th he wrote again, 'I have heard by a round about route that Herschel says my book is "is the law of higgledy-piggledy". What this exactly means I do not know, but it is evidently very contemptuous. If true this is great blow & discouragement.'[3]

Two years later Herschel sent Darwin a signed copy of the second edition of his own book *Physical Geography*. In the first edition Herschel had stated his belief that species change had occurred 'by a series of overlappings, leaving the last portion of each in co-existence with earlier members of the newer series'; Herschel noted that this had been written 'previous to the publication of Mr Darwin's work' but had not, he thought, been disproved by it.

In the second edition he included a new footnote saying that 'We can no more accept the principle of arbitrary and casual variation selection as a sufficient account *per se*' than 'the Laputan method of composing books' (by assembling random words) would give a sufficient explanation of 'Shakespeare and the Principia'.[4] Without some kind of 'intelligent direction', which he doesn't think Darwin means to deny, he was unable to conceive how any law of selection could have led to such results.

[1] Letter to Sir John Herschel, 11 November 1859, Darwin Correspondence Database, http://www.darwinproject.ac.uk/entry-2517, accessed 6 May 2015.

[2] Letter to Charles Lyell, 23 November 1859, Darwin Correspondence Database, http://www.darwinproject.ac.uk/entry-2543, accessed 6 May 2015.

[3] Letter to Charles Lyell, 10 December 1859, Darwin Correspondence Database, http://www.darwinproject.ac.uk/entry-2575, accessed 6 May 2015.

[4] J. F. W. Herschel, *Physical Geography* (1861), 12.

On the other hand, he does not want to deny that 'intelligent direction' may 'act according to law', and that 'Such law, stated in words would be no other than the actual observed law of organic succession . . . including all the links in the chain that have disappeared'. 'But', he argues, 'the one law' ('intelligent direction') was 'a necessary supplement to the other' ('arbitrary . . . variation selection') and 'ought in all logical propriety to form a part of its enunciation'. 'Granting all this', he concludes, 'and with some demure as to the genesis of man, we are far from disposed to repudiate the view taken of this mysterious subject in Mr Darwin's book'.[5]

What did this mean?

In a letter to Lyell Darwin complains that although Herschel argues 'that the higher law of providential arrangement shd always be stated', in fact 'astronomers do not state that God directs the course of each comet & planet'. Herschel himself had argued that in all his works God appeared to operate 'through a series of intermediate causes', and in this case 'the view that each variation has been providentially arranged seems to me to make natural selection entirely superfluous, & indeed takes whole case of appearance of new species out of the range of science'.[6]

He did not wish to say 'that God did not foresee everything that would ensue', but was aware that 'here comes very nearly the same sort of wretched embroglio as between free-will & pre-ordained necessity'. Would Herschel say 'that you ought always to give the higher providential Law, & declare that God had ordered . . . that certain mountains should arise.' It seemed to him that views like that of Herschel or his friend Asa Gray 'merely show that the subject in their minds is in Comte's theological stage of science'.[7]

Most Beautiful and Wonderful

Over the course of their lives the religious convictions of Darwin and Herschel moved in somewhat different directions. When William Herschel had suggested to his son that he might be ordained as a clergyman, the young man had vigorously protested. Although his father had argued that he shouldn't be worried by intellectual scruples, John had replied by describing the Church as 'a system of organised self-deception'. As he grew older, and particularly after his marriage, Herschel's religious feelings seem to have grown warmer and deeper.

[5] J. F. W. Herschel, *Physical Geography*, 12.
[6] Letter to Charles Lyell, 1 August 1861, Darwin Correspondence Database, http://www.darwinproject.ac.uk/entry-3223, accessed 6 May 2015.
[7] Letter to Charles Lyell, 1 August 1861.

Darwin, by contrast, had accepted his father's plan that he go to Cambridge to take holy orders. Looking back he found this 'somewhat ludicrous' but insisted that since he 'did not then doubt the strict and literal truth of every word in the Bible' it wasn't at the time insincere. In the autobiography that he wrote for his family in 1876, however, he described how in the years after returning from his travels he had gradually come 'to disbelieve in Christianity as divine revelation'.[8]

These years had taken a heavy toll on both men. Darwin had suffered continual ill health (though finding some relief in a hydrotherapeutic cure which he described in a letter to Herschel). It was, though, the shattering death of his beloved daughter Annie in 1851 which perhaps most profoundly affected his outlook on life, and by the time he came to write his autobiography he was aware of a change that had taken place within himself.

In his *Beagle* journal he had written that while standing in the grandeur of a Brazilian forest it was not possible 'to give an adequate idea of the higher feelings of wonder, devotion and admiration which fill and elevate the mind', but now, 'the grandest scenes would not cause any such convictions and feelings to rise in the mind. It may be that I am like a man who has become colour blind'.[9] Whereas poetry, pictures, and music had once caused him intense delight, now he had suffered a 'curious and lamentable loss of all the higher aesthetic tastes'.[10] Reading Shakespeare seemed 'so intolerably dull that it nauseated me'.[11]

Herschel too had not been well. In 1850 for the first time in his life since he had taken up astronomy, he had followed Newton in accepting a salaried public post as Master of the Mint. Although he turned out to be a very able administrator, the separation from his family and from his scientific work, together with the strain of managing the Mint personnel, rapidly undermined his health. In 1854 a nervous breakdown forced him to resign. His peace of mind gradually returned, but his health never fully recovered.

Unlike Darwin, however, he experienced no loss of his religious or aesthetic sensibilities. In 1840 when his father's great 40-foot telescope had been finally dismantled he marked the occasion by assembling his entire family inside the tube to sing a requiem he had composed for the old instrument. Now in his spare time between sitting on scientific committees, and

8 C. R. Darwin, *The Autobiography of Charles Darwin* (1958), 86.
9 C. R. Darwin, *The Autobiography of Charles Darwin*, 91.
10 C. R. Darwin, *The Autobiography of Charles Darwin*, 139.
11 C. R. Darwin, *The Autobiography of Charles Darwin*, 138.

writing articles and books on astronomy, geology, and other scientific subjects, he embarked on translations of Schiller, Dante, and Homer, and took an interest not only in the science of photography but in the developing art of the process. Julia Margaret Cameron was a regular visitor and her portrait of Herschel in old age is one of her greatest photographs [Fig. 37.2]. His face never quite lost the 'distressed expression' which Darwin had noticed at the Cape, but his faith seems to have stood firm. 'Whatever God sends is welcome', he wrote in his diary, in response to increasing ill health.

Despite their different religious journeys, Herschel and Darwin still shared a large area of common ground. In his letter of thanks to the older man, Darwin, while insisting that there was no evidence of intelligent design of individual organisms (unless one was prepared to admit that God had 'designed the feathers in the tail of the rock-pigeon to vary in a highly peculiar manner', just so that man 'might select such variations and make a Fan-tail'), also admitted that 'one cannot look at the Universe with all living productions & man without believing that all has been intelligently designed'.[12]

This was not a pious pose. In his candid 1876 autobiographical sketch Darwin could still write of 'the extreme difficulty or rather impossibility of conceiving this immense and wonderful universe, including man with his capacity for looking backwards and far into futurity as a result of blind chance or necessity', and 'when thus reflecting . . . I deserve to be called a Theist'.[13] This sense with 'many fluctuations' had, though, over the years gradually 'become weaker'. Could the mind of man, developed from that of the animals, be trusted 'when it draws such grand conclusions'?[14] Although as he wrote three years later 'in my most extreme fluctuations I have never been an atheist',[15] faced with the mystery of the beginning of things he felt that 'I for one must be content to remain an agnostic'.[16]

This questioning of human capacities did not, however, extend to scientific capacities (as logically it might have done). In a letter of appreciation written nine months before his death, to the author of a book called *The Creed of Science*, he states that 'you have expressed my inward conviction, though far more vividly and clearly than I could have done, that the Universe is not the result of chance',[17] and in his autobiography he remembered that at the time of writing *The Origin of Species* a

37.2 Julia Margaret Cameron's photograph of John Herschel.

[12] Letter to Sir John Herschel, 23 May 1861, Darwin Correspondence Database, http://www.darwinproject.ac.uk/entry-3154, accessed 6 May 2015.

[13] C. R. Darwin, *The Autobiography of Charles Darwin*, 92.

[14] C. R. Darwin, *The Autobiography of Charles Darwin*, 93.

[15] Letter to John Fordyce, 7 May 1879, Darwin Correspondence Database, http://www.darwinproject.ac.uk/entry-12041, accessed 6 May 2015.

[16] C. R. Darwin, *The Autobiography of Charles Darwin*, 94.

[17] Letter to William Graham, 3 July 1881, Darwin Correspondence Database, http://www.darwinproject.ac.uk/entry-13230, accessed 6 May 2015.

theistic interpretation of this conviction had been 'strong in my mind'.[18]

This is confirmed both by the 1842 sketch of the theory and by the notebooks in which he felt his way towards the idea of natural selection.

In one notebook he contrasts the notion of special individual creations to the more 'magnificent' and 'grander' idea of God creating through 'astronomical causes ... which cause changes in geography & changes of climate ... then superadded to changes of form in organic world', as being more worthy of him 'who is supposed to have said "let there be light and there was light" '.[19] In another he points out that

Astronomers might formerly have said that God ordered each planet to move in its particular destiny.—In the same manner God orders each animal created with certain form in certain country, but how much more simple & sublime powers—let attraction act according to certain law[s], such are inevitable consequen[ces]—let animals be created, then by the fixed laws of generation, such will be their successors.[20]

This perspective was also articulated by a Scottish landowner, Patrick Matthew, who (anticipating Darwin and Wallace by 30 years) published an account of natural selection in *Naval Timber and Arboriculture* (1831). In it he argues that

There is a law universal in nature, tending to render every reproductive being the best possibly suited to its condition that its kind, or that organized matter, is susceptible of, which appears intended to model the physical and mental or instinctive powers, to their highest perfection, and to continue them so. This law sustains the lion in his strength, the hare in her swiftness, and the fox in his wiles.[21]

Reviewing *The Origin of Species* in *The Farmer's Magazine* Matthew wrote,

I challenge anything of Bridgwater prize origin, or of any other higher origin, as showing grandeur of design— means to end—display of infinite wisdom equal, or to be compared to the great self-modifying-adaptive scheme

[18] C. R. Darwin, *The Autobiography of Charles Darwin*, 94.
[19] C. R. Darwin, *Notebook D: Transmutation* (1838), Cambridge University Library, DAR 123.
[20] C. R. Darwin, *Notebook B: Transmutation* (1837–1838), Cambridge University Library, DAR 121.
[21] P. Matthew, *Naval Timber and Arboriculture* (1831), 364.

of Nature which I many years ago pointed out in 'Naval Timber and Arboriculture', and which Mr. Darwin has in his recent work so ably brought forward.[22]

In Darwin's 1842 sketch he writes similarly how 'it accords with what we know of the law impressed on matter by the Creator, that the creation and extinction of forms, like the birth and death of individuals should be the effect of secondary [laws] means'.[23]

Both this and a passage that follows found their way into the famous final 'Recapitulation and Conclusion' of *The Origin of Species*, in which Darwin (in language which recalls Herschel's picture of the natural philosopher walking amid wonders) invites the reader to contemplate 'an entangled bank, clothed with many plants of various kinds, with birds singing on the bushes, with various insects flitting about, and with worms crawling through the damp earth, and to reflect that these elaborately constructed forms . . . have all been produced by laws acting around us' and concludes that 'There is a grandeur in this view of life, with its several powers, having been originally breathed into a few forms or into one; and that, whilst this planet has gone cycling on according to the fixed laws of gravity, from so simple a beginning endless forms most beautiful and most wonderful have been, and are being, evolved.'[24]

Two Truths

In his letter to Herschel, Darwin describes how many younger scientists in different branches of science had (on Herschelian principles) accepted his theory 'because they find they can thus group and understand many scattered facts', and ends (almost pathetically) by saying that he mentions this only because 'I shd. value your partial acquiescences of my views, more than that of almost any other human being.'[25]

In the second edition of *The Origin* he had gone some way towards Herschel's idea of stating 'the providential law' by inserting three words into the final sentence. This now talked of life 'with its several powers having been originally breathed by the Creator into a few forms or into one'.[26] By 1863 he had begun to think better of this insertion, writing to his friend Joseph Hooker, 'I have long regretted that I truckled to public opinion & used [the] Pentateuchal term of creation, by which

[22] M. E. Weale, 'Patrick Matthew's Law of Natural Selection', *Biological Journal of the Linnean Society* (2015), doi: 10.1111/bij.12524.

[23] C. R. Darwin, *The Foundations of the Origin of Species* (1909), 51.

[24] C. R. Darwin, *On the Origin of Species* (1902), 491.

[25] Letter to Sir John Herschel, 23 May 1861.

[26] C. R. Darwin, *The Origin of Species*, 490.

I really meant "appeared" by some wholly unknown process',[27] and to *The Athenæum* that 'in a purely scientific work I ought not have used such terms'.[28]

The quotation from Bacon's *Advancement of Learning* that Darwin used as an epigraph to *On the Origin* which stated that 'no man . . . can be too well studied in the book of God's word or in the book of God's works' had gone on to warn against an 'unwise mixing' of these two disciplines. Such mixings, however, in one form or another, were not easy to avoid.

Darwin in his autobiography writes that already on the *Beagle* he had 'come . . . to see that the Old Testament from its manifestly false history of the world, with Tower of Babel, the rainbow as a sign etc, etc . . . was no more to be trusted than . . . the beliefs of any barbarian',[29] and in later life contributed money to the defence of Bishop Colenso when he was tried by an ecclesiastical court for his critical book on the Pentateuch.

Herschel, at the same period, was much less dismissive of the Old Testament. In his letter to Lyell from the Cape, he argued that there was scope to reconcile scriptural chronology with the findings of geology since 'the lives of the patriarchs may as reasonably be extended to 5,000 or 50,000 years apiece as the days of creation to as many thousands or millions of years'.[30] His central principle was that 'we must not impugn the scripture chronology but we *must* interpret it in accordance with *whatever* shall appear on fair inquiry to be the *truth* for there cannot be two truths'.[31]

This question of whether the scriptural stories were simply the same kind of thing as pagan creation myths—'the beliefs of any barbarian' as Darwin has it—had for centuries been a critical issue at the heart of the development of science within the Abrahamic faiths in general and within Christianity in particular. Many insisted that they were not: that the Scriptures did not set out to answer physical questions but pointed instead to a God standing outside the circle of nature guaranteeing its meaning and unity. That has been the thread that we have been following all the way from Philoponus and Roger Bacon to Galileo and Newton. Yet in the light of increasing knowledge, could such assertions still be supported?

At almost exactly the same moment that the vision of the prehistoric past, which Darwin had derived from his voyage on the *Beagle*, was giving a new urgency to this question, a succession of intrepid Victorian travellers in another part of the world

[27] Letter to Joseph Hooker, 29 March 1863, Darwin Correspondence Database, http://www.darwinproject.ac.uk/entry-4065, accessed 19 August 2015.

[28] Letter to *The Athenæum*, 18 April 1863; see N. Spencer, *Darwin and God* (2009), 82.

[29] C. R. Darwin, *The Autobiography of Charles Darwin*, 85.

[30] W. F. Cannon, 'The Impact of Uniformitarianism: Two Letters from John Herschel to Charles Lyell, 1836–1837', *Proceedings of the American Philosophical Society* 105 (1961), 305.

[31] W. F. Cannon, 'The Impact of Uniformitarianism: Two Letters from John Herschel to Charles Lyell, 1836–1837', 305.

had begun to uncover a possible answer. Their results would challenge many traditional assumptions but could also be seen as supporting the central claim of these assertions.

Following the story of these discoveries might appear to take us off at a tangent to the direction we have so far been travelling, and readers impatient to continue the scientific story (and with no interest in biblical archaeology) may wish to skip to Part X. They will, however, have missed something important and might care to read the summary at the end of Chapter Forty-Three. This is because what (among other things) these finds began to reveal were the origins of the religious concept of penultimacy. They provide a record of the time and place where the critical idea that God could not be identified with any part of the universe first began to be clearly articulated.

The discoveries that slowly began to uncover the development of this idea involved the patient unearthing of many ancient objects from the sands of the Middle East. The crucial moment, however, was the deciphering of a script.

PART IX
In the Beginning II

CHAPTER THIRTY-EIGHT

The Literary Inquest

On the 20th of May 1857 a small group, which included Herschel's friend William Whewell, began to gather at the rooms of the Royal Asiatic Society in New Burlington Street to determine the results of a dramatic experiment. The task of the group was to compare the contents of four sealed packets which had been deposited with the society's secretary.

The first packet to arrive had been sent by William Henry Fox Talbot, the same man who had profited from Herschel's contribution to the development of photography. Fox Talbot had many strings to his bow, and it was he who had proposed the experiment.

A second had been deposited by Dr Julius Oppert, a German-born scholar working in France, who had happened to be present at the meeting when the proposal for the experiment had been discussed, and who had asked if he could take part.

A third packet, and the last to arrive, had been contributed by an Irish clergyman called Dr Edward Hincks. Hincks, who to his considerable annoyance had been the last person to be notified about what was going on, had had little more than a fortnight in which to supply his contribution.

The fourth packet had been provided by the famous imperial soldier Sir Henry Creswicke Rawlinson. Rawlinson had returned to England after almost a quarter of a century spent in the east, and it was a discovery made during his travels there that had made the whole experiment possible.

A Passage to India

Thirty years earlier, in 1827 (four years before Darwin embarked on the *Beagle*), the 17-year-old Henry Rawlinson had sailed on the *Neptune* to Bombay to take up a commission as a soldier in the East India Company. One of his fellow passengers on the four-month voyage was Sir John Malcolm, a diplomat and oriental scholar, on his way to take up the governorship of Bombay. According to Rawlinson's brother George, Henry's conversations with Malcolm planted the seeds of an interest in Persian language and history that were in time to produce 'the most momentous change' in his life. When, six years later, Rawlinson, by then fluent in Urdu, Marathi, and Persian languages, was sent by the company to train the army of the Shah of Persia, these seeds began to germinate.

En route to Tehran, Rawlinson rode out to visit the ruins of Persepolis. For centuries European travellers had been perplexed by the strange wedge-shaped markings that they had found in the ruins of the ancient Persian capital. One visitor persuaded himself that they had been formed by worm casts. In 1700 a professor of Hebrew at Oxford called Thomas Hyde coined the term 'cuneiform', literally 'wedge form' (from the Latin *cunus*—wedge) to describe them. Hyde (working from very inadequate drawings) thought they could not be a script and must be purely decorative.

A rather similar puzzlement had been caused by the hieroglyphics that travellers had found in Egypt, but in the case of cuneiform the puzzle turned out to be even harder to solve. This is because whereas hieroglyphics were only used to represent the Egyptian language, cuneiform had been adapted over centuries to write down a whole succession of different languages. Like the Roman script we use today, which was first used to write down Latin and is now used to write down languages as different as Finnish and Swahili, cuneiform was a writing system that was used for more than 3000 years to write down languages as various as Sumerian, Akkadian, Elamite, and old Persian. To decipher cuneiform you not only had to understand the way it was used to write down different languages, you also had to reconstruct each of these different languages. It was a formidable task.

In 1765 a Danish explorer, Carsten Niebuhr, had made accurate drawings of inscriptions at Persepolis, and had recognized that some of the inscriptions contained three different scripts, and therefore possibly three different languages. At the turn of the nineteenth century, Silvestre de Sacy, an oriental scholar in Paris, began to work out a language known as 'middle Persian' from some parallel inscriptions in Greek and middle Persian that had been found near Persepolis at Naqsh-e Rustam.

Some of these inscriptions contained the names and titles of kings. In 1802 a young German schoolteacher, Georg Friedrich

Grotefend, decided to try and see whether he could identify any of these names and titles in the trilingual cuneiform inscriptions recorded by Niebuhr. He guessed they might include the formula 'Xerxes, great king, son of Darius, great king, king of kings, son of Hystaspes'; and he was right. He managed to identify the cuneiform signs for 'Xerxes', 'Darius', 'Hystaspes', and 'king'. And from that he began to try to work out a cuneiform alphabet for old Persian.

Rawlinson began to make his own copies of the inscriptions he found as he wandered round the ruins and visited surrounding sites. He soon realized that Grotefend had been correct, and that a number of trilingual inscriptions—where the same text had been carved in three different ancient languages—offered the best hope of trying to work out what they meant. When he was sent by the Shah of Persia to train Kurdish troops from Kermanshah, Rawlinson discovered that he was in the vicinity of by far the most extensive of these trilingual inscriptions: a huge series of panels 25 foot high and 70 foot long, carved 200 feet above the plain in the mountain of Behistun.

The Behistun Rosetta

After Darius' sculptors had finished this huge monument recording the victories of the king, the mountain path leading to it was quarried away, leaving the inscriptions inaccessible from below [Fig. 38.1]. A French artist, Eugène Flandin, had attempted to reach it in 1840. After a terrifying climb, 'fearing every moment that I would be hurtled to the bottom', he eventually reached the inscriptions with 'bloody feet and hands', only to discover that 'the narrowness of the ledge on which I found myself forced against the rock without being able to move a single inch'[1] made it impossible to take a copy [Fig. 38.2].

Rawlinson, who (in addition to his linguistic gifts) was a fine athlete with a good head for heights, breezily remarked that he did 'not consider any great feat in climbing to ascend to the spot where the inscriptions occur', and later described how he 'used frequently to scale the rock three or four times a day without the aid of rope or ladder'. Once on the ledge, he admitted,

ladders were indispensable in order to examine the upper portion of the tablet; and even with ladders there is considerable risk, for the foot-ledge is so narrow, about

[1] E. Flandin, *Voyage en Perse de MM Eugène Flandin, peintre, et Pascal Coste, architecte*, vol. I (1851), 450–1.

38.1 The Behistun inscription.

38.2 The ledge of Behistun.

eighteen inches or at most two feet in breadth . . . the upper inscriptions can only be copied by standing on the topmost step of the ladder, with no other support than steadying the body against the rock with the left arm, while the left hand holds the notebook, and the right hand is employed with the pencil.

In this position, he wrote, 'I copied all the upper inscriptions, and the interest of the occupation entirely did away with any sense of danger.'[2]

Danger though there was. On one later occasion he tried to cross a gap in the ledge by laying a ladder over it, only for one side of the ladder to come away and go 'crashing down the precipice', leaving Rawlinson hanging by his hands until he managed to scramble back. On his last visit to Behistun in 1847, when he set out to copy the Babylonian portion of the inscription, even he was defeated. The inscription was in a spot where the rock was scarped and projected some feet over the recess so that it could not be approached by any of the normal means. It was, he admitted, 'quite beyond my power of climbing'. Even the local cragsmen who tracked mountain goats over the entire face of the mountain said that this spot was unapproachable.

The problem was solved in the end by 'a wild Kurdish boy' who volunteered to make the attempt, and was promised a considerable reward if he succeeded.

The boy's first move, as Rawlinson described it,

was to squeeze himself up a cleft in the rock a short distance to the left of the projecting mass. When he had ascended some distance above it, he drove a wooden peg firmly into the cleft, fastened a rope to this, and then endeavoured to swing himself across to another cleft at some distance on the other side; but in this he failed, owing to the projection of the rock. It then only remained for him to pass over to the cleft by hanging on with his toes and fingers to the slight inequalities on the bare face of the precipice, and in this he succeeded, passing over a distance of twenty feet of almost smooth perpendicular rock in a manner which to a looker-on appeared quite miraculous. When he had reached the second cleft, the real difficulties were over. He had brought a rope with him attached to the first peg, and now, driving in the second, he was

[2] H. Rawlinson, 'X—Notes on Some Paper Casts of Cuneiform Inscriptions upon the Sculptured Rock at Behistun Exhibited to the Society of Antiquaries', *Archaeologia* 34 (1851), 73–6.

enabled to swing himself over the projecting mass of rock. Here with a short ladder, he formed a swinging seat, like a painter's cradle, and, fixed upon the seat, he took under my direction the paper cast of the Babylonian translation of the records of Darius which is now at the Royal Asiatic Societies rooms and which is of almost equal value for the interpretation of the Assyrian inscriptions as was the Greek translation of the Rosetta stone for the intelligence of the hieroglyphics of Egypt.[3]

Getting these paper casts was only the beginning. The task of interpretation was to prove equally difficult. Back in the stifling heat of Baghdad, Rawlinson laboured on the texts he gathered, sheltering in a spot 'where a waterwheel turned by the Tigris poured a continuous stream of water over a summerhouse at the extreme of the residency garden and hanging over the river'. He was accompanied in his work by a tame mongoose, a tame leopard named Fahed, and a pet lion 'found as a cub whose mother had been shot in a bed of rushes and flags near the Tigris' and which 'would follow him about all over his house and garden like a dog and never be happy unless he could be with him'.[4]

A continual spur to these efforts of decipherment was the astonishing discoveries being made on the banks of the Tigris 200 miles upstream, by a young Englishman called Austen Henry Layard.

The Buried Cities

Layard had been working as a solicitor's clerk in London when in 1839, after passing his law exams, he set out for Ceylon to work as a lawyer. Travelling overland through Persia he became fascinated with what he found there. Visiting Mosul he had crossed the Tigris to explore the huge heaps of earth that were supposed to mark the site of ancient Nineveh. There was little to see. The whole area was covered with grass and flowers and a few Arabs who had pitched their tents were grazing flocks. 'But even then', Layard later wrote, 'as I wandered over and among these vast mounds, I was convinced that they must cover the vestiges of the great capital, and I felt an intense longing to dig into them.'[5]

The French consul at Mosul, Paul-Émile Botta, had made some remarkable finds in a mound at Khorsabad, 12 miles

[3] A. G. Rawlinson, *A Memoir of Major-General Sir Henry Creswicke Rawlinson* (1898).
[4] A. G. Rawlinson, *A Memoir of Major-General Sir Henry Creswicke Rawlinson.*
[5] A. H. Layard, *Sir A. Henry Layard—Autobiography and Letters* (1903), 306–7.

DISCOVERY OF THE GIGANTIC HEAD.

38.3 Discovery of human-headed lion at Nimrud, 1846.

north-east of Mosul, and finally in November 1845, with the backing of the British ambassador, Sir Stratford Canning, and the assistance of Hormuzd Rassam (an Assyrian Christian who was later to take over the excavations and make important discoveries of his own), Layard began his own dig some miles to the south at the ancient site of Nimrud.

Within days of beginning Layard was greeted one morning with the news that an enormous head had appeared in the ground. The terrified diggers thought this must be Nimrud himself, but by nightfall the excavations had uncovered a 12-foot-high human-headed winged lion (which Layard promptly shipped off to the British Museum) [Fig. 38.3].

The following year an even more remarkable object was uncovered: a black obelisk sculpted on four sides and covered with extensive inscriptions [Fig. 38.4].

On Christmas day 1846 Layard despatched 23 cases, including one containing the obelisk, on a raft to float down the Tigris to Baghdad. On the 6th of January Rawlinson, having had the cases hauled up 'to a high open space in front of our

38.4 Black limestone obelisk of Shalmaneser III from Nimrud.

naval depot', got his first glimpse of it. 'The monument', he wrote to Layard, 'is ... the most noble trophy in the world and would alone have been well worth the expense of excavating Nimrud.' Throwing aside all other cuneiforms he immediately 'set to work to work tooth and nail on the Assyrian'.[6] Although the text could not yet be read, progress was being made, and in a letter to Layard, Rawlinson reported that 'they write to me from England that Assyrian antiquities were exciting great interest and that the clergy had got perfectly alarmed at the idea of there being contemporary annals whereby to test the credibility of Jewish history. A brother indeed of mine, a fellow of Exeter ... , protests most vehemently against the prosecution of the enquiry. Did you ever hear such downright *rot*?'[7] Whatever the alarm among the clergy at Oxford it was not shared in a vicarage at Killyleagh on the shores of Strangford Lough.

Breaking the Code

The Revd Edward Hincks, the vicar of Killyleagh, was the son of the Professor of Oriental Languages at the Belfast Academical Institution. He inherited his father's linguistic gifts and having studied Hebrew and Arabic at Trinity College, Dublin, had contributed to the deciphering of hieroglyphics and become expert in the field before turning his attention to cuneiform in the 1840s.

Edward Hincks had been elected as a Fellow of Trinity College in 1813 but had left after six years to work as a Protestant minister in the north of Ireland. From 1825 till his death in 1866 he served as the vicar of Killyleagh in County Down. Unlike Rawlinson (who on his return from the east was lionized by London society), Hincks worked all his life in isolation, hardly leaving the village (where as a strong supporter of Catholic emancipation he was not always popular) and publishing in relatively obscure Irish journals. He maintained, nevertheless, a correspondence with scholars around the world.

By 1850 Rawlinson and Hincks, working independently, had each made enormous strides in the deciphering of Assyrian and Babylonian cuneiform. On 29 December 1849 Rawlinson arrived back in London after 22 years in the east. At the beginning of February 1850, with the paper casts of Behistun inscriptions suspended around the walls, he addressed a meeting of the

[6] L. Adkins, *Empires of the Plain* (2004), 229.
[7] L. Adkins, *Empires of the Plain*, 241.

Royal Asiatic Society, chaired by the Prince Consort, at which he read his translation of the inscription on the black obelisk.

He was able to discern that the inscription began with an invocation to the gods of Assyria to protect the empire but admitted that he could not follow the sense of the complete invocation. This was in part because 'considerable difficulty still attaches to the pronunciation of proper names'. The obelisk seemed to have been erected, he thought, by a king called 'Temen-bar II' and the main body of the text concerned the events of his reign. The second row of reliefs on one side shows the figure of a man prostrating before the Assyrian king. Rawlinson mistakenly read the inscription below this as saying that offerings have been sent by 'Yahua, son of Hubiri', remarking that this was 'a prince of whom there is no mention in the annals and of whose native country I am therefore ignorant.'[8]

The following year Hincks pointed out that this should in fact be 'Ya.ua' the son of Kh'umum r.ii', that is 'Jehu the son of Omri'—the usurping king of Israel described in the biblical Book of Kings. Hincks' identification was not only one of the first independent confirmations of the existence of someone mentioned in the Old Testament, it also gave previously unknown information about him (his submission to an Assyrian king is not mentioned in the Old Testament) and provided a picture: the relief sculpture is the earliest known image of an Israelite king.

Some months earlier Rawlinson had been able at last to work out the names of the kings who had built the palaces that Layard and Botta had been excavating. 'Temen-Bar II' was Shalmaneser II, and in the inscription on an enormous bull that Layard had shipped back he was able to identify the names 'Judah', 'Hezekiah', and 'Jerusalem'. In 1851 he published his complete translation of the inscription which was, he claimed, Sennacherib's report of his attack on Judah and Jerusalem in the reign of King Hezekiah.

At the same moment that Rawlinson was translating the inscription on the bull, Edward Hincks was translating a very similar text on a prism that had been acquired back in 1830 by the British Resident in Baghdad, Colonel Taylor. When Layard published Hincks' translation (which, like Rawlinson's, caused immense excitement) in a book describing the excavations at Nineveh, he concluded that 'there can be little doubt the campaign against the cities of Palestine recorded in the inscriptions

[8] L. Adkins, *Empires of the Plain*, 279.

of Sennacherib at Kouyunjik [Nineveh] is that described in the Old Testament. The events agree with considerable accuracy.'[9]

Finding the Library

Layard had made his first excavations at Nineveh in 1847 when he had uncovered a room of bas reliefs depicting what was later realized to be Sennacherib's siege of the Judean city of Lachish. In 1849 when he returned to the site, with modest funding from the British Museum, he made an even more remarkable discovery. In the course of their excavations the diggers uncovered two small rooms in Sennacherib's palace filled 'to the height of a foot or more from the floor' with clay tablets. A visitor to the site remarked that 'many of them resembled cakes of Windsor soap, except instead of "Old Brown Windsor" they were covered with most delicately cut arrow-head hieroglyphs'.[10]

Four years later, in December 1853, Layard's assistant Homuzd Rassam had just begun excavating Assurbanipal's palace when he discovered, in the centre of a room lined with reliefs of a royal lion hunt, another immense library of clay tablets. Together with the reliefs, the whole mass of these tablets from Assurbanipal's great library (some 25,000 in all) were shipped back to the British Museum.

The wealth of information and interest for biblical studies that these texts represented was immense, but could the translations of them be trusted? The rivalry and perpetual disagreements between Hincks and Rawlinson and other scholars in the field meant that public confidence in their conclusions was at a very low ebb. The question of how such trust could be established needed somehow to be answered.

---

[9] A. H. Layard, *Discoveries in the Ruins of Nineveh and Babylon* (1853), 144.
[10] L. Adkins, *Empires of the Plain*, 291.

CHAPTER THIRTY-NINE

Breaking the Seals

William Henry Fox Talbot had first become fascinated with cuneiform at the beginning of the 1850s. He had corresponded on the subject with Edward Hincks, and in 1856 the British Museum had sent him a lithographed copy of an inscription by Tiglath Pileser I on a clay cylinder to translate [Fig. 39.1]. Rawlinson was also making a translation of this, which the museum intended to publish. On March 17th Fox Talbot sent his completed translation to the Royal Asiatic Society with the request that the packet containing his manuscript might not be opened until the publication of Rawlinson's translation when the two could be compared.

Fox Talbot's packet was laid on the table at a meeting of the society four days later and his request read out. Dr Julies Oppert, who happened to be present, announced that he also was working on the cylinder. He therefore asked that when his translation was finished he also 'might be allowed to deposit his version . . . with the secretary for the object of more fully carrying out the views of Mr Talbot, by affording three independent versions of the same document'.[1]

This was immediately agreed to and it was decided that rather than wait for the British Museum publication, both Rawlinson and Hincks should be invited to submit translations. This would be a powerful test of their system. If the translations did not agree it would show 'the decipherment had broken down and no confidence could be placed in the translation'. If, on the other hand, the translations turned out to be identical or nearly identical it would vindicate the scholars' claims, because 'it would be against calculation that three or four independent inquirers could possibly read and understand a long inscription of 1,000 lines in the same way unless they were working in the right path'.[2]

The secretary of the society 'engaged to keep the sealed packet in safe keeping', a committee of scholars 'whose names it was thought would command general respect' was set up,

39.1 Clay prism of Tiglath-Pileser I.

[1] W. H. F. Talbot et al., 'Comparative Translations', *Journal of the Royal Asiatic Society of Great Britain and Ireland* 18 (1861), 150–219.

[2] W. H. F. Talbot et al., 'Comparative Translations', 150–219.

and a date was fixed for a special meeting on May 20th, when the packets would be opened and the 'literary inquest' could take place.

Whereas the others had had the inscription for months, Hincks received it on April 26th, leaving him only time to translate the parts that seemed to him of most importance. Nevertheless the results were decisive.

When the seals were broken and the manuscripts compared, three of the translations were more or less identical, 'while it appeared to be merely owing to Dr Oppert's very imperfect acquaintance with the English language that a difficulty was found in bringing his version into unison with the others'. The committee concluded that there was a 'sufficient variety of words to test ... the extent of the knowledge claimed by the translators of the sound of the words, and of the language to which the words are supposed to belong',[3] and noted that 'the two principle scholars in the field', Rawlinson and Hincks, often agreed word for word.

Though doubters still remained, this cleared the way for the task ahead. A massive quantity of texts had been accumulated. What would the examination of them now reveal?

---

[3] W. H. F. Talbot et al., 'Comparative Translations', 150–219.

CHAPTER FORTY

The Intellectual Picklock

The man who began the immense task of sorting through this mass of material started his working life as an apprentice bank note engraver and died at the age of 36 in the British consulate in Aleppo. By the time of his death George Smith, 'deluge Smith' the 'intellectual picklock', had become a nationally known figure whose discoveries were eagerly followed in the pages of the *Daily Telegraph*.

Bird Tracks and Bibles

His short but remarkable career had begun from a childhood fascination with accounts of the exploits of Rawlinson and Layard. During his apprenticeship he left his work bench at Bradbury and Evans each day at 12 o'clock and spent his lunch hour at the British Museum and all his spare earnings on books about Assyria. On one occasion he had heard a museum assistant remarking what a shame it was that nobody took the trouble to read 'them bird tracks' and discover what they might reveal, and out of this he developed an abiding interest in 'those parts of Assyrian history which bore upon the history of the Bible'.[1]

When he learnt that George Smith had taught himself to read cuneiform, Rawlinson gave him permission to examine the store of paper casts in his work room at the British Museum. Smith began to take his own paper impressions of the clay texts to study at night, and seemed to be able to remember every fragment that he had seen. One of the first discoveries he made among these was an inscription that enabled him to date Jehu's payment to Shalmaneser described on the black obelisk.

On Sir Henry's recommendation Smith was then appointed assistant in the Assyriology Department in 1867. He began searching through the many boxes full of fragments of terracotta tablets covered in dirt that Layard and Rassam had sent back from Nineveh, and ordering them into categories:

[1] G. Smith, *Assyrian Discoveries* (1875), 9.

40.1 Fragment of the *Epic of Gilgamesh* with the flood story.

religious and other, historical, and mythological. It was among this last category that in 1872 he came across a fragment of an account of the flood [Fig. 40.1].

Smith's reaction to his discovery was closer to that of a twenty-first-century footballer scoring a goal than to the usual picture of a Victorian scholar. According to museum archives, 'he jumped up and rushed about the room in a state of great excitement and to the astonishment of those present began to undress himself'. After this outburst he began searching for more fragments and soon discovered that the first tablet he had found was the eleventh in a series of twelve. In the account of the flood, however, 17 lines were missing.

He made a translation of the whole sequence and read it to a packed meeting of the newly formed Biblical Archaeology Society on the 3rd of December 1872. The reading was received with huge acclaim and the vote of thanks was delivered by William Gladstone, who ended by warning 'that we must be on our guard against travelling too fast in these matters', but suggesting that through the labours of Rawlinson and Smith, 'we shall be permitted to know a great deal more than our forefathers in respect of the early history of mankind—perhaps the most interesting and most important of all the portions of the varied history of our race, with reference to the weighty interests that are involved as regards science or religion.'[2]

In the course of his reading Smith had drawn attention to the 17 missing lines, and this had had its effect. Sir Edwin Arnold, the editor of the *Daily Telegraph*, offered a thousand guineas for anyone who could find the missing segment. Smith was granted leave by the trustees of the museum to take up this challenge and set off on 20 March 1873.

Three Expeditions

After crossing the Syrian Desert on a horse he reached Nineveh on May 2nd. The prospect of finding anything did not look good. The ruins of Nineveh measured three miles across, and when he discovered Layard's excavations he found that 'the pit had been used since the close of the last excavations for a quarry' and 'the bottom of the pit was now full of massive fragments of stone . . . all in utter confusion'. The thought of needles and haystacks cannot have been far from his mind.

[2] *The Times*, 4 December 1872.

On May 21st, however, the *Daily Telegraph* published a telegram from Smith announcing a new discovery. On the fifth day of the excavation he had, he later wrote,

> sat down to examine the store of fragments of cuneiform inscriptions from the day's digging, taking out and brushing off the earth to read their contents. On cleaning one of them I found to my surprise and gratification that it contained the greater portion of seventeen lines of inscription belonging to the first column of the Chaldean account of the deluge and fitting into the only place where there was a serious blank in the story.[3]

Further expeditions followed. In 1874 Smith went again to Nineveh, this time on an expedition financed by the museum. Such was the public interest in these discoveries that he was despatched again in 1876, but this expedition was to end in disaster.

There was plague in the area. Smith's colleague Walter Eneberg died. Smith himself, hundreds of miles from medical treatment, rode 350 miles across the desert following old caravan trails and eventually collapsed in a village on the Turkish border. A dentist called John Parsons who had been sent out to look for him brought him in a cart 60 miles to Aleppo where he died in the consulate. In the last entry in his journal he wrote, 'I have done my duty thoroughly—I do not fear the change.'

Short though his career had been, it had, almost miraculously, opened a new door into the world of some of the early chapters of the Book of Genesis, which might conceivably answer some of the questions that now gathered round them.

---

[3] G. Smith, *Assyrian Discoveries*, 97.

CHAPTER FORTY–ONE

In a Strange Land

The prehistoric world that was being uncovered in the caves of the Dordogne and the world of giant reptiles that was being revealed in fossil discoveries were each entirely new and unexpected. To walk through the door which had been opened by Smith, on the other hand, was to encounter a landscape that, for Victorian readers of the Bible, seemed both deeply strange and oddly familiar.

The parallel account of the Flood that he had discovered was quite recognizable, but it appeared at the end of a magnificent and previously entirely unknown poem: *The Epic of Gilgamesh*. This remarkable work tells the story of the adventures of Gilgamesh, the king of Uruk, and his friendship with the wild man Enkidu.

In the second half of the poem, after the death of Enkidu, Gilgamesh goes on a quest to find a man named Utnapishtim. This Utnapishtim (the name may mean 'he found life') is the only man to have survived the Flood, and he holds the secret of eternal life. When Gilgamesh after many struggles and terrifying journeys at last finds him, Utnapishtim recounts for Gilgamesh the story of the Flood: how he was warned by the god Ea to build a giant boat, how he took into it the seed of all living things and cattle and wild beasts, how when the Flood subsided the boat came to rest on the mountain Nimush, and he sent out, first, a dove and then a swallow which both returned, and finally a raven which did not.

The version of *Gilgamesh* that Smith discovered dates from the seventh century BC and was apparently written by a 'master scribe and incantation priest' called Sin-leqe-unnini. Smith conjectured that it was a retelling of a much older story, and some 90 years later he was proved right.

Atra-hasis

The quantity of tablets unearthed by Layard and Hormuzd Rassam was so great (and the number of scholars capable of reading them so small) that translating them took many years. When some tablets, which had been dug up in Nineveh and had been in the British Museum since 1879, were finally translated in 1960, they turned out to contain another version of the Flood story.

In this version (where the author is named as Nur-Ayu junior scribe) the survivor of the Flood is called Atra-hasis—'extra wise' and the story begins with a time 'when the gods instead of man | did the work'. To avoid the back-breaking labour of digging out canals to irrigate the land, mankind is created. After a time, however, the god Enlil complains that 'the noise of mankind has become too much. I am losing sleep over their racket'. After a series of attempts to get rid of this nuisance he sends the Flood. Atra-hasis fortunately is warned by the dissident god Enki to build a boat and put on it '(birds) that fly in the sky, cattle [of Shak] and wild animals of open country';[1] to the fury of Enlil he manages in this way to survive.

As the Atra-hasis tablets were pieced together it became clear that the 17 lines that Smith had discovered in Nineveh in 1873 belonged to this version of the story rather than to *Gilgamesh*. It also became clear that the story of Atra-hasis had been written around 1700 BC—a thousand years earlier than the seventh-century version of *Gilgamesh*. Other versions of *Gilgamesh* were found that did date to around 1700 BC, but neither *Gilgamesh* nor the Atra-hasis epic seems to have been the earliest version of the Flood story.

Zi-ud-sura

In the Ashmolean museum in Oxford is a small clay block, 20 centimetres high known as the 'Weld-Blundell prism'. On the sides of the block is a list of Sumerian kings. The early kings rule for enormous lengths of time (28,800 and 36,000 years are mentioned), then we are told 'the flood swept over. After the flood swept over and kingship descended from heaven for a second time.'[2] After that the lengths of reigns gradually begin to become more moderate, reducing to 900 or so years and later as few as 3 (among these latter rulers are intriguing characters like 'Kug-Bau the woman tavern keeper').

According to this list the last king before the Flood was 'Ubara-Tutu who took the kingship to Shuruppag'. The prism is currently dated to around 1800 BC but these names occur

[1] S. Dalley, *Myths from Mesopotamia* (1989), 31.
[2] 'Electronic Text Corpus of Sumerian Literature: The Sumerian King List', http://etcsl.orinst.ox.ac.uk/index1. htm, accessed 20 August 2015.

on older tablets elsewhere. Among the earliest Sumerian fragments that we can read is a tablet dating from 2600 BC: 'The
instructions that Šuruppak the son of Ubara-Tutu gave to his
son Zi-ud-sura'. In some fragments of a Sumerian story found
at Nippur, which dates to around 1700 BC, this Zi-ud-sura is
named as the man that the Sumerian god Enki warns of the
Flood, and who survives it with his family and 'the animals and
seed of mankind'.

These extraordinary discoveries presented Victorian scholars
with a question which continued to engage their successors
in the following century. What was the relationship between
these Mesopotamian Flood stories and the account in Genesis?
As more tablets were discovered and translated, the picture that
began to emerge was one of considerable fluidity. Different
elements of stories seemed to have passed from one nation to
another, and to have been told and retold in different contexts
and with different meanings. Was the relationship between the
Mesopotamian stories and the Genesis story another example
of this kind of retelling?

Fragments of the *Epic of Gilgamesh* have been found at Me-
giddo in Israel and at Ras Shamra on the coast of modern-day
Syria, but it was not necessary to suppose that the writer of
Genesis had actually read *Gilgamesh* or even heard of it. The
telling and retelling of the Flood story in different languages
and in different contexts suggested that the story must have
been at least as well known in oral culture as in written texts.
Given the antiquity and widespread diffusion of the Sumerian
traditions it is difficult to imagine that Genesis could have been
written in ignorance of them. Was there any evidence within
Genesis that suggested a knowledge or awareness of Mesopotamian stories and religions?

Noah's Ark

Something like such a conscious rejection of earlier ideas
seemed to emerge when the Genesis account of the Flood was
laid alongside these earlier versions of the story.

In Babylonian mythology mankind is described as having
been created from the blood of the god Kingu, after the defeat of his lover, the sea god Tiamat, in order to 'serve the gods
without remission', to 'work their lands, build their houses'.[3]
The Atra-hasis epic, the Babylonian version of the Flood,

[3] N. K. Sandars, *Poems of Heaven
and Hell from Ancient Mesopotamia*
(1971), 101.

echoes this theme. Having been created to save the gods from having to dig canals, humanity is then destroyed because of the noise they are making, but saved, in the person of Atra-hasis, by Enki, the fresh water god, who subverts the actions of his fellow divinities.

In the Genesis stories there are no warring gods. The one God is creator of all. Far from being a slave race, humanity is created 'in the image of God, male and female', and appointed to rule the Earth on God's behalf. It is this new theology which flows through the rewritten Flood story. The Flood in Genesis becomes not an act of arbitrary despotism, but a judgement against a world that has 'ruined itself'. The ark (far from being the subversion of a rival god) is depicted as God's own provision to rescue a righteous man and bring the ship of humanity to safety.

Meanings and Motives

These discoveries began gradually to introduce a new basis of historical evidence into the long tradition of interpreting the Flood.

Lacking any such basis the early church fathers had tended to look for spiritual allegories in Genesis in something of the same way that stoic philosophers had looked for allegories in the Greek myths. Thus writing in the fourth century AD John Chrysostom had described the story of the deluge as 'a figure of things to come'.[4] For other church fathers the ark was 'primarily a symbol of the church, with those inside destined for salvation, and those without doomed to perish'.[5] Medieval commentators following this lead found symbolic significances in every last detail of the ark, from its three levels (symbolizing it was thought faith, hope, and love), to its 50 cubit length (symbolizing it was supposed the 50 days of Pentecost).

Reformation exegetes who rejected such allegorical approaches to Genesis began instead to look at these stories as a source of straightforward historical information that couldn't be had from anywhere else. 'We are utterly dis-provided of any history of the world's creation', wrote John Donne, 'except we defend this Book of *Moses*, to be Historical and therefore literally to be interpreted.'[6] Hence, as Peter Harrison has pointed out, the details of stories like the Flood 'were now related to more mundane questions of science and logistics. Where did

[4] P. Harrison, The Bible, Protestantism, and the Rise of Natural Science (1998), 128.
[5] P. Harrison, *The Bible, Protestantism, and the Rise of Natural Science*, 128.
[6] J. Donne, *Essays in Divinity* (1952), 18.

the waters come from and where did they eventually go? What mutations of the earth took place as a result of the Deluge?'[7]

It was against a background of two centuries of such literal interpretations that Darwin (who having read Lyell's *Geology* was not as 'dis-provided' as Donne) dismissed the Old Testament's 'manifestly false history of the world with the Tower of Babel the rainbow as sign etc etc'[8] as not to be trusted.

The cuneiform inscriptions discovered by Smith and others began to suggest a way of reading these biblical texts that involved neither allegorizing them nor reading them as literal history. When a story is retold, the point and meaning of the retelling can sometimes be found in the changes that are made from the original. It was reasonable to hope that by putting ancient versions of these stories alongside each other it might be possible to gain some sense of the original intention of the Genesis writer (or writers) and some insight into what these narratives might have meant in their original context.

How far could that context be recovered? As excavations at Babylon continued in the years after Smith's death, the answer to that question began to emerge.

---

[7] P. Harrison, *The Bible, Protestantism, and the Rise of Natural Science*, 128.
[8] C. R. Darwin, *The Autobiography of Charles Darwin* (1958), 85.

CHAPTER FORTY-TWO

By the Waters of Babylon

By the waters of Babel
We sat down and wept
When we remembered Sion
On the willows between them
We hung up our harps
For those that led us captive
Asked us for songs
And those who made us weep asked us to rejoice
'Sing us one from "The songs of Sion" '.
How shall we sing the Lord's song
In a strange land?[1]

The Book of Kings and the Book of Chronicles both end with descriptions of the fall of Jerusalem and the forced transport of its people to Babylon. First, as punishment for having rebelled against Nebuchadnezzar II, the young king Jehoiachin and his court were taken in shackles the 700 miles to Babylon; then eleven years later his uncle Zedekiah, together with the bulk of the population, trod the same bitter path leaving behind the smouldering ruins of their city and temple.

The story of the exile and the seemingly miraculous return of the exiles to their homeland was the apparent context for the writing and editing of a large percentage of the biblical writings, but had the events actually happened? While late-nineteenth-century evangelicals were committed to reading every word of Scripture as literal truth, late-nineteenth-century sceptics were equally committed to reading it as wholly myth-ological. Could archaeology adjudicate? As it turned out, the archaeological traces of the Babylonian captivity were dug up in reverse order.

Captives and Exiles

The first positive evidence was discovered in 1879 when Hor-muzd Rassam, Layard's erstwhile assistant, digging at Babylon, in

[1] R. Wagner, *In a Strange Land* (1988), Psalm 137:1–4.

42.1 Cyrus cylinder.

42.2 Ration tablet mentioning Jehoiachin.

42.3 Babylonian Chronicle.

² D. J. Wiseman, *Chronicles of Chaldaean Kings* (1956), 33.

what turned out to be the ruins of the Esagila temple, unearthed and sent back to the British Museum the so-called 'Cyrus cylinder' [Fig. 42.1]. This appeared to be a public monument on which, after his conquest of Babylon, Cyrus proclaims his policy of returning captured images of gods to their sanctuaries and captured peoples to their settlements. Although the Cyrus cylinder made no reference to the Israelites and relates specifically to the local cults around Babylon, such a policy is consistent with what is described in the biblical books of Ezra and Chronicles.

A specific reference to the Jewish captivity was discovered during the 14-year excavation of Babylon undertaken by the German archaeologist Robert Koldewey at the turn of the century. In his excavations near the Ishtar gate, Koldewey discovered the royal archive room of Nebuchadnezzar II containing tablets dating from 595 to 570 BC. Among these were tablets which, when they were translated by the German Assyriologist Ernst Weidner in the 1930s, turned out to list rations of oil and barley for various individuals [Fig. 42.2]. These included '10 [sila of oil] to the king of Judah Yaukin [Jehoiachin]; 2 and a half sila [oil] to the offspring of Judah king'. This was particularly striking because the last sentences of the Book of Jeremiah and the Book of Kings both record that after being released from imprisonment in Babylon 'day by day the king gave Jehoiachin a regular allowance'.

In 1956 D. J. Wiseman published a translation of fragments of the Babylonian Chronicle found in the British Museum collections [Fig. 42.3]. This described how Nebuchadnezzar II 'besieged Jerusalem (literally the city of Judah) and seized it on the second day of the month of Adar'.² The immense wealth of materials in the British Museum have continued to the present day to throw up discoveries which add to or confirm these details.

A recent such discovery was made in 2007 by Michael Jursa, a visiting Viennese professor of assyriology, when he was examining a tablet discovered at Sippar in 1870 and acquired by the museum in the 1920s [Fig. 42.4]. This was an apparently mundane receipt for a quantity of gold dated 'Month XI day 18 year 10 of Nebuchadnezzar king of Babylon', the property of 'Nabu-Sharrussu-ukin the chief eunuch', but Jursa recognized that the name and title were the same as 'Nebo-Sarsekim the chief eunuch' described in Jeremiah 39:3, a minor character in the narrative who was present at the fall of Jerusalem

when Zedekiah the king was blinded and taken in shackles to
Babylon.

If these catastrophic events had actually occurred, how had
the Israelites managed to survive their encounter with the formidable civilization of Babylon? What strategies had enabled
them to maintain their own cultural identity when everything
that had shaped it had been removed? What reply did they
find to the psalmist's desperate question 'How shall we sing the
Lord's song in a strange land?'

The answer that many scholars have suggested is that they
produced a literature.

42.4 Record of Nabu-sarrusu-
ukin's gold deposit.

Is It Not Written?

The exiled court of Jehoiachin may well have carried their
dynastic scrolls with them into exile. The Book of Kings frequently asks rhetorically, 'is it not written in the annals of David?' or 'the annals of Solomon' or 'the annals of the Kings of
Israel'. The books of Chronicles additionally refer to 'the records of Nathan the prophet', 'the records of Gad the seer', 'the
prophecy of Ahijah the Shilonite', and 'the visions of Iddo the
seer concerning Jeroboam son of Nebat'. None of these ur-
texts survive, but it has been conjectured that the task of editing
and using them to formulate the great theological histories of
Kings and Chronicles may have begun during the Babylonian
exile and been inspired by the challenge of that situation.

Something similar might have happened with temple and
prophetic literature carried to Babylon by exiled priests like
the prophet Ezekiel.

What, though, of the stories of humanity's origins?

The Book of Daniel, looking back on the exile, describes
how young Israelites from the nobility were taught 'the language and literature of the Babylonians ... they were to be
trained for three years and after that they were to enter the
King's service'.[3] In a recent book Irving Finkel, the curator of
the British Museum's cuneiform collection, has argued for the
plausibility of this account.[4]

That 'the month names used today in the modern Hebrew
calendar preserve the ancient names used in Nebuchadnezzar's
capital' is an 'eloquent measure of permanent Babylonian influence'.[5] Finkel describes a new tablet which mentions the
staged release of birds and animals going in 'two by two', and a

[3] Daniel 1:4–5.
[4] Y. I. Finkel, *The Ark before Noah* (2014).
[5] Y. I. Finkel, *The Ark before Noah*, 259.

round shape for the ark. More specifically Finkel draws attention to Babylonian school texts from this period which contain extracts from 'The great ages of man', 'the Sargon legend', and 'the epic of Gilgamesh'. The close verbal similarities between the Gilgamesh and the Genesis account of the Flood become easier to account for since 'the three works which best exemplify the process of borrowing were *on the school curriculum*. The trainee Judaeans would have encountered these texts in their palace classroom.'[6]

The production of the history of the captive nation, Finkel conjectures,

> must have been carried out by a group of specific individuals who had access to all existing records, under an agreed editorial authority. One must envisage a Bureau of Judaean History. It is against this backdrop that the incorporation of particular Babylonian traditions becomes intelligible. Perhaps there was a shortfall of native ideas among Hebrew thinkers about the beginning of the world and civilisation. Whatever the case certain powerful Babylonian narratives were taken up but, crucially, not adopted wholesale. The beginning of the Book of Genesis would be unrecognisable without the cuneiform substratum but the stories were given *a unique Judaean twist that allowed them to function in a wholly new context.*[7]

Evidence of the radical nature of this twist and the way that it turned Babylonian religion on its head had first appeared in a book that George Smith, published in 1876, the year of his death: *The Chaldean Account of Genesis.*

Enuma Elish

In 1874 Smith had begun piecing together and translating a long poem, covering seven tablets, which begins with the words 'Enuma élish la nabu shamamu'—when there was no heaven. The first part of the poem tells the story of the birth of the gods and the war in heaven in which Marduk (sometimes called Bel—'Lord') becomes king of all the other gods, defeats Tiamat, the primeval chaos monster, and establishes order in the universe. This leads to the construction of a tower and temple at Babylon, built according to Marduk's specific instructions by the 'Anunnaki gods' who for a 'whole year long . . . set

[6] Y. I. Finkel, *The Ark before Noah*, 254.
[7] Y. I. Finkel, *The Ark before Noah*, 249.

bricks in moulds'. In the second year 'they had raised its head ESAGLIA it towered, the earthly temple the symbol of infinite heaven'. Marduk and the Anunnaki then take up residence, mankind is created to serve their needs, and 'Great Babylon' is established as the 'dear city of God'.

When systematic excavations of Babylon and other sites began to be organized in the years following Smith's death it quickly became clear that the great stepped ziggurats of Mesopotamia had been as much a feature of that landscape as were the pyramids in Egypt [Fig. 42.5]. The mud bricks from which they were constructed though had not lasted so well: the tower of Babylon today looks like 'a bank of mud in a watery depression'.[8] These extraordinary buildings would have been unlike anything the exiled Judaeans had seen before.

The description of the tower of Babel in Genesis 11 ('Babel' being the Hebrew word for 'Babylon') reveals that the writer of that story was aware of the method of ziggurat construction— 'come let us make bricks and bake them thoroughly', and of the contrast with other methods—'They used brick instead of stone and tar instead of mortar' (Gen. 11:3). Whether or not the author had any specific knowledge of the Babylonian creation poem, the idea of 'a tower that reaches to the heavens' (Gen. 11:4) suggests some awareness of the religious meaning that the Babylonians attributed to their ziggurat.

In fact no resident in the city could have failed to be aware of it. As excavations at Babylon proceeded, several tablets were found that indicated the importance of the poem Smith had translated in the national cult. These describe the liturgy of the

42.5 Ziggurat of Ur.

[8] N. K. Sandars, *Poems of Heaven and Hell from Ancient Mesopotamia* (1971), 40.

42.6 Map of Babylon.

new year festival in which 'after the second meal of the late afternoon . . . the urgalla-priest . . . shall recite [while lifting the hand] to the god Bel [the composition entitled] *Enuma Elish*'.[9] In *Enuma Elish* (as it seems we should call it) the building of the ziggurat which establishes 'Great Babylon' as the home of the god Marduk makes it a source of both religious and political power.

The remarkable Babylonian clay map in the British Museum illustrates this religious and political claim by depicting Babylon as the centre of the world [Fig. 42.6]. Outsiders coming into the city might have been able to detect the fragility of the claim.

Babylon was not the only place to claim this status. Dunnu was another important city around 1700 BC, and a poem known as the *Theogony of Dunnu* tells how 'At the very beginning plough married earth' and describes 'how they built Dunnu forever as his refuge'.[10] Since rival creation stories and rival gods were associated with different places, they could be rewritten by political events.

When in 689 BC the Assyrian king Sennacherib conquered Babylon, the statue of Bel/Marduk was carried off to the Assyrian capital, Assur. Assur was also the name of the god of the Assyrian nation, and an Assyrian version of *Enuma Elish* has been found, dating from this time, in which the name of Assur replaces that of Marduk.

The story of the tower of Babylon in Genesis, however, involved a far more radical kind of rewriting than simply replacing the name of one god with that of another. Instead of seeing the ziggurat as a 'home for God', which establishes Babylon (or some rival) as the centre of religious and political power, the thrust of the Genesis story is to subvert what it portrays as the arrogant striving 'to make a name for ourselves'.

Babel Babble

There is nothing in the Babylonian poem about speaking in a single language, but some tablets of a Sumerian saga have been found which do foresee this. These describe how Enmerkar, the king of Kulaba, tries to force the submission of the king of Aratta and demands that he should obey the command of the goddess Inanna and build a great tower as a shrine, 'the mountain of the shining me'. Here a prophecy will be spoken that

[9] N. K. Sandars, *Poems of Heaven and Hell from Ancient Mesopotamia*, 47.
[10] S. Dalley, *Myths from Mesopotamia* (1989), 279.

'the many-tongued lands' will 'address Enlil together in a single language'.[11]

Genesis turns this kind of prophecy on its head. When Babel is built, instead of uniting humanity in a single language the opposite takes place. According to the Babylonians, 'Babel' means 'gate of god'. In Genesis, in what seems to be a deliberate parody, it is interpreted to mean 'confusion'.

The polemical thrust of this retelling of the Babylonian foundation myth is paralleled in the latter sections of the Book of Isaiah, where the religion of the Jewish captives is directly contrasted with the mythologies that they found enacted in the great annual religious rituals of Babylon.

I Am He

The tablets which describe the liturgy of the Babylonian new-year festival recount how on the third day of the festival a goldsmith and woodcarver are called to make statues of two gods: Bel/Marduk and Nebo, his son. On the tenth day (after *Enuma Elish* had been read) these statues were to be carried in procession under the Ishtar gate into boats on the canal and then into the temple house (a relief on the now destroyed temple of Bel at Palmyra illustrates a procession somewhat like this) [Fig. 42.7].

Isaiah Chapter 44 begins by satirizing this idea of making a god out of a piece of wood, from part of which a man makes a fire to cook his meal while 'from the rest he makes a god . . . he prays to it and says, "Save me you are my god"'. Chapter 46

42.7 Procession of Bel from the temple of Bel at Palmyra.

[11] 'Electronic Text Corpus of Sumerian Literature: Enmerkar and the Lord of Aratta,' http://etcsl.orinst. ox.ac.uk/index1.htm, accessed 20 August 2015.

of Isaiah describes both the procession of Marduk and the hu-
miliating parody of the annual ceremony when Sennacherib
conquered Babylon and carried off these statues to the Assyrian
capital:

> Bel bows down, Nebo stoops low
> Their idols are borne by beasts of burden
> The images that are carried about are burdensome
> A burden to the weary . . .
> They themselves are carried into captivity.

In contrast to these man-made gods, the God who made man
turns this whole picture upside down:

> Listen to me O house of Jacob
> All you who remain of the house of Israel
> You whom I have upheld since you were conceived
> And have carried since your birth
> Even to your old age and grey hairs.
> I am he, I am he who will sustain you
> I have made you
> And I will carry you.[12]

There is some evidence that there was a movement in a mono-
theistic direction even within Babylonian thought. A remarkable tablet of Marduk theology has been found which sets out
the attributes of lesser gods as aspects of Marduk himself:

> Nabu is Marduk of accounting
> Sin is Marduk as illuminator of night
> Shamash is Marduk of justice
> Adad is Marduk of rain[13]

The Jewish critique of Babylonian religion went far beyond
this. The opening of Genesis, while still evidencing a cuneiform substratum, proposes an understanding of the character of
God and his relationship to the created world that was unique
in the ancient world, and uniquely influential. In the fullness
of time it was to have a profound impact on the penultimate
curiosities of science.

---

[12] Isaiah 46:1–4.
[13] Y. I. Finkel, *The Ark before Noah*, 243.

CHAPTER FORTY–THREE

Adam and Adapa

In 1887, five years after the death of Darwin and just at the moment that the discovery at La Mouthe was about to confirm the reality of prehistoric painting, a number of tablets written in Akkadian cuneiform began to appear on the antiquities market.

These, it transpired, had been dug up by local Egyptians from the ruined city of el-Amarna: the short-lived capital that the Pharaoh Akhenaten had built for himself. The tablets had come from a building that archaeologists later identified from inscriptions as 'The Bureau of the Correspondence of Pharaoh' and mostly consisted of a remarkable diplomatic correspondence—'the Amarna letters'. Among them though was a tablet, dated to 1400 BC, that told the story of Adapa—the primal man, the first of the seven sages and the son of Eridu, the first human city. A version of the same story from the seventh century BC, translated into Babylonian, was later found on tablets in the library of Assurbanipal in Nineveh.

The story tells how Adapa is summoned to heaven because while out fishing he has 'broken the wing of south wind'. The god Ea tells him how to behave and advises him that when he is offered the bread of death he must not eat, and when offered the water of death he must not drink. Adapa follows this instruction but then discovers he has actually refused the bread of life and the water of life and thus lost the opportunity of immortality.

What was the relationship between this and the Genesis account of Adam and Eve?

Adapting Adapa

The Adam and Eve story contains many of the same elements of the Adapa story—the first human being losing the prospect of immortality by not eating a special food (the fruit of the Tree of Life)—but apparently arranged to suggest a radically different meaning.

In Genesis instead of seeing human beings as victims of divine trickery or fatal mischance (it is not clear in the story of
Adapa which is to blame), men and women are seen as tra-
gically responsible for their own destiny: subject to temptation
but making their own choice. The reactions of God to human
choices are consequently portrayed not as arbitrary tricks or the
actions of fate, but as moral judgements. Yet even humanity's
disobedience cannot defeat God's beneficence: as they are expelled from the garden Adam and Eve are clothed by God and
given a cryptic promise of ultimate victory over their tempter.

If the writers of Genesis were adapting the form of the
Adapa story, they did so by jettisoning almost all of the Babylonian mythology out of which the story comes.

In other Babylonian texts Adapa appears as one of the Apkallu: the seven sages. These were fish-men, with the body of
fish and the heads of men, who lived by night in the sweet
waters of Apsu, and came out by day to teach mankind all the
skills of civilization: writing, divination, kingship, agriculture,
mathematics, and the building of cities [Fig. 43.1]. They were
thought to have lived in the golden age before the Flood where
men might live for 36,000 years in a world without disease or
death in childbirth (these lifespans were recorded in the Sumerian king lists which, with a Babylonian interlinear translation,
formed part of the curriculum for those learning cuneiform[1]).

43.1 Apkallu fish-men.

[1] Y. I. Finkel, *The Ark before Noah*
(2014) .

In Genesis this idea of a golden age has been almost wholly excised (the only traces that remain are the much more moderately long-lived antediluvian patriarchs and a single mention of the mysterious Nephilim[2]). The world before the Flood is represented as 'full of violence'. Even in the Garden of Eden the idea of a world without death, toil, and the pains of childbirth appears only as an unrealized potentiality: a road not taken.

In all the various Mesopotamian stories mankind is represented as a slave race created by the gods to do their work for them. In the complementary account of Creation in the first chapter of Genesis human beings are created male and female 'in the image of God'. Their creation (seen by God as 'very good') comes as the climax of a process which is described in language that seems carefully phrased to distinguish it from rival accounts.

Shamash

In the description of the creation of the Sun and the Moon we are told that 'God made two great lights; the greater light to rule the day, and the lesser light to rule the night, and also the stars' (Gen. 1:16). There is something a bit convoluted about these phrases. Why talk about 'the greater light' rather than 'the Sun', the normal term that is used throughout the rest of the Bible?

The answer which many scholars have suggested is that the Hebrew word for sun, *shemesh*, is the same word as the name of the Mesopotamian sun god Shamash. In the *Epic of Gilgamesh* it is all-seeing Shamash who helps the hero defeat the monster Humbaba. There was a great temple of Shamash (known as e-barra: 'the shining house') in two Babylonian cities, Sippar and Larsa, with smaller temples at Babylon, Nineveh, Ur, Mari, and Nippur.

The Moon, Genesis seems to tell us, is not 'Marduk as illuminator of night', neither is the Sun the great god Shamash, 'Marduk as justice' [Fig. 43.2]. Both are merely 'lights' created by God. The stars, which in Babylonian thought were the controllers of human destiny, are referred to almost as a divine afterthought. This careful circumlocution is not accidental: it reflects a strategy that pervades the whole biblical literature.

43.2 Shamash giving the law to Hammurabi from the Code of Hammurabi.

[2] Genesis 6:4.

Idols and Idolatory

The full implications of a faith in an all-powerful creator God may have only become fully visible when Jewish and Babylonian religious traditions appeared side by side, but the faith itself had not originated in Babylon. Though some have argued the contrary, all the evidence we have suggests that the Jewish exiles brought it with them. While the dating of biblical literature is notoriously difficult, the prophetic books of Hosea, Amos, Jeremiah, and the early chapters of Isaiah, which explicitly address situations before the exile took place (and which the weight of scholarly opinion locates in the pre-exilic period), robustly proclaim God as sole creator.

A repeated accusation is that 'Israel has forgotten his maker' (Hos. 8:14) and turned to idols. While the nations worship 'the work of men's hands, wood and stone', it is he 'that builds his lofty palace in the heavens and sets its foundations on the earth; he calls for the waves of the sea' (Amos 9:6); it is he alone 'who appoints the sun to shine by day, who decrees the moon and stars to shine by night, who stirs up the sea so that its waves roar' (Jer. 31:35).

What follows is the insistent prophetic demand that to live in tune with this God the people must care for the poor in their economic lives and abandon idols in their worship.

According to the tradition recorded in Exodus this did not absolutely exclude the use of visual imagery. In the tabernacle (and later the temple) there are images of cherubim representing the supernatural world and images of almond flowers representing the natural creation. This visible imagery, however, is used to point towards an invisible reality that cannot be represented. Thus, while the cherubim stand on either side of the mercy seat, the seat itself remains empty (Exod. 25:17–20).

For the Jews in Babylon who started to create a national literature, what was the literary equivalent of this repudiation of visible idols?

The Appearance of the Likeness

In later Judaism the tradition developed that while the name of God could be written it could not be spoken (only murmured by priests in the temple on the Day of Atonement). When the exiled priest Ezekiel received his revelation 'among the captives' by the waters of Babylon, he describes his 'visions of God' with a similar caution. Amid all the temple imagery of cherubim and fire that comes rolling towards him is that which at best can be described as 'the appearance of the likeness of the glory of the LORD' (Ezek. 1:28).

This vision of the glory of God in Babylon prefigures Ezekiel's vision of the glory leaving the temple in Jerusalem, which is a prelude to its physical destruction. That cataclysm, however, which might have extinguished Jewish religion, had the paradoxical effect of reinforcing the conviction that the God worshipped by the Jews was not merely a tribal deity but the creator of everything. Hence the psalm which describes how 'They burned your sanctuary to the ground, they defiled the dwelling place of your name' moves to affirming that 'God is my king from long ago', the one who 'established the sun and the moon'.[3]

This enlarged sense of God as creator, which pervades the exilic and post-exilic literature, is expressed in language that is just as circumspect and allusive as that which is used to describe God's own person. In striking contrast both to the detailed accounts of Mesopotamian and Egyptian creation myths and to the mechanical conjectures of contemporary Greek philosophy, the Hebrew writers evoke the process of creation through shifting and provisional metaphors.

Myth, Mechanism, and Metaphor

Enuma Elish describes heaven and earth as being formed from the body of the sea god Tiamat. Egyptian myths describe how a mound or egg emerged from the primeval waters. The early Greek philosophers of this time describe the world beginning from water, air, or fire.

In the later chapters of Isaiah, by contrast, Creation is sometimes compared to putting up a tent, where God 'stretches out the heavens like a canopy and spreads them out like a tent to live in'[4] and at other times likened to making pottery where 'we are the clay, you are the potter'.[5] In the Book of Job God compares himself to a builder, asking Job 'where were you when I laid the foundations of the earth? . . . On what were its bases sunk? Or who laid its cornerstone?'[6] Psalm 104 again uses the simile of putting up a tent, but also describes the creation as like putting on clothes with God wrapping 'himself in light as

[3] Psalm 74:7, 12, 16.
[4] Isaiah 40:22.
[5] Isaiah 64:8.
[6] Job 38:4–6.

with a garment' and covering the earth 'with the watery depths as with a garment'.[7]

The 'days' of Genesis similarly suggest an ordering of Creation but not a mechanism (these are 'days' which exist even before the creation of the Sun and the Moon). The occasional fragments of mythologies or physical theories that appear (such as mentions of 'Rahab', the world being 'without form and void', or 'the waters above and below the earth'[8]) are passing references that are not elaborated on.

None of these shifting images suggests a commitment to any particular physical theory, but neither does this allusive metaphorical language imply a detached Epicurean concept of divinity. On the contrary when Ezekiel saw in his vision that the presence of God from the holy of holies in Jerusalem had appeared by the river Kebar in Babylon, the first thing that he records is that 'I heard the voice of one speaking'.[9] God, who was mysteriously and continuously responsible for all created things and somehow present always and everywhere, was also a God who spoke and listened to men.

In consequence while the affirmation that 'the LORD our God is one God' is a consistent theme across the whole of the Hebrew Scriptures, the far-reaching implications of this radical doctrine are worked out in what appears at times like a continuously evolving dialogue between God and man.

The distinctive characteristics of this dialogue began to appear in ever more high relief as further aspects of Near Eastern religion continued to be uncovered, and prominent among these was the idea whose winding course we have been following throughout the preceding chapters. The concept is of a transcendent unity which cannot be identified with anything in the physical world but to which nevertheless all things point.

---

[7] Psalm 104:2, 6.
[8] Genesis 1:2, 7.
[9] Ezekiel 1:28.

CHAPTER FORTY-FOUR

Ariadne's Thread

The 1887 discovery of the Amarna letters soon led to further excavations in Akhenaten's abandoned capital by Flinders Petrie, Alessandro Barsanti, and others. The letters though were not the only significant finds that had been made there. In 1833 when the Scottish Egyptologist Robert Hay visited the site he had found a number of tombs filled with pottery and rubbish. In 1883 the French sent a Mission Archelogique to Amarna which began to clear these out. In their first season there an Egyptologist called Urbain Bouriant found and transcribed on the west wall of the tomb of Ay an extraordinary poem. This was *The Great Hymn to the Aten*, possibly written by Akhenaten himself, which represents the closest parallel so far discovered to the perspective of biblical monotheism.

The Works of God

One of the most radical consequences of monotheism was the realization that if the one God is the creator of all, then God is necessarily responsible for everything in creation that is inimical to mankind as well as everything beneficial.

In Mesopotamian and Egyptian thought, the struggle between good and evil took place on a cosmic level with a host of gods and demons all involved. The complex syncretistic nature of these religious traditions precluded anything as straightforward as dualism. Nevertheless, within the Mesopotamian pantheon an evil goddess like Lamaš-tu could be held responsible for miscarriages and cot deaths, while the demon god Pazuzu, who describes himself on a statue as 'king of the evil wind demons', could be a protective against her [Fig. 44.1].

In the more straightforwardly dualistic Zoroastrian thought of Persia, Ahura Mazda the creator 'made every land dear (to its people)' while Angra Mainyu the evil one 'counter-created the serpent in the river . . . the locust which brings death unto cattle and plants'.[1]

44.1 The demon-god Pazuzu, 'king of the evil spirits'.

[1] L. H. Mills and J. Darmesteter, *The Zend-Avesta* (1880), 2–3.

In Akhenaten's heretical reformulation of Egyptian religion by contrast, everything is brought to a great simplicity [Fig. 44.2]. In *The Great Hymn of Aten*, the Sun god Aten is the 'sole god' who moulds to his wish 'All men, cattle, and wild beasts, | Whatever is on earth, going upon (its) feet, | And what is on high, flying with its wings'.[2] All traces of the cosmic drama, in which every night the solar barque Apophis survives attempts to sink it, have disappeared. Nevertheless, even here, it is while 'Darkness hovers | earth is silent | Their Maker rests in light-land' that 'Every lion comes from its den | All serpents bite'.[3]

In Psalm 104, which otherwise closely parallels Akhenaten's hymn (so closely that some scholars have suggested a relationship), the heavenly bodies are mere creatures of God and obey his command. Likewise the beasts who hunt in the dark remain in God's presence and are explicitly described as part of his good creation: 'He made the moon to mark off the seasons, and the sun knows when to go down . . . the lions roar for their prey and seek their food from God . . . How many are your works, O Lord! In wisdom you made them all'.[4]

44.2 Akhenaten worshipping the Aten with Nefertiti and their two daughters.

[2] J. B. Pritchard et al., *The Ancient Near East* (1958), 227–30.
[3] M. Lichtheim, *Ancient Egyptian Literature* (1976), 96–9.
[4] Psalm 104:19–24.

The same perspective is found in Genesis. While the prophecy spoken in *Enmerkar and the Lord of Aratta* looked forward
to a time when there would be 'no snake',[5] in the goodness of
the Garden of Eden the presence of the snake is included and
is explicitly described as part of God's creation.[6]

In the Book of Job, God's speech 'out of the whirlwind'
comes to its climax with a description of the great monsters
Behemoth and Leviathan. Victorian writers who celebrated
'all things bright and beautiful' were shocked by the Darwinian picture of 'monsters of the prime tearing each other in the
slime'. Job is not so squeamish. Behemoth's bones are 'tubes
of bronze, its limbs like rods of iron' and yet Job is invited to
'look at Behemoth, which I made along with you; . . . it ranks
first among the works of God' (Job 40:15, 18, 19). Leviathan's
mouth is 'ringed about with fearsome teeth', yet God delights
in him and declares that 'Nothing on earth is its equal' (Job
41:14, 33).

William Blake's poem 'The Tyger' asks 'Did he who made
the lamb make thee?' The answer from the Book of Job is an
emphatic 'yes'.

The Problem of Pain

This collision between the belief in God's goodness and an
insistence that he is the creator of everything is a paradox that
is explored throughout the biblical literature. The puzzle and
problem of 'nature red in tooth and claw' were not a Victorian
discovery.[7] In Genesis neither the presence of the snake nor
God's cryptic promises are explained, while elsewhere the Bible contains a kind of interrogation of God that outside some
examples in Greek drama had few equivalents in the ancient
world.

In the psalms the questions 'Why O LORD?', and 'How long
O LORD?' are constant refrains. Particular answers are sometimes mooted, but no attempt is made to propose a solution
that covers all cases. In the Book of Job, every explanation of
Job's sufferings brought forward by his comforters is rejected
by Job, who angrily accuses his friends of speaking wickedly on
God's behalf: 'will you speak deceitfully for him? Will you show
him partiality?'[8] At the end of the book God himself joins in,
accusing the comforters of having 'not spoken about me, as my
servant Job has'.[9]

[5] 'Electronic Text Corpus of Sumerian
Literature: Enmerkar and the Lord of
Aratta', http://etcsl.orinst.ox.ac.uk/
index1.htm, accessed 20 August 2015.
[6] Genesis 3:1.
[7] A. Tennyson, *The Poems of Tennyson* (1969).
[8] Job 13:7–8.
[9] Job 42:8.

The cognitive dissonance between a belief in God's kindness and the experience of nature's indifference could only be resolved by an emphasis (which again has few ancient parallels) on the need to 'trust in the LORD': 'Though he slay me yet will I hope in him'.[10] 'Though the fig tree does not bud | and there are no grapes on the vines, | though the olive crop fails and the fields produce no food, . . . yet I will rejoice in the LORD, | I will be joyful in God, my saviour'.[11] The basis for such trust is rooted in the revelation of a God who redeems: who brings light out of dark and can even use the terrors of the ocean 'the mighty waters . . . his wonderful deeds in the deep'[12] to bring about his good purposes. The requirement for such trust is grounded in a sense of the limitations of human knowledge which the contemplation of Creation itself exposes.

What God Has Done

Since God in Hebrew thought cannot be identified with any aspect of nature, the scale of Creation becomes daunting to contemplate:
'When I look at your heavens
The work of your fingers
The moon and the stars which you created,
What is man that you remember him
The son of man that you visit him?[13]
The dimensions of the universe not only defeat any human attempts to imagine them, but render absurd any attempt to comprehend the mind of God or sit in judgement on his actions: 'Who has measured the waters in the hollow of his hands or with the breadth of his hand marked off the heavens? . . . Who can fathom the spirit of the LORD or instruct the LORD as his counsellor?'[14]

The Book of Job constitutes a remarkable example of this evolving Hebrew dialogue between God and man. The 179 or so questions with which the book concludes begin by asking, 'Where were you when I laid the earth's foundation?' and reiterate this demand in respect of every aspect of Creation, from the movements of 'the beautiful Pleiades' to the flapping of an ostrich's wings. Job's response, 'surely I spoke of things I did not understand',[15] echoes the psalmists realization that 'Such knowledge is too wonderful for me, too lofty for me to attain'.[16]

[10] Job 13:15.
[11] Habbakuk 3:17, 18.
[12] Psalm 107:23–4.
[13] R. Wagner, *The Book of Praises* (1994), Psalm 8.
[14] Isaiah 40:12–13.
[15] Job 38:4, 42:3.
[16] Psalm 139:6.

Despite these limitations, human being are far from being the ignorant pawns of the gods depicted in Mesopotamian myths. Where Genesis has men and women 'made in the image of God' to rule over the created world, the psalmists see humanity as 'little less than God . . . Lord of the work of your hand'.[17] Human beings are 'fearfully and wonderfully made' and the recognition that God's works 'are wonderful' is an intrinsic part of our nature, something which the 'soul knows abundantly'.[18] Both the God-given reach and the limitations of the human condition are eloquently summed up in the formulation of Qoholeth, 'the teacher': 'He has made everything beautiful in its time. He has also set eternity in the human heart, yet no one can fathom what God has done from beginning to end.'[19]

Following the Thread

Qoholeth's concise summary was quoted by Galileo in his *Letter to the Grand Duchess Christina*. It is perhaps the closest the biblical writers come to describing what at the beginning of the book we referred to as 'ultimate curiosity'. It brings us back to where we began.

Our argument in Part I suggested that it was 'the Garden of Eden moment' (the period when early humans were first propelled into a cognitive world that required some kind of priority scheme to navigate) which promoted the capacity for ultimate curiosity to centre stage in human life. To steer a consistent path a moral GPS needs its moral satellites. Hence the integrative struggle to unite our internal mental world—'to act as a whole'—becomes, we argued, a larger struggle to comprehend the entire external world as some sort of meaningful whole.

This larger cognitive struggle employs every sense and capacity in the attempt to create an integrated picture of reality, and the synergies that result produce in time radical new ways of perceiving and describing the world.

Ultimate curiosity, the impulse to see beyond the rim of the physical world, becomes in this way a continuous driver for new discoveries within the physical world.

The resulting entanglement of what we now call 'religion' and 'science' has then been the basis of the Ariadne's thread that we have been following through the rest of the book. Thus while the circumstances in fifth-century Greece that gave

[17] R. Wagner, *The Book of Praises*, Psalm 8.
[18] Psalm 139:14.
[19] Ecclesiastes 3:11.

rise to an idea of 'the divine' as a reality behind the traditional stories of the gods opened both the need and the possibility of a new kind of natural philosophy, the Socratic search for an underlying moral order provided a motive which energized the Platonic and Aristotelian science that came in its wake.

The Abrahamic vision of a God who could not be identified with the created realm presented an overwhelming and in the end irresistible challenge to the idea that the heavens were divine, which had effectively been fossilized in the ambiguous relationship between Greek philosophy and religion.

'The vast ontological presence that is the Hebrew God'[20] encountered by all who followed in Abraham's footsteps caused the pagan myths to be rewritten in such a way that they became signs and pointers to a new way of reading the world: a way of conceiving an order which lies beyond the horizon of human comprehension, but could nevertheless be traced (as it was in the centuries that followed) in everything from the furthest reaches of the stars and galaxies to the minute intricacies of Darwin's 'entangled bank'.

The discoveries of late-nineteenth-century explorers and scholars may have provided hard evidence of such rewriting, but they were not the first people to notice that the biblical writings pointed in a different direction to the pagan mythologies. The idea of Creation as a framing hypothesis, a rubric written over the whole physical universe, has been a thread which (though frequently becoming snagged on literalistic interpretations of Scripture and coercive attempts to police religious thinking) has run through Jewish, Islamic, and Christian thought for roughly two thousand years.

In the light of this long story it might seem less unexpected than it appeared at the outset to find religious invocations inscribed over the entrances of two buildings that stand at the threshold of the modern scientific world.

On closer inspection of the histories of these two buildings, however, it might still come as a surprise to discover the extent to which a strong slipstream of religious motivation was responsible for pulling individuals, professions, and even whole universities over that threshold.

---

[20] S. May, *Love* (2011), 255.

PART X

Through the Laboratory Door

CHAPTER FORTY–FIVE

Science in a Time of Cholera

The 1854 Oxford cholera epidemic began in the height of summer. On August 6th a butcher's wife in Walton Street showed the first symptoms of the disease. Ten hours later she was dead. The following week a charwoman from Gas Street and a prisoner in the county gaol were both showing symptoms, and on August 30th six new cases were reported from different areas through the city. By the end of October when the epidemic had run its course, there had been nearly two hundred cases of which more than half had died.

The epidemic in Oxford took place at the same time as the Soho epidemic in London, which over the same months saw more than 600 deaths. In Soho Dr John Snow was able to trace the source of the epidemic to a single infected pump in Broad Street. In Oxford a large part of the city's water supply was compromised. Most of it was taken out of the Isis and Cherwell downstream from the town's various sewage outlets, while the supply pipe for the county gaol sucked up water from a millstream branch of the Isis just ten feet from where the gaol's own drain discharged.[1]

Snow had published his first essay on 'The Mode of Communication of Cholera', in which he proposed the idea that cholera was a water-borne disease, in 1849. It had been ignored by most of his London colleagues who stuck to the traditional miasma theory, which held that the disease was spread through foul air. In Oxford it was not only scientific medicine that was ignored: science as a whole barely registered on the university's agenda.

When a young anatomy lecturer, Henry Acland, took up his post in 1845 he had found that '[t]he science studies of the University were for various causes almost extinct'. There were no laboratories, no apparatus, no books. 'I felt the work before me desperate and hopeless.... All physical science was discountenanced.'[2]

He had attempted to remedy the situation by proposing that the university should build a fully equipped centre exclusively

[1] J. Parfit, *The Health of a City* (1987), 21–32. From 1949 to 1984 the Parfit family owned and lived in the house where much of this book was written.

[2] J. B. Atlay, *Sir Henry Wentworth Acland* (1903), 133.

devoted to the sciences, but his proposal had fallen on deaf ears. Undaunted he had written 'hundreds and thousands of letters', and in 1849 a new proposal had been launched. Opinion in the university was sharply divided; the crucial vote for the new scheme was due to be taken on the 11th of December 1854. By then Acland had little time for campaigning. At the beginning of September he had been appointed as consulting physician to the Board of Health and found himself in charge of the city's attempts to combat the outbreak of cholera.

Dr Acland

Henry Wentworth Acland had been born into an evangelical family. His father, Sir Thomas, was an associate of William Wilberforce and the anti-slavery campaigners of the 'Clapham sect'. According to his biographer, Henry Acland never departed from 'the ... simple belief which he had learned under his father's roof'.[3] Attracted to science from a young age he had, from the outset, conceived this vocation in religious terms.

As a young man Acland had talked with the great experimental scientist Michael Faraday about what he might do in the future. Faraday after a long silence had said, 'that which I know best and anticipate most is that I shall go to be with Christ'.[4] Acland was so impressed that to the end of his life he kept these words written on the flyleaf of his Bible.

When he was about seventeen he wrote to his sister, 'you know my old idea of being a physician ... I am sure nothing is more calculated than this study to show us the frail nature and composition of man and the great and incomprehensible wisdom of God in His works'.[5] Like Charles Darwin, he at first found anatomical studies a distressing experience and had composed a prayer that he used daily on reaching the hospital, asking God to be with him: 'in these my fearful studies'.[6]

Acland's piety, like that of the Clapham sect, was expressed in practical action. When Shaftesbury and other evangelicals were campaigning against child labour, the Aclands befriended a little chimney sweep. Every Sunday they turned their rambling house in Broad Street into a shelter for the little sweeps from Oxford and the surrounding towns. As a doctor in the Oxford area he soon built up a huge practice among both rich and poor (it was said that 'no one of any respectability thought of dying before seeing Dr Acland'), often travelling up to 70 miles a day

[3] J. B. Atlay, *Sir Henry Wentworth Acland*, 499.
[4] J. B. Atlay, *Sir Henry Wentworth Acland*, 499.
[5] J. B. Atlay, *Sir Henry Wentworth Acland*, 28.
[6] J. B. Atlay, *Sir Henry Wentworth Acland*, 85.

on his rounds after finishing his work at the university. This superabundant energy, together with his contacts throughout the community, made Dr Acland a natural leader in the crisis the city was now facing.

A Memoir of the Cholera

Acland's organizational skills were soon in evidence. He divided the city into districts with a medical attendant responsible for each area, arranged messengers who could transport blankets, food, and medicines to where they were needed, supervised a temporary hospital and laundry on a field in the north of Jericho, and began recruiting Oxford ladies as volunteer nurses.

At the same time, like Snow in London, he documented the precise distribution and spread of the disease from day to day, publishing the results two years later as *Memoir on the Cholera at Oxford* with detailed maps and statistics. Unlike Snow he did not rule out the idea that cholera could be air-borne as well as water-borne, but it was evident to him that 'the application of combined observations to medicine' would in time 'establish ... the truth of some observations, and eliminate others to absurdity'.[7] The association between sanitary conditions and the spread of the disease was already evident and his text included maps of districts of the city left undrained and parts of the river still contaminated with sewage.

Snow, who died in 1859, had little time to act on his discovery, but Acland became a passionate sanitary campaigner; overseeing the complete reconstruction of the water supply and sewage in Oxford, serving on committees, and going up and down the country advising on drainage schemes.

The experience of the cholera epidemic had also brought home to him the need for professional nursing and the importance of hygiene in hospitals. When his wife died, 'The Sarah Acland Home for Nurses' (subsequently the Acland Hospital) was founded in her memory. Not long before his own death, Florence Nightingale wrote to Acland that whereas she could remember when pyaemia generated *in* hospitals was a common thing, 'Now we shout so loud when there is a case they can hear us all over London. This is an amazing change. And we owe it principally to you and what you have taught us.'[8]

Acland devoted much of his life to raising the standards of his profession (he went on to become President of the General

[7] H. W. Acland, *Memoir on the Cholera at Oxford* (1856), 72.

[8] J. B. Atlay, *Sir Henry Wentworth Acland*, 481.

Medical Council). His broader ambition was to promote scientific understanding as a whole, and this for him was ultimately a religious issue.

Our Local Barnum

In 1845 when Acland was first appointed as Dr Lee's Reader in Anatomy the so-called 'Oxford Movement', which sought to renew Catholic tradition within the Church of England, was at its height. It seemed to Acland that 'the intellect of the university was wholly given to ecclesiastical and theological questions'.[9] Having heard a rumour that Keble, Pusey, and the other leaders of the movement discouraged the study of the natural sciences because they encouraged arrogance, Acland went to call on Pusey, who was one of the Lee trustees.

He began by asking if the rumour was true, and was told that it was. Acland then asked whether this meant that the more he fulfilled the duties to which Pusey had appointed him, the more he would be held up by Pusey as a dangerous and mischievous member of society? The theologian had the grace to laugh. He would admit that the desire for knowledge and the ability to get it were both gifts of God and were to be used as such, and promised that 'while you discharge your duties in that spirit you may count on my assistance whenever you need it'.[10] This was to prove crucial in the events that followed.

The committee gathered to launch the 1849 museum proposal consisted of 20 members, all but 3 of whom were clergy. Acland had a gift for bringing people together (Benjamin Jowett the Master of Balliol called him 'our local Barnum'[11]) and this was soon expanded to a committee of 60, including the Bishop of Oxford and the professors of divinity and ecclesiastical history.

The vision for the building had also expanded. It would now 'include adequate room for the reception of zoological, geological, mineralogical, anatomical, and chemical collections . . . a series of apparatus for experimental philosophy . . . lecture rooms, laboratories for the use of the professors and students of these several departments of science . . . a general scientific library'.[12] The estimated cost stood at £30,000 and the opposition was intense.

In the end the vote on December 11th was 68 to 64 in favour. It was Pusey and the other Tractarians who turned the scale. Their commitment to the project was inspired by

[9] J. B. Atlay, *Sir Henry Wentworth Acland*, 133.
[10] J. B. Atlay, *Sir Henry Wentworth Acland*, 141.
[11] Although best known for his circus, in 1841 P. T. Barnum had opened 'Barnum's American Museum' on Broadway and St Anne's in Manhattan, which included stuffed animals and models of cities as well as hot air balloon rides from the strolling garden on the roof.
[12] J. B. Atlay, *Sir Henry Wentworth Acland*, 201.

Acland's insistence on its religious purpose. A collect written for the laying of the foundation stone in 1855 asked God to grant that the building to be erected on that spot might 'foster the progress of those sciences which reveal to us the wonders of thy creative powers'.

The laying of the foundation stone did not, however, end the opposition, which now concentrated on starving the project of resources.

Nisi Dominus

This Babylon of a new museum is before us
Have we any students in Natural History? No
Do we require this new museum? No
Have we one farthing wherewith to build? No
Have we one farthing wherewith to endow? No
Are these the times for setting about such a folly? No

So read one of the election leaflets which appeared whenever monies needed to be voted to the project. Two designs had been proposed for the museum, one in a classical style and one in 'Rhenish gothic'. Acland favoured the latter, referred to by the motto *Nisi Dominus* (from Psalm 127:1 'Unless the LORD build the house the labourers labour in vain').

A skilled amateur artist who took painting lessons from Samuel Palmer, Acland was a close friend of the art critic John Ruskin (it was he who suggested the idea for Millais' famous portrait of Ruskin which hung for many years in Acland's own house in the Broad). When by a narrow margin the convocation voted for the gothic design, Ruskin was jubilant. As he wrote in a letter to Acland,

I hope to be able to get Millais and Rossetti to design flower and beast borders—crocodiles and various vermin— such as you are particularly fond of—Mrs Buckland's 'dabby things'– and we will carve them and inlay them with Cornish serpentine all about your windows—I will pay for a good deal myself, and I doubt not to find funds. *Such* capitals as we will have! [Fig. 45.1][13]

The intention was that all the decoration 'should illustrate the Kosmos, as religious histories or allusions for the most part are represented in ecclesiastical edifices'.[14]

[13] J. B. Atlay, *Sir Henry Wentworth Acland*, 241.
[14] J. B. Atlay, *Sir Henry Wentworth Acland*, 217.

45.1 The Oxford University Museum under construction.

In the event when Rossetti came up to Oxford he fell in love with the New Union building and the pre-Raphaelites got diverted there. Ruskin though took an interest in every detail of the museum, even building with his own hands one of the brick columns in the interior (it was later demolished and rebuilt more securely by a professional bricklayer).

The university's commitment to the gothic ideal was rather less wholehearted, particularly when the costs began to escalate. The first version of the ambitious and technically innovative roof of the central court, constructed as Acland put it from 'railway materials—iron and glass', proved to have been built with pillars too narrow to support its weight and the whole thing had to be taken down and redesigned, at an extra cost of £5000 [Fig. 45.2].

A later rumpus involved two red-bearded stonemasons from Ballyhooly, County Cork. James and John O'Shea had been brought over to carve the capitals of every pillar into plants representing the different botanical orders. In 1859 when funds began to run out, O'Shea had begged to be allowed to continue 'for the sake of art alone', but the university refused, and

45.2 The glass roof of the Oxford University Museum.

after a contretemps in which he had first been accused of destroying university property by carvings cats around a window [Fig. 45.3], and subsequently dismissed, Acland found him carving parrots and owls round the porch to symbolize members of convocation. He made the Irishman stop—leaving a stumpy trail that can still be seen today [Fig. 45.4].[15]

Although the museum was incomplete and the university had, as Ruskin put it, 'carved on its façade the image of her parsimony', it powerfully expressed the ambition to unite art, science, and religion in a single enterprise.

At their last meeting as very old men, Ruskin asked Acland to carry back his benediction: 'Say to my friends at the Oxford Museum from me "May God bless the reverend and earnest study of nature and of man to his glory and the good of all mankind" ' [Fig. 45.5].[16] The popular legend of what happened

[15] The accuracy of Acland's account has been questioned by B. J. Gilbert, 'Puncturing an Oxford Myth', http://oxoniensia.org/volumes/2009/gilbert.pdf, accessed 22 August 2015. The chronology of the events does seem to have become confused but the suggestion that Acland made up the whole story is unconvincing.

[16] J. B. Atlay, *Sir Henry Wentworth Acland*, 476.

45.3 James O'Shea carving the windows of the Oxford University Museum.

45.4 Mutilated sculptures on the doors of the Oxford University Museum.

45.5 Ruskin and Acland.

at the museum suggests that this pious ideal had been under-mined even before the building had been completed. The reality was more complicated. It would be truer to say that it represented a new phase of an age-long integrative struggle: a struggle at the centre of which Acland now found himself.

CHAPTER FORTY–SIX

A Visit to the Museum

In late June of 1860 the British Association for the Advancement of Science (BA) held its third meeting in Oxford. Acland was the local secretary of the association and although the university museum was still unfurnished, he arranged for all the events to be held there.

On the afternoon of Saturday, June 30th, Dr John William Draper of New York University was scheduled to deliver his paper and it was understood that the Bishop of Oxford Samuel Wilberforce would also speak. This was the first time that *On the Origin of Species* had been debated in a public forum and interest was intense. Feeling the stress of the moment in his stomach, Darwin had been admitted to Dr Lane's Hydropathic Clinic, but his great supporter T. H. Huxley decided at the last moment that he would attend. The crowd that gathered was so large (one author has put it at more than 1000) that the lecture theatre was too small and everyone adjourned to the still bookless library.

The nature of the debate that followed has itself become a subject of debate.

Writing nearly 40 years later in *Macmillan's Magazine*, Isabella Sidgwick recalled how, after Dr Draper's 'somewhat dry' address,

> the Bishop rose, and in a light scoffing tone, florid and fluent . . . assured us there was nothing in the idea of evolution . . . Then turning to his antagonist with a smiling insolence, he begged to know, was it through his grandfather or his grandmother that he claimed descent from a monkey. On this, Mr. Huxley slowly and deliberately arose . . . and spoke those tremendous words, words which no one seems sure of now. . . . He was not ashamed to have a monkey for an ancestor, but he would be ashamed to be connected with a man who used great gifts to obscure the truth. The effect was tremendous. One lady fainted and had to be carried out: I for one jumped out of my seat.[1]

[1] 'A Grandmother's Tales', *Macmillan's Magazine, 1859–1907* 78 (1898), 425–35.

According to a letter written by John Richard Green, how-ever, just four days after the event, Wilberforce's joke had arisen from some earlier remark of Huxley's. He has the Bishop say-ing, '"he had been told that Professor Huxley had said that he didn't see it mattered much to a man whether his grandfather were an ape or no! Let the learned Professor speak for himself" and the like'. Green then has Huxley replying to the Bishop, 'If there were an ancestor whom I should feel shame in recall-ing, it would be a *man*, a man of restless and versatile intellect, who, not content with an equivocal success in his own sphere of activity, plunges into scientific questions with which he has no real acquaintance'.

Five weeks before the Oxford meeting Wilberforce had written a long review on *On the Origin of Species* for *The Quar-terly Review*, addressing genuine scientific questions raised by Darwin's book and arguing that 'To oppose facts in the nat-ural world because they seem to oppose Revelation, is . . . but another form of . . . lying for God'. When it was published in July, immediately after the Oxford debate, Darwin wrote to his friend Joseph Hooker that 'It is uncommonly clever; it picks out with skill all the most conjectural parts and brings forward well all the difficulties. It quizzes me splendidly'.[2]

Only in the last paragraphs of his review did the bishop move beyond science. Pointing out that 'Mr. Darwin writes as a Christian', he concluded by asking how far Darwin's the-ory was compatible with Christianity, and asserts that the bib-lical doctrine of man, his fall and redemption, the incarnation, the gift of the spirit—'all are equally and utterly irreconcilable with the degrading notion of the brute origin of him who was created in the image of God'.[3]

Many of the clergy present at the Oxford meeting seem to have strongly disagreed with this last idea (which Wilberforce seems to have repeated). Hooker (who had also spoken) re-ported that in coming out of the debate he had been 'con-gratulated and thanked' by numerous Oxford clergymen.[4] The next morning when Frederick Temple (a former pupil of Wil-berforce and a future Archbishop of Canterbury) got up in St Mary's to preach the university sermon on the subject of 'The Present Relations of Science to Religion', he argued that as science discovers that aspects of nature once thought to be mysterious were actually governed by laws 'we must look for the finger of God' in the laws of nature themselves [Fig. 46.1].

[2] C. R. Darwin, *The Life and Letters of Charles Darwin* (1887), vol. II, 324–5.
[3] S. Wilberforce, 'Review of Darwin's *On the Origin of Species*', *The Quarterly Review* (Jul. 1860), 225–64.
[4] L. Huxley, *Life and Letters of Sir Joseph Dalton Hooker* (1918), 527.

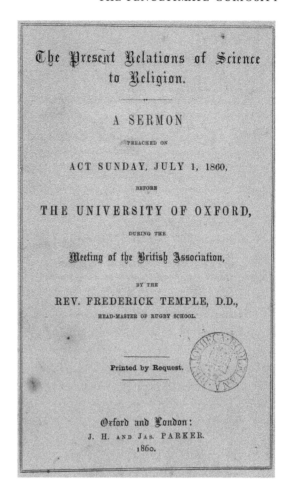

46.1 Frederick Temple's sermon on The Present Relations of Science to Religion (1860).

Worshippers coming out of church concluded that he had 'espoused Darwin's ideas fully'.[5]

Acland's concern, however, was the damage that this kind of dispute caused to the reputation of science, and from his perspective the real problem had begun earlier in the week.

The Eirenicon

At one of the so-called 'section D' meetings on Thursday June 28th, the great anatomist Richard Owen had caused consternation among his colleagues by attempting to show that there were structures of the brain in man (in particular the hippocampus minor) different to that in all other animals. This idea which Owen thought would enable people to 'come to some

[5] J. H. Brooke, *Science and Religion* (1991), 274.

conclusions . . . of the truth of Mr. Darwin's theory' had already been discredited, and Huxley, who was present, had not been slow to point this out.

Owen had been staying with the Bishop at Cuddesdon that weekend, and in a letter to the Archbishop of Canterbury written in 1862, Acland argued that it had been this issue that had led to the dispute 'in which the Bishop of Oxford charged Professor Huxley to the affect that his assertions unwarranted by facts had an irreligious tendency'.[6] Acland, according to his biographer, 'felt a deep repugnance at anatomical and technical matters being discussed in such a temper before a mixed audience'. The only result, he believed, would be to destroy people's confidence in the 'records of science' and 'the calmness and candour of scientific men'.[7]

Two years later, at the BA meeting in Cambridge, there was an even more bitter exchange between Owen and Huxley, when Huxley dissected in front of an audience a series of specimens which demonstrated that all of Owen's unique structures, including the hippocampus minor, could be found in other primates. Owen, however, had refused to concede.

Acland said nothing in the meeting, but the next day wrote a long letter to Owen, trying to help him to extricate himself from an evidently false position. He noted that he had not actually heard Owen repeat his claim that the hippocampus was peculiar to man, although that was the impression people had come away with, and asked therefore whether Owen would tell him, and allow him to publish 'in what respect Mr. Huxley has misunderstood your opinion, or misinterpreted your words', and in what way 'the discussion of these two years' had 'modified your original views'.

Acland thought that the argument played into the hands of narrow-minded people who were suspicious of science when there was nothing worth fighting about. The real question was one of technical zoology which 'the public confound . . . in a misty manner with the essential nature of man'. 'Whatever views or hypotheses, or guesses Mr. Huxley, or you, or Mr. Darwin or the Bishop of Oxford may have as to the origin of man', Acland argued, 'you are all agreed that however he so became, he is in some manner made in the image of God—a spiritual being', the only creature on Earth that worshipped God. 'However near he be to other material natures, nothing can change, nor anything but himself degrade his higher nature'.[8]

[6] J. B. Atlay, *Sir Henry Wentworth Acland* (1903), 306.
[7] J. B. Atlay, *Sir Henry Wentworth Acland*, 303.
[8] J. B. Atlay, *Sir Henry Wentworth Acland*, 305.

Owen responded positively but in the end it proved impossible to agree terms for publishing an 'Eirenicon'. It was at this point that Acland had written to the Archbishop of Canterbury.

Acland begged Archbishop Longley not to get involved in the controversy. Either Owen was right in his facts or Huxley was right in his. The Bishop of Oxford had 'unintentionally given an unfortunate turn to the debate by resting sentiments dear to all men on at least questionable facts'. Owen had 'accidentally mis-stated certain differences upon which afterwards great issues were supposed to hang, and now he does not like to retract'. But despite all the fuss there was nothing of a religious kind at issue.

'Suppose no difference of a material kind could be found to exist between Man and Apes, should we be brutes? Or would brutes be men?' He implored the archbishop to discourage the clergy from taking sides in scientific disputes. What was to be feared? 'Scientific inquiry, ever shifting, can attain only to what? To a further knowledge of the facts which he ordained who made the world and us with them'.[9]

The furore over this issue seems to have persuaded Acland that some public statement of scientific principles might be useful. The Anglican clergy were not the only ones muddying the scientific waters. In 1853 Auguste Comte's massive six-volume *Positive Philosophy*, first published between 1830 and 1842, had appeared in English—'freely translated and condensed by Harriet Martineau'.

The central thesis in Comte's great work was that the only authentic kind of knowledge was the scientific knowledge which came through a positive affirmation of theories by strict scientific method. The human race, having passed through a 'theological' or 'fictitious' phase and a 'metaphysical' phase, was now in the 'scientific' or 'positive' phase, in which science was the guiding light and supernatural beliefs must be discarded.

Comte's ideas were becoming increasingly fashionable. When an opportunity arose to put a different point of view Acland took it.

The Harveian

In 1865 Acland had been invited to give the Harveian Oration to the Royal College of Physicians. This lecture, in memory of William Harvey, the discoverer of the circulation of the blood,

9 J. B. Atlay, *Sir Henry Wentworth Acland*, 307.

had traditionally been delivered in Latin on a medical topic to a naturally somewhat restricted audience. Acland chose to break with tradition by using the opportunity to launch a polite but powerful critique of the French philosopher, and to give his lecture in English before an invited audience which included the Prince of Wales (to whom he had served as personal physician) and William Gladstone, the Prime Minister.

Harvey was significant for Acland because like most of the seventeenth-century pioneers of modern science, he believed in what were called 'final causes'. This was a term which went back to the philosophy of Aristotle, who had distinguished four kinds of causation: material, formal, efficient, and final. If you boil a kettle, the metal that the kettle is made from and the water inside it would, in Aristotle's scheme, be the material causes of what happens. The laws which determine how electricity heats the element and the element heats the water would be the formal causes. You would be the efficient cause of the kettle boiling, while the final cause might be your intention to make a cup of tea.

In respect of science then, a belief in final causes was 'the conviction that every arrangement in the Natural World is the result of *Design*—that every effect is *intended and has a Purpose*'.[10] William Whewell, the great Cambridge historian of science, had argued that this conviction was a critical ingredient in Harvey's work. Comte, on the other hand, had argued that modern science abolished the idea that there was any kind of purpose in nature. Who was right?

Acland begins by considering Comte's argument that a belief that there is a purpose in nature gives rise to a kind of religious wonder that gets in the way of science. Focussing on 'the pretended wisdom of nature' prevents us, according to Comte, from noticing nature's flaws and defects. Comte's central example of what he describes as a 'puerile affectation' was people who admired what, in his view, was 'the fundamental uselessness' of the lens in the eye. This was not a good example, because, as Acland explained to his audience, the lens is far from useless (Comte might have done better to have cited vestigial organs like the appendix or the male nipple). After the lecture Acland underlined the point by inviting the Prince of Wales, Mr Gladstone, and other luminaries into a small darkened room to examine each other's eyes with Hermann von Helmholtz's ophthalmometer.

[10] H. W. Acland, *The Harveian Oration* (1865), 4.

Acland does not deny that disease and 'physical evil' can seem to challenge the idea that there is any purpose behind nature, but he replies to Comte's specific point by quoting Darwin's account of the evolution of the eye. This is full of a wonder that 'in no way derogates from the splendour of the instrument, but only attempts to account for the mode of its construction'. Admiring nature and studying it are, Acland argues, in no way contradictory. Darwin himself had written that the more he studied nature, the more he was impressed with the conclusion that 'the contrivances and beautiful adaptations . . . transcend in an incomparable degree the contrivances and adaptations which the most fertile imaginations of the most imaginative man could suggest with unlimited time at his disposal'.[11]

Acland goes on to point out that there was a fundamental non sequitur at the heart of Comte's argument. He had no difficulty in agreeing with Comte that final causes cannot be used as a means of scientific investigation. For one thing we do not know 'the intermediate steps by which they operate'; for another, 'there may be several final causes for the same condition'.[12] Your intention in boiling a kettle might be to make a cup of coffee as well as a cup of tea. This was simply restating the view of seventeenth-century pioneers of science like Francis Bacon who described introducing final causes in science ('intelligent design' and 'dumb design' would both come into this category) as 'impertinent'.

Yet just because we were not able to scientifically investigate God's purposes, it did not follow, as Bacon had equally insisted, that there were no such purposes—that there were no ultimate reasons why things exist. Nor did it follow that such metaphysically central and existentially crucial questions were in any way illegitimate. In fact they could act as drivers that shaped the whole way we see the world and act in it.

The contrast between Harvey and Comte, he concludes, was striking. Harvey 'was not guided in his discovery of the Circulation of the Blood by any metaphysical speculation or religious dogma whatever'[13]—he relied on observation, reasoning, and experiment. But while Comte despising religious wonder manages, 'notwithstanding his knowledge and genius', to see 'fundamental uselessness' where there is in fact important function, Harvey '[b]elieving . . . that there is purpose as well as harmony in the material world . . . acted in this faith'[14] and in doing so laid the basis of scientific medicine.

[11] H. W. Acland, *The Harveian Oration*, 40.
[12] H. W. Acland, *The Harveian Oration*, 6.
[13] H. W. Acland, *The Harveian Oration*, 53.
[14] H. W. Acland, *The Harveian Oration*, 72.

The Theologian and the Scientist

Acland made no further public statements about religion and science, but in private continued to warn against confusing the two. In 1890 he began receiving letters from Gladstone (for the moment out of office) who was preparing a book entitled *The Impregnable Rock of Holy Scripture*. Acland tried to persuade the ex-premier against trying to make Genesis 'square' with science in 'its details'. It seemed to be true that 'the Light and Sun came: the earth cooled, the vegetables began to come, and grew; the animals began, and evolved; and in the end the blessed boon of Faith and Prayer and Reason came', but it was unwise to go beyond this. The science of today would not be the science of tomorrow and 'No one supposes that the present state of scriptural interpretation or of natural science is final'.[15]

Acland, like Temple, had 'espoused Darwin's ideas fully'. Twenty years earlier in 1870 there had been a proposal to confer an honorary Oxford degree on Darwin, which Pusey had opposed. Acland had had a long interview with his theologian friend. Eventually, once again, he had managed to win Pusey round. In a subsequent letter to him he pointed out that there was not 'a single irreverent passage or uncharitable passage in Darwin's writings'. It was his 'exceeding eminence and his character as a working man that justify and required me to beg you to pause before bringing about his rejection here'. Men, he argued, must be allowed to 'state without fear the facts they discover'. The student of nature 'takes the universe as he finds it. He learns what scraps of it he can; and if morally tender, like Harvey, Newton or Hunter he seeks and adores'.[16] There was no reason why this should compromise religious faith.

Acland sent copies of his Harveian lecture to many of his acquaintances, including agnostic scientists and philosophers like John Tyndall and Herbert Spencer. The warmest response came from Darwin, who recalled 'with much pleasure our short acquaintance at Oxford'. He went on to confess to Acland 'the hopeless confusion of mind' that he had been left in after corresponding about divine purposes in nature with his friend the American botanist Asa Gray. On the one hand, Darwin felt that 'it grates against common sense to look at this world with all its inhabitants as originating without express design', but on the other hand, 'I cannot believe that any one structure is expressly designed in the common meaning of the word'. As a result, he

[15] J. B. Atlay, *Sir Henry Wentworth Acland*, 485–6.
[16] J. B. Atlay, *Sir Henry Wentworth Acland*, 348.

concludes, 'looking at the subject from two opposite points of view, I am driven to two opposite conclusions'. Writing about the overall nature of science he says, 'I believe we entirely agree that purpose or design is one of the surest & simplest roads to discovery in Natural History.'[17]

This larger picture was crucial for Acland. The kind of questions posed by the scientist and the theologian were very different, but, he argued, 'they are correlative one of the other, and together make the sum of that portion of human experience by which man strives to work his way in the labyrinth of his present state'.[18] The empirical methods of the early pioneers of science like William Harvey had been inspired by the belief that there was indeed a divine order in the universe and it seemed to Acland that 'the idea of comprehensive plan or unity of design . . . in the works of nature' was one 'with which, in some form or other, the philosophic observer . . . cannot safely dispense, without sacrificing all hope of attaining any conception at all of nature as a whole'.[19]

In suggesting this, Acland was writing as a student of science. For all his contributions to scientific progress he was not himself an original scientist and had not experienced the discovery of this kind of unity at first hand. The same could not be said for his counterpart in Cambridge who had been responsible for the design of the Cavendish Laboratory.

[17] C. R. Darwin, Letter to H. W. Acland, 8 December (1865) Bodleian, MS Acland d 81:63.
[18] H. W. Acland, *The Harveian Oration*, 75.
[19] H. W. Acland, *The Harveian Oration*, 12–13.

CHAPTER FORTY–SEVEN

Experiments of Thought

On 1 November 1879 the Revd William Guillemard, the rector of the Cambridge church of St Mary the Less, called at a house in Scroope Terrace not far from the Cavendish Laboratory. He had come to bring communion to a bed-ridden parishioner who at the age of 48 was dying from advanced abdominal cancer. He was told that the patient had had a bad night (he was to die a few days later) and he did not expect there to be much conversation. As the rector robed himself (Father Guillemard followed all the rituals of the Oxford Movement) he was astonished to hear from the bed a Scottish voice, with the trace of a Galloway accent, reciting from memory all five verses of George Herbert's poem about the robing of Aaron:

> Holiness on the head,
> Light and perfection on the breast ...
> My doctrine tun'd by Christ, (who is not dead,
> But lives in me while I do rest)
> Come people; Aaron's dressed.

James Clerk Maxwell, the man in the bed, had been surprising people all his life.

When as a little boy, with a then strong Galloway accent, he had first appeared at the Edinburgh Academy dressed in a tweed tunic, frilly collar, and square-toed shoes (all designed by his father), his classmates had nicknamed him 'dafty'. Their opinion was confounded when at the age of 14 he had produced a scientific paper that was read to the Royal Society of Edinburgh.

A few years later it was followed up by a second paper: 'On the Equilibrium of Elastic Solids'. The starting point of this was an experiment carried out in an improvised laboratory above the washhouse at home, where he had shone a beam of polarized light into a twisted cylinder of gelatin. The then 18-year-old Maxwell had discovered that as the light shone through the stressed jelly, the strain patterns became visible to the naked eye, and he had thereby invented a technique that would be widely used by engineers until eventually being superseded by computer modelling.

The constant stream of original discoveries that flowed from him over the next 30 years ranged over an astonishing variety of fields. In some instances he initiated whole new subjects: writing the founding papers on what became the science of cybernetics and the science of global analysis. In others he transformed existing fields out of all recognition. Even the greatest of his contemporaries sometimes failed to understand the significance of his ideas. Unlike the experiment that confirmed Einstein's theory of general relativity which was performed during Einstein's own lifetime and resulted in immediate fame, the experiment by which Heinrich Hertz proved the existence of electromagnetic waves, which Maxwell's most important theory had predicted, took place eight years after Maxwell's early demise. In consequence it was others like Hertz who enjoyed the practical implementation of Maxwell's predictions, and those like Marconi who exploited them technologically and commercially.

This time lag of understanding, together with his utter lack of interest in self-promotion, meant that Maxwell never achieved the kind of celebrity accorded to Newton or Darwin (both of whom were buried in Westminster Abbey). Yet while the significance of what he had done may have been obscure to his contemporaries at the time, and has largely remained unrecognized by the general public, it was not lost on his successors.

Twentieth-century physicists have spoken of this discovery 'as one of the greatest leaps ever achieved in human thought' and described him as among 'the most penetrating intellects of all time'.[1] To Albert Einstein (whose own theories, as he acknowledged, were based on Maxwell's) it seemed that 'one scientific epoch ended and another began with the work of James Clerk Maxwell'.[2] The great American physicist Richard Feynman wrote that 'from a long view of the history of mankind—seen from, say, ten thousand years from now—there can be little doubt that the most significant event of the nineteenth century will be judged as Maxwell's discovery of the laws of electrodynamics'.[3]

On Saturday November 1st, when Father Guillemard called, he may have been only dimly aware of the intellectual distinction of his parishioner, though he was certainly impressed that

[1] B. Mahon, *The Man Who Changed Everything* (2003), 17.
[2] B. Mahon, *The Man Who Changed Everything*, 1.
[3] R. P. Feynman, R. P. Leighton, and M. Sands, *The Feynman Lectures on Physics* (1965), II–1–8.

Maxwell seemed to know the whole Bible by heart. What had struck him was the depth of faith that he encountered. He told Maxwell's friend and biographer Lewis Campbell that Maxwell 'had gauged and fathomed all the schemes and systems of philosophy and found them utterly empty and unsatisfying, "unworkable" was his own word about them—and he turned with a simple faith to the Gospel of the Saviour'.[4]

Victorian deathbed scenes need to be treated with caution, but there is an element here that rings true. Maxwell (like Darwin) had always been conscious of the limitations of human knowledge, but (unlike him) never became resigned to ignorance or agnosticism. Even where absolute knowledge was impossible, the empirical criterion of 'workability' could, in both science and religion, provide a means of avoiding error and advancing towards truth.

The Rings of Saturn

As a small child James Clerk Maxwell had constantly interrogated his parents about the world around him. If the answer to the question 'what's the go o' that?' was unsatisfactory, it would be followed up by a supplementary 'but what's the *particular* go of it?' Told about a blue stone he would ask, 'but how d'ye *know* it's blue?' One means of finding answers had been mathematics, and as soon as he discovered geometry he had begun writing equations to describe geometrical propositions. These were not always right. In later life he told a friend that 'I am quite capable of writing a fancy formula' (meaning a wrong one). His deepest interest was in finding formulas that not only worked as pure mathematics but accurately described phenomena in the real world, and at the age of 14 he had found one that did.

Starting from the schoolboy exercise of first drawing a circle with a pin stuck in a piece of paper tied to a pencil with a piece of string, and then drawing an ellipse by enclosing two pins and a pencil in a taut loop of string, Maxwell had invented a variation. He had first tied the string between the pin and the pencil and then looped it round the other pin before pushing the pencil against the string to produce a new curve. It was at that point that he began writing the series of equations for multifocal ovals which formed the basis for the paper that was read to the Royal Society of Edinburgh, and turned out to have a practical application in optics [Fig. 47.1].[5]

[4] L. Campbell, *The Life of James Clerk Maxwell* (1882), 389.
[5] J. C. Maxwell, *The Scientific Letters and Papers of James Clerk Maxwell* (1990).

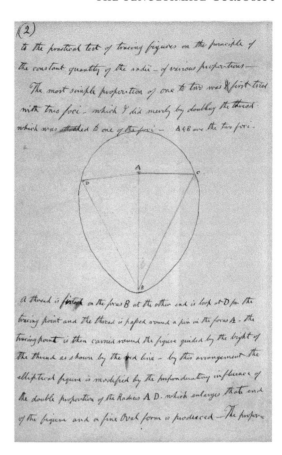

47.1 James Clerk Maxwell's figure showing how to draw an oval.

Ten years later this mathematical aptitude produced what the Astronomer Royal described as 'one of the most remarkable applications of mathematics to physics that I have ever seen'. His starting point this time had been a problem set for the St John's College, Cambridge, Adams prize (established to commemorate John Couch Adams' prediction of the existence of the planet Neptune by mathematical calculation). The question set had concerned the mysterious rings of Saturn. Under what conditions would they be stable if they were (a) solid, (b) fluid, or (c) composed of many pieces?

In an extraordinary series of calculations which took him two years to complete, Maxwell was able to show that neither solid nor fluid arrangements would be stable, and that the rings by elimination must consist of 'flights of brickbats', a conclusion confirmed by the photographs taken by *Voyagers 1* and *2* in the 1980s.

While these calculations were dramatic examples of the power of mathematics to provide knowledge about otherwise inaccessible phenomena, Maxwell never supposed that calculation was sufficient in itself. In his inaugural lecture at Marischal College, Aberdeen (his first teaching post), he made it clear that he had 'no reason to believe that the human intellect is able to weave a system of physics out of its own resources without experimental labour'. Whenever the attempt had been made, he argued, 'it has resulted in an unnatural and self contradictory mass of rubbish'.[6] Maxwell's childhood question for instance, 'but how d'ye *know* it's blue?', could not be answered by pure mathematics. Indeed the need for experimentation was nowhere more evident than in the vexed question of the perception of colour.

The Colour Box

Ever since Ibn al-Haytham had written his *Book of Optics* it had been apparent that the understanding of sight must involve both the physics of light and the biology of the human eye. Maxwell's first step, therefore, had been to design and build the world's second ophthalmometer (the previous year Hermann von Helmholtz had independently constructed the first: a prototype of the instrument with which Acland later encouraged Gladstone and the Prince of Wales to examine each other's eyes). This gave interesting insights but didn't itself reveal the process of colour perception.

While Newton had shown that white light was a mixture of all the colours in the rainbow, there were some colours like brown which didn't appear in the solar spectrum. These Newton supposed must arise from a mixing of the spectral colours, but how did this process work?

A hundred thousand years before Newton, the painter in the Blombos Cave had made orange from mixing red and yellow pigments. Her artistic successors had found that apart from black and white, any colour could be made from mixing three primary colours: red, yellow, and cyan. It was on the basis of this artistic practice that Thomas Young, a nineteenth-century English doctor, suggested that the human eye might have three corresponding colour receptors which could combine to produce a single perceived colour in the mind.

One of Maxwell's teachers at Edinburgh University, James Forbes, had tried to test this idea by spinning a disc marked

[6] B. Mahon, *The Man Who Changed Everything*, 70.

with different colours so fast that they would appear to blur into a single colour. The results were puzzling. When he spun a disc that was half-yellow and half-blue the colour that appeared was not green but pink.

Forbes was forced by ill health to give up experimentation, but Maxwell continued his work and found the answer to Forbes' puzzle. A coloured pigment like ultramarine blue absorbs all the light of the spectrum except the blue light, which it reflects back. Thus, when artists mix their different light-absorbing pigments together, they are creating a *subtractive* process in which the eye only receives whatever rays from the spectrum have not been absorbed. When coloured light is mixed by a spinning disc or a prism, this is an *additive* process in which the wavelengths recombine towards white (a painter who starts working with stained glass soon discovers these complexities).

Using first an improved version of Forbes' spinning disc [Fig. 47.2], and later prisms in a large coffin-shaped colour box (which alarmed their neighbours when James and his wife, Katherine, placed it beside an attic window to test their finds), Maxwell was able to show that the three kinds of colour receptor in the eye are not red, yellow, and cyan, but red, green, and blue.

In 1861 when Maxwell was invited to give a Royal Institution talk about colour vision, he devised a way of demonstrating this to a large audience. First, photographing a tartan ribbon with red, green, and blue filters, he then projected all three photographs simultaneously onto a screen. The result (though some luck was involved) was the first ever colour photograph. Maxwell went on to construct a geometrical method of representing the relationship between colours—the Maxwell colour triangle—which forms the standard chart for colour reproduction and digital colour images that is still used today.

For all their ingenuity, the mathematical methods that Maxwell had employed here and elsewhere were basically conventional ones. In the late 1850s, however, he started using a technique that enabled him to describe phenomena that had hitherto seemed to be beyond the scope of human knowledge.

47.2 James Clerk Maxwell with his colour top.

The Demon in the Gas

In 1859 Maxwell read a paper 'On the Mean Length of the Paths Described by the Separate Molecules of Gaseous Bodies' by the

German physicist Rudolf Clausius, which raised in his mind a formidable problem. The kinetic theory of gases—which proposed that gases consisted of huge numbers of molecules moving in all directions—offered a good explanation of the laws of pressure, volume, and temperature, but seemed unfeasible to be described mathematically. Newton's laws of motion might tell you the answer in principle but it was impossible to calculate the position of each molecule.

Nine years earlier, however, Maxwell had read an article by John Herschel in the *Edinburgh Review* describing the work of the Belgian statistician Adolphe Quetelet. Using similar techniques Maxwell was able to show that a gas would settle down to a statistical scatter of velocities—now known as the Maxwell distribution. This was the first ever statistical law in physics (the model for future developments of statistical and quantum mechanics), and on the basis of this new law Maxwell was able to make the startling prediction that the viscosity of a gas should be independent of its pressure. This seemed to run in the face of common sense, but a few years later James and his wife, Katherine, in their Kensington attic, were able to perform the experiment that proved it was correct. The temperature was controlled by Katherine stoking a fire in the room.

It was 11 years later that he developed some of these ideas in a book on *Theory of Heat*. In this he set out what have become known as 'the Maxwell relations' (differential equations expressing the relationships between pressure, volume, temperature, and entropy) and introduced the public to what his friend William Thomson described as 'Maxwell's demon'.

Suppose, he suggested, a tiny creature ('the demon') were to operate a shutter separating two compartments of gas that were both at the same temperature. Whenever the demon saw a fast-moving molecule approaching in compartment A he would let it through into compartment B, and whenever he saw a slow-moving molecule approaching in compartment B he would let it through into compartment A. The heat—in the rapidly moving molecules—would gradually flow from compartment A as it got colder to compartment B as it got warmer, thus contradicting the second law of thermodynamics (which states that heat cannot flow from a colder to a hotter body). In doing so it would create a perpetual motion machine. Why, he asked, wouldn't this work?

'Maxwell's demon' baffled physicists for 60 years.[7] It was a textbook example of what some years previously the Danish physicist Hans Christian Ørsted had described as a 'Gedankenexperiment', or 'thought experiment'. Trying to solve the puzzle it posed was to lead to a deeper understanding of the statistical nature of the second law of thermodynamics, and to the development of information theory, in something of the same way that Einstein's thought experiment of riding on a light beam led to the development of relativity theory.

Statistical laws and thought experiments could in these different ways both provide workable methods of advancing towards truth even where absolute knowledge was impossible. It was, however, Maxwell's use of mathematics to describe his deep intuition of the order and unity of the universe that was most dramatically to expand the reach of human understanding.

[7] O. Maroney, 'Information Processing and Thermodynamic Entropy', in *The Stanford Encyclopedia of Philosophy* (Fall 2009 edn), ed. E. N. Zalta, http://plato.stanford.edu/archives/fall2009/entries/information-entropy/. Accessed 27 August 2015.

CHAPTER FORTY-EIGHT

The Unity of Nature

In 1820 while lecturing to his students at Copenhagen, Hans Christian Ørsted, the same Danish physicist who had coined the term 'gedankenexperiment', switched on an electric current. As he did so he noticed that the needle of a magnetic compass that happened to be lying on the bench beside him suddenly jerked. Some months afterwards he published the results of a series of experiments undertaken in the wake of this observation, which showed that electric current produces a circular magnetic field as it flows through a wire.

Eleven years later, working in his laboratory at the Royal Institution in Albemarle Street, the British scientist Michael Faraday tried another experiment. Reasoning that if electricity could induce magnetism in this way the converse might also be true, he moved a magnet in and out of a coil of wire and succeeded in demonstrating that magnetism could indeed induce an electric current.

The vast potential of these discoveries began to unfold when Faraday first used them to build the world's first dynamo, the forerunner of all electric power generators, and then by reversing the process back to Ørsted's observation, to construct the first ever electric motor. The manifold uses of such devices took time to emerge. The immediate question that faced scientists was the nature of the relationship between magnetism and electricity.

Lines of Force

When Acland referred to the impossibility of dispensing with 'the idea of comprehensive plan or unity of design . . . in the works of nature . . . without sacrificing all hope of attaining any conception at all of nature as a whole'[1], he might well have had Michael Faraday in mind. Faraday was a member of what he himself described as 'a very small and despised sect of Christians, known, if known at all, as *Sandemanians*',[2] and this

[1] H. W. Acland, *The Harveian Oration*, 12–13.
[2] Faraday, Letter to Ada Lovelace, in H. B. Jones, *The Life and Letters of Faraday* (1870), vol. ii, pp. 195–6.

religious commitment, as Acland had discovered as a young man, was at the centre of everything Faraday did.

For Newton the idea that the laws of motion might have a universal extent was a natural inference from the idea of a universal being. For Faraday the theological overtones of what he called 'unity in one' seems in a similar way to have shaped his inferences about the relationships between the forces of nature (so that having established the connection between the electrical and magnetic forces, he continued in old age to struggle unsuccessfully to find a unity between these forces and that of gravity).

Faraday thought of electricity and magnetism as different forms of the same force exerted at right angles to each other, so that while mathematical physicists (like the French scientists Poisson and Ampère) were content to talk in Newtonian terms of 'action at distance', he had developed a theory of how these two forces worked together. Both magnets and electric charges, he concluded, were surrounded by regions of influence—what he described as 'lines of force', which he thought could actually be seen in the distinctive pattern that iron filings make when scattered on a piece of cardboard placed over a magnet.

Faraday had no mathematical training (his father was a blacksmith and he had been an assistant bookbinder before being taken on as an assistant by Humphry Davy). His theoretical ideas tended for that reason to be discounted by most physicists. After Maxwell had graduated from Cambridge, however, it was Faraday's ideas that began to interest him.

'Before I began the study of electricity', he wrote,

> I resolved to read no mathematics on the subject till I had first read through Faraday's "Experimental Researches in Electricity". I was aware that there was supposed to be a difference between Faraday's way of conceiving phenomena and that of the mathematicians, so that neither he nor they were satisfied with each other's language. I had also the conviction that this discrepancy did not arise from either party being wrong.[3]

At the end of 1855 and the beginning of 1856 Maxwell read a two-part paper to the Cambridge Philosophical Society, 'On Faraday's Lines of Force', in which he attempted 'to simply state the mathematical methods by which I believe that

[3] M. Goldman, *The Demon in the Aether* (1983), 137.

electrical phenomena can be best comprehended and reduced to calculation'.[4] He had done this by developing a remarkable analogy between electric and magnetic force and the flow of an imaginary weightless and incompressible fluid.

It was Maxwell's older compatriot William Thompson who, while still an undergraduate at Cambridge, first noticed that equations for the strength and force of static electricity took the same form as equations which described the flow of heat through a solid material. Using similar equations Maxwell had found he was able to derive mathematical formulae for the action of Faraday's 'lines of force' which merged them into a single 'flux' across what he described as a 'field' (an idea that was to become a crucial concept in twentieth-century physics).

The following year he sent the paper to Faraday and received a delighted reply:

Albemarle Street W.
25th March 1857

My DEAR SIR,

I received your paper and thank you very much for it. I do not venture to thank you for what you have said about 'Lines of Force' because I know you have done it for the interests of philosophical truth; but you must suppose it is most grateful to me and gives me much encouragement to think on. I was at first almost frightened when I saw so much mathematical force made to bear upon the subject, and then wondered to see that the subject stood it so well.[5]

Faraday found that Maxwell, unlike other mathematicians, had given him a 'perfectly clear idea of your conclusions which though they may give no full understanding of the steps of your process, gave me the results neither above nor below the truth and so clear in character that I can think and work from them.' In his reply he goes on to say that he is hoping to make some experiments on 'the time of magnetic action'.

The analogy of flowing liquid applied only to the action of static electricity and magnetism. To describe what happened when a moving magnet makes a current flow in a loop of wire Maxwell found that he needed to resort to a differential equation, which was eventually to lead him to the prediction of electromagnetic waves. Meanwhile Faraday's suggestion that

[4] J. C. Maxwell, *The Scientific Papers of James Clerk Maxwell* (1890), vol. I, 187.
[5] L. Campbell, *The Life of James Clerk Maxwell* (1882), 200–1.

electromagnetic effects were not instantaneous but involved some sort of travelling fluctuation struck Maxwell like a revelation and spurred him on to further efforts.

Wheels and Cogs

In 1860 Maxwell had been appointed as Professor of Natural Philosophy at King's College London, and it was during his time at King's that he set himself the task of trying, as he put it, 'to form a mechanical conception' of what Faraday had called 'the electrotonic state'.

Faraday had argued that in his experiment with a magnet and a loop of copper wire, it appeared that the magnet was inducing a kind of tension in the wire ('the electrotonic state') so that when the magnet moved, an electric current flowed in the wire. How could this be physically imagined?

Maxwell came up with a picture of closely packed spinning cells whose centrifugal force would cause them to expand along one axis as they rotated and contract along the other [Fig. 48.1]. The contraction would cause a tension along their axis of spin and a pressure as they expanded at right angles to the angle of spin, in the same way that magnets exert an attraction along their length and a sideways repulsion.

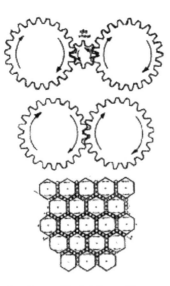

48.1 James Clerk Maxwell's drawings of idler wheels.

In between these spinning cells he imagined there could be 'idle wheels': counter-rotating little wheels that engineers put between cogs. On Maxwell's model these idle wheels would be equivalent to particles of electricity. If a line of idle wheels rotate—an electrical current—then the cogs move in another direction—a circulating magnetic field around the current. If there is a rotation of the cogs—a magnetic field—then the idle wheels move—an electric current.

Maxwell could describe all this mathematically. He wrote it up in a two-part paper which appeared in instalments in the *Philosophical Magazine* from March to May of 1861. He may have intended to stop there. Having written up his results he and Katherine left King's College for their summer vacation at the family home in Galloway.

Let There Be Light

Over the course of the summer in Galloway he began to reflect more deeply about his model of cogs and wheels. Were they rigid or elastic? How did they share their space with ordinary matter? Why did some materials conduct electricity and others not? If they were elastic and their elasticity could be modified by different kinds of material (conductors and non-conductors) then much could be explained. Electrostatic energy could be conceived as a kind of potential energy like that in a spring, with magnetic energy like the rotational energy in a flywheel.

If, he argued (going back to his model of electricity as a kind of fluid), 'a conducting body may be compared to a porous membrane', a non-conductor would be 'like an elastic membrane which may be impervious to the fluid but transmits the pressure of the fluid on one side to that on the other'.[6] This would give rise to a novel kind of physical phenomenon that he referred to as a 'displacement current'.

It followed from all this that these energies were not inherent in matter, they existed as it were around ordinary matter, and they could exist in empty space.

When Maxwell worked out the mathematics, adding in this new 'displacement' phenomenon, it all fitted beautifully. Before he had time to develop his equations further, however, he noticed something else.

All elastic materials transmit waves. Even the smallest change in an electric or magnetic field would send out an electric ripple in one direction and a magnetic ripple in another. The whole universe would be filled with the surge of *electromagnetic* waves.

Somewhat like the waves of the sea, in which there is motion of the water at right angles to the direction of propagation of the wave, these would be *transverse* waves (as opposed to *longitudinal* waves like sound waves in a gas where the molecules move in the same direction as the sound). Since it was already known that the speed of waves in an elastic medium is given by the square root of the elastic modulus divided by the density, their speed could be calculated. This suddenly raised an extraordinary possibility.

Light was known to travel in transverse waves; and since some recent optical experiments by the French physicists Hippolyte Fizeau and Léon Foucault, its speed had been accurately determined. Could light, magnetism, and electricity all be related? Maxwell had left all his reference books in London. Communication and transport in the 1860s were a slow

[6] J. C. Maxwell, *The Scientific Papers of James Clerk Maxwell*, vol. I, 490–1.

business, so that it was not until the end of summer when he returned to King's that he was able to check his findings. The results were worth waiting for.

Twelve years earlier Faraday had given a private lecture on magnetism to Prince Albert in which he had told the Prince that 'I cannot doubt that a glorious discovery in natural knowledge, *and of the wisdom and power of God in creation*, is awaiting our age.'[7] At the end of the summer of 1861 he received a letter which must have seemed like a fulfilment of that prophecy. Maxwell informed him that he had 'worked out the formulae in the country before seeing Weber's number, which is in millimetres, and I think we now have strong reason to believe, whether my theory is a fact or not, that the luminiferous and electromagnetic medium are one.'[8]

He had brought light, electricity, and magnetism into a single unified theory. In 1862 he published a third part to his paper, announcing that 'The velocity of transverse undulations in our hypothetical medium, calculated from the electromagnetic experiments of M. M. Kohlrausch and Weber, agrees so exactly with the velocity of light calculated from optical experiments of M. Fizeau, that we can scarcely avoid the inference that *light consists in the transverse undulations of the same medium which is the cause of electric and magnetic phenomena*.'[9]

This, however, was not the end of his investigations. Maxwell had been anxious from the outset to stress the 'provisional and temporary' nature of his wheel and cog hypothesis. He had not brought it forward 'as a mode of connexion existing in nature' or even as an 'electrical hypothesis'.[10] It was rather a stepping stone for thought—what today would be called a 'model'. The next step would take him out beyond what could be pictured through mechanical images into a realm where it was necessary to walk by numbers, not by sight.

[7] G. N. Cantor, *Michael Faraday* (1991), 67.
[8] Maxwell to Faraday, 19 Oct. 1861, in L. Campbell, *The Life of James Clerk Maxwell*, xii.
[9] J. C. Maxwell, *The Scientific Papers of James Clerk Maxwell*, vol. I, 500.
[10] B. Mahon, *The Man Who Changed Everything* (2003), 104.

CHAPTER FORTY–NINE

The Works of the Lord

In his inaugural lecture at King's College London, delivered the year before his discovery of electromagnetic waves had been made, Maxwell had stated his conviction that 'As physical science advances we see more and more that the laws of nature are not mere arbitrary and unconnected decisions of Omnipotence, but they are essential parts of one universal system in which infinite Power serves only to reveal unsearchable Wisdom and eternal Truth'. These advances not only revealed a deep underlying rationality in the universe, but pointed to a corresponding echo of that rationality in the human mind.

Kepler and Galileo had thought that it was our ability to discern the 'shining in the mind of God' which proved 'that the human mind is the work of God'. In much the same way Maxwell argued that

> When we examine the truths of science and find that we can not only say 'This is so' but 'This must be so for otherwise it would not be consistent with the first principles of truth' . . . we should think what a great thing we are saying, when we pronounce a sentence on the laws of creation, and say they are true, or right, when judged by the principles of reason. Is it not wonderful that man's reason should be made a judge over God's works, and should measure, and weigh, and calculate, and say at last "I understand I have discovered—It is right and true'.[1]

This did not lead him to soft pedal his insistence that the model of cogs and wheels should not be thought of as a literal description of underlying reality. It did, however, encourage the hope that it might be possible to gain a workable grasp of 'the go' of things and that mathematics might be the best means of doing so. The method that most helped him was one he discovered in a work called *Méchanique Analytique*, by an Italian-born, eighteenth-century mathematician, Giuseppe

[1] R. V. Jones, 'James Clerk Maxwell at Aberdeen, 1856–1860', *Notes and Records of the Royal Society of London* 28 (1973), 57–81.

Luigi Lagrange, which set out to describe the dynamics of a mechanical system in entirely abstract terms.

The Mysterious Belfry

The critical aspect of Lagrange's system was that it used what the twentieth century called black box analysis—calculating the outputs of a system (the black box) from a knowledge of the inputs, and of what they characteristically produce without needing to know what goes on inside.

Maxwell compared this to a mysterious belfry. In an ordinary belfry the bell ringers' ropes go up through a hole in the floor to a simple unseen mechanism which rings the bells as the ropes are pulled. Suppose though that instead of each rope being connected to a single bell there was an elaborate cat's cradle of rope connections in the unseen belfry above the ringers' heads. Suppose further, says Maxwell, 'all this machinery is silent and utterly unknown to the men at the ropes who can only see as far as the holes above them'.[2] As long as we know which rope is being pulled and which bell characteristically sounds, we can calculate the result without needing to have any knowledge of the belfry's hidden mechanism.

Hence, discarding his whole elaborate model of cogs and wheels, he set about writing a sequence of what turned out to be four (to physicists wonderfully beautiful) equations which describe the way that the magnetic and electric fields circulate around one another. These miraculously give as a solution a form of equation that was known to represent a transverse wave travelling at speed c, equal to the reciprocal of the square root of the product of the electric constant and the magnetic constant (the quantity that had been shown to correspond to the experimentally measured speed of light) [Fig. 49.1].

Because they define the way we transmit both energy and information Maxwell's four equations form the basis, in one way or another, of every electronic technology in the modern world. Within 50 years of his death they were beginning to be seen by physicists as '*the* most basic laws of the physical world'.[3]

In 1887 the famous nul result of Albert Michelson and Edward Morley's experiment showed that light appeared to travel at the same speed whatever the travelling speed of the observer. Starting not so much from the experiments as from Maxwell's equations, this became one of the postulates of Einstein's

$$\nabla \cdot \mathbf{E} = \frac{\rho}{\varepsilon_0}$$

$$\nabla \cdot \mathbf{B} = 0$$

$$\nabla \times \mathbf{E} = -\frac{\partial \mathbf{B}}{\partial t}$$

$$\nabla \times \mathbf{B} = \mu_0 \left(\mathbf{J} + \varepsilon_0 \frac{\partial \mathbf{E}}{\partial t} \right)$$

49.1 Maxwell's equations.

[2] B. Mahon, *The Man Who Changed Everything* (2003), 121.
[3] B. Mahon, *The Man Who Changed Everything*, 181.

theory of relativity, which held that the speed of light c was an absolute universal constant (the equation $E = mc^2$ was to come later).

Maxwell's equations thus revealed what has been described as 'the fundamental characteristic of the universe, nature's gearing between space and time'.[4] They had done so furthermore by a new methodology of science which involved leapfrogging over the unknown (and perhaps unknowable) physical realities to arrive at a mathematically workable kind of knowledge.

Hamlet's Ghost

This paradoxical ability of mathematics to obtain a workable knowledge of reality without actually knowing how it worked, which Eugene Wigner was later to dub 'the unreasonable effectiveness of mathematics',[5] chimed with Maxwell's larger sense that what philosophers call 'the epistemological conditions' of knowledge—our ways of knowing the different kinds of things we know—were not straightforward.

While T. H. Huxley (with whom Maxwell was joint scientific editor of the *Encyclopaedia Britannica*) could breezily assert that 'there is but one kind of knowledge and one means of acquiring it', insisting as a corollary that 'Science has no intention of signing a treaty of peace ... nor of being content with (less than) absolute victory and uncontrolled domination over the whole realm of intellect', Maxwell's view was more profound.

In the year of his death he wrote a review in *Nature* of a book called *Paradoxical Philosophy* in which the authors had tried to use what they called 'psychophysics' in support of spiritualism. Maxwell had little time for this, but in the course of his review reflects on the scope of scientific methodology. 'One of the severest tests of a scientific mind', he suggests, 'is to discern the limits of the legitimate application of scientific methods'. Adapting the line in *Hamlet* he argues that 'there are many things in heaven and earth which by the selection of our scientific methods have been excluded from our philosophy'.[6]

The viewpoint of the detached observer (typically that of the scientist) would, however valuable and necessary, inevitably screen out the kinds of knowledge available to the committed participant. That reality had more aspects than could be captured by a study of physical causes was revealed in the fundamental choices of human existence: 'either be a machine

[4] B. Mahon, *The Man Who Changed Everything*, 181.

[5] E. P. Wigner, 'The Unreasonable Effectiveness of Mathematics in the Natural Sciences. Richard Courant Lecture in Mathematical Sciences Delivered at New York University, May 11, 1959', *Communications on Pure and Applied Mathematics* 13, 1–14.

[6] J. C. Maxwell, *The Scientific Papers of James Clerk Maxwell* (1890), 759.

and see nothing but "phenomena" ', Maxwell once wrote to a friend, 'or else try to be a man, feeling your life interwoven, as it is, with many others, and strengthened by them whether in life or death'.[7]

The relationship of religion and science seems, as he conceived it, to have involved a similar kind of interweaving of knowledge of objects and persons. A prayer found in a drawer after his death begins with a meditation on Psalm 8: 'When we consider the heavens the work of thy fingers, the moon and the stars which thou has ordained, teach us to know that thou art mindful of us . . . showing us the wisdom of your laws', before going on to ask, 'that looking higher than the heavens may we see Jesus'.

The personal relationship with an unseen deity that was opened to the worshipper through prayer was always central for Maxwell.[8] Like Faraday, he used to visit the sick to pray with them when this was welcome. Visitors to the Maxwell home, Glenlair, were often struck by the extempore daily prayers conducted by the master of the household. Among these visitors in 1879 was his biographer William Garnett, who for the past eight years had been Maxwell's demonstrator at the Cavendish Laboratory, and had observed how closely his religious and scientific thinking wrapped themselves around one another.

The Cavendish Doors

Maxwell had not been the University of Cambridge's first choice to run the laboratory. The move to build a laboratory had begun in 1860, the same year Oxford had completed the University Museum. In Cambridge, while other sciences had acquired their own buildings, physics was still being taught in poorly equipped college laboratories. The syndicate appointed to look into the question in 1860 did not report until 1869, and although it recommended appointing a professor of experimental physics and building a laboratory in which he could teach, there were some who could see little point in an undergraduate performing experiments that had already been carried out by his betters.

'If he does not believe the statement of his tutor—probably a clergyman of mature character, recognised ability and blameless life,' wrote one don, 'his suspicion is irrational and manifests a want of power of appreciating evidence'.[9] Others thought that

[7] B. Mahon, *The Man Who Changed Everything*, 72–3.
[8] L. Campbell, *The Life of James Clerk Maxwell* (1882).
[9] A. MacFarlane, *Lectures on Ten British Mathematicians of the Nineteenth Century* (1916), 140–1.

the estimated cost of £6000 was excessive. This situation was resolved by the intervention of the Chancellor of the University, the Duke of Devonshire, who in 1870 offered £6300 to cover the entire cost of building and equipping the new laboratory. This generous offer overwhelmed all objections and in its wake the university senate established a new chair of experimental physics and set about the business of appointing the first Cavendish Professor.

The first approach was made to William Thomson—later Lord Kelvin. When he turned it down, Thomson was asked if he would approach the German physicist Hermann von Helmholtz. Helmholtz, however, was happy in Berlin, where he had just been appointed head of the new institute of physics. Only then were approaches made to Maxwell. 'Cambridge', as Martin Goldman, another of Maxwell's biographers, remarks, 'was luckier than it deserved'; of three very great physicists, 'it got the greatest'.[10]

When Maxwell arrived back in Cambridge in 1871 to take up his new post, he found himself a professor of experimental physics with no laboratory: 'I have no place to erect my chair but must move about like the cuckoo, depositing my notions in the Chemical Lecture room 1st term, in the Botanical in Lent and in the Comparative Anatomy in Easter'.[11] The new laboratory, to be sited in Free School Lane, could be designed entirely to his specifications. His father, John Clerk Maxwell, had designed his own house (and his son's clothes) and Maxwell, following in these paternal footsteps, visited both Glasgow and Oxford to gather his own ideas.

In London he had found that magnetic disturbances caused by Thames steamers had interfered with some of his measurements. These were not a problem in Cambridge but traffic vibrations could be, so a special magnetic room was built with three monolithic piers set on their own foundations through holes in the floor, on which sensitive equipment could be mounted. Other devices included a large water tank set on a 50-foot tower to drive a big vacuum pump with vacuum lines running to various rooms in the building, and 18-inch window sills on which heliostats—mirrors following the Sun—could be set to direct light into the laboratory (this was still in the days before electric light) [Fig. 49.2].

All the features that Maxwell had incorporated were minutely described in an article in *Nature*, which appeared on

10 M. Goldman, *The Demon in the Aether* (1983), 11.
11 L. Campbell, *The Life of James Clerk Maxwell*, 389.

49.2 The interior of the Cavendish Laboratory.

25 June 1874, a week after the building was formally presented to the university by the Duke.[12] Among the features which the writer particularly notes is the entrance to the laboratory: 'The doors, which are very massive, are beautifully carved in oak and bear, in old English letters, the inscription "Magna opera Domini exquisita in omnes voluntates ejus," which is the Vulgate version of Psalm cxi. 2.' [Fig. 49.3].

Although according to Garnett, 'the drawings were ably prepared and the builder's work supervised by Mr W. M. Fawcett,

49.3 Inscription on Maxwell's Cavendish doors.

[12] 'The New Physical Laboratory of the University of Cambridge', *Nature* 10 (1874), 139–42.

M.A., of Jesus College, the Architect appointed by the University, the Laboratory and its equipment . . . sprang from the mind of Maxwell as complete as Athene from the brain of Zeus', and this quotation, 'perfectly expresses Maxwell's mental attitude to the researches which he hoped would be carried out by himself and his students in the laboratory he had designed'.[13]

Maxwell's teaching methods were never prescriptive. He thought that people should 'not be led into other men's ways of thinking under the pretence of studying science',[14] and he encouraged his most able students to pursue their own ideas. While he thought it crucial that the student of science should not be withdrawn or cut off from the wider 'study of man' (and was himself involved in a small discussion group that included Joseph Lightfoot, Brooke Westcott, and Fenton Hort— the most distinguished New Testament scholars of the period), he was wary of attempts to link particular scientific theories to specific metaphysical or theological ideas.

When, for instance, in 1876 Charles Ellicott, the scholarly Bishop of Gloucester and Bristol, wrote to ask for his scientific opinion about a verse in Genesis, Maxwell was extremely cautious in his reply. The bishop asked him whether he agreed that the concept of the 'aether'—the medium through which light was thought to move—could help explain why in Genesis 1 the creation of light precedes the creation of the Sun and the stars. Maxwell, like every other physicist at the time, assumed that light must travel through some medium and his whole theory was formulated in terms of what he calls the 'luminiferous and electromagnetic medium' (only later when Einstein's general theory of relativity reformulated these ideas was it recognized that the concept of the aether was no longer tenable).

In his reply to Ellicott, Maxwell warned that 'if it were necessary to provide an interpretation of the text in accordance with the science of 1876 (which may not agree with that of 1896) it would be very tempting to say that the light of the first day means the all-embracing aether, the vehicle of radiation, and not actual light, whether from the sun or any other source'. But whether or not his own assumptions were valid he could not 'suppose this was the idea meant to be conveyed by the author of the book to those for whom he was writing', and would be very sorry if an interpretation 'founded on a most conjectural scientific hypothesis were to get fastened to the text of Genesis'. 'The rate of change in scientific hypotheses',

[13] J. J. Thomson, *James Clerk Maxwell* (1931), 109.
[14] B. Mahon, *The Man Who Changed Everything*, 70.

he pointed out, 'is naturally more rapid than that of biblical interpretations, so that if an interpretation is founded on such a hypothesis it may help to keep it above ground long after it ought to be buried and forgotten'.[15]

Maxwell was acutely aware of the provisional nature of scientific theories. It might be as Blaise Pascal had written that 'we burn with desire to find a firm foundation, an unchanging solid base on which to build a tower rising to infinity', but science could not provide that foundation. Even long-accepted theories might need to change to accommodate new data, 'the foundation splits and the earth opens up to its depths'.[16] However powerful the predictive power of science, it could not predict its own discoveries.

This meant that a degree of humility was always appropriate. It might be that it was impossible to see anything without 'the focussing glass of theory', sometimes adjusted to one pitch of definition, sometimes to another, 'so as to see down into different depths through the great millstones of the world',[17] but we should not mistake human theory for absolute truth. 'The true logic of this world', he suggested in an echo of Pascal, 'is the calculus of probabilities'.[18]

Pascal's conclusion was that man, though 'a thinking reed', must nevertheless concentrate on 'thinking well', which for Maxwell meant trying to achieve a degree of unity in both life and thought. At the age of 23 he wrote a reflection that captures the continual search for coherence that has been our theme throughout this book: 'Happy is the man who can recognise in the work of Today a connected portion of the work of life, and an embodiment of the work of Eternity'.[19]

He appeared to have achieved this aspiration. Father Guillemard was not the only visitor to his bedside to have been impressed by what they found. Maxwell's local doctor in Galloway included a remarkable tribute in the medical notes he sent to Maxwell's Cambridge physician, stating that quite aside from his patient's scientific achievements, 'I must say he is one of the best men I have ever met'. Professor Hort recalled the dying man saying, 'what is done by what is called myself is, I feel, done by something greater than myself in me . . . The only desire I can have is like David to serve my own generation by the will of God and then fall asleep'.[20]

For Maxwell, however, the quest to comprehend the world as a meaningful whole could never be considered an achieved

[15] L. Campbell, *The Life of James Clerk Maxwell*, 323.

[16] B. Pascal, *Pensées and Other Writings* (1995), 70.

[17] L. Campbell, *The Life of James Clerk Maxwell*, 237.

[18] J. C. Maxwell, *The Scientific Letters and Papers of James Clerk Maxwell* (1990), vol. I, 197.

[19] B. Mahon, *The Man Who Changed Everything*, 47.

[20] B. Mahon, *The Man Who Changed Everything*, 173.

state. Science might swim in the slipstream of ultimate questions but the gap between the two could never be finally
closed. The integrative struggle must (while we see through a
glass darkly) always take the form of a dynamic process.

This was the reason that he gave for refusing an invitation to
join a newly formed society, the Victoria Institute, which was
established to promote the harmony of science and religion.
In an incomplete draft found among his papers after his death
he writes:

> I think Christians whose minds are scientific are bound
> to study science that their view of the glory of God may
> be as extensive as their being is capable of. But I think
> that the results that each man arrives at in his attempts to
> harmonise his science with his Christianity ought not to
> be regarded as having any significance except to the man
> himself and to him only for a time, and should not receive
> the stamp of a society. For it is the nature of science es-
> pecially of those branches of science which are spreading
> into unknown regions to be continually . . . ,[21]

and here the manuscript (rather appropriately) ends.

[21] L. Campbell, *The Life of James Clerk
Maxwell*, 404–5.

PART XI
Epilogue

Epilogue

The Oxford University Museum stands near the corner of Parks Road and South Parks Road where in an engraving and an 1860 photograph (taken by Henry Wagner) it rises in splendid isolation [Figs. 50.1, 50.2]. A photograph taken near the same spot today shows it virtually hidden, and a succession of impressive new laboratories now line these two roads on either side for almost a quarter of a mile [Fig. 50.3].

In Free School Lane the Cavendish Laboratory had no room for such expansion. During the late 1960s (by which time 20 members of the laboratory had won a Nobel Prize, with 9 more to follow in the remainder of the century) the over-crowded structure increasingly appeared to be inadequate to the demands that were being made upon it, and the decision was taken to abandon Maxwell's original building and construct a new laboratory on a green field site to the west of Cambridge (though still within cycling distance of the centre).

The physical expansion of laboratories in Oxford and Cambridge is testament to the massive intellectual achievements of science during the twentieth century, and a demonstration of

50.1 The Oxford University Museum, 1860.

50.2 The Oxford University Museum photographed by Henry Wagner in 1860.

50.3 The Oxford University Museum (just visible on far left), photographed by Roger Wagner in 2014.

the vastly increased resources that have been poured into science over the past 70 years. Encouraged by the contribution of science to two world wars, governments and private companies all over the world have invested massively in the scientific enterprise. The fabulous return from their investments has been a continual stream of discoveries that are transforming human society in ways never before seen, and raising questions never before faced.

The question that 16 years ago had seemed to us to be posed by the Cavendish Laboratory and the University Museum was

a simple one: 'How did the religious invocations over the entrances of these two original scientific buildings come to be
there?' The fundamental answer we have argued is that the penultimate curiosity of science has throughout human history
swum in the slipstream of an ultimate metaphysical curiosity
rooted in the human need to make sense of the world as a
whole.

Following the strongest effects of that slipstream has revealed
a number of features in the configuration of that metaphysical
curiosity which seem to have contributed to its strength (or
otherwise):

- The idea of a single, beneficent, rational agency whose rationality could be both expressed in mathematics and read
  in the humblest aspect of Creation.
- The idea that this agency could not be identified with
  anything within the universe but gave to the whole a law-
  like character.
- The idea that truth is not the exclusive property of any
  single civilization.
- The idea that truth could not be imposed by force but involved the right, even the duty, of individual investigation
  and experiment.

All of these features appeared to us to have been significant
factors. The first two were concerned, so to say, with '*what* God
is': the metaphysical character of divinity and how it related
to the natural world. The last two were concerned more with
'*who* God is': the moral nature of divinity and the relationship
between divinity and humanity. When at the beginning of the
seventeenth century Francis Bacon highlighted the two leading
motives for scientific enquiry as 'the glory of God and the relief
of man's estate', they were directly related to these twin aspects
of ultimate curiosity.

At the beginning of the twenty-first century while altruism and pure curiosity are still operative, it is hard to deny
that money, power, and the pursuit of national, corporate, and
personal advantage have become much more visible drivers of
scientific discovery.

How might this relate to our slipstreaming metaphor?

Although the flock of geese we observed flying over the
Cherwell 16 years ago appeared to be flying in a stable V

formation, this we now know was not the case.[1] Geese and other birds circulate within their formations taking turns to fly at the front, rather like riders in the 'peloton'—the group of leaders in the Tour de France.

The circulating peloton of motives that drives drug companies to develop new products is an easily observable phenomenon. Does it follow (pursuing the metaphor) that the slipstream generated throughout human history by what we have called 'ultimate curiosity' has either sunk to the back of the peloton or fallen behind the front runners and effectively ceased to operate?

By the time this book was nearing its conclusion we had been living in different parts of Oxford for more than a decade. Our discussions now often took place in the context of a termly forum of scientists, philosophers, and theologians (and a lone artist) which met over dinner in different colleges to discuss areas of common interest. These meetings continually testified to the breadth and complexity of the issues that now arise where scientific discovery intersects with other areas of thought (issues which, for the reason Maxwell gave, can never be finally resolved). Here, though, was one issue which could not be separated from the initial question we had asked.

The vast expansion of scientific modes of thinking, evident in the new laboratories of Oxford and Cambridge, has certainly introduced new material and interested motives into the practice of science, but has it brought about a deeper more philosophical change? Has the long entanglement of religious and scientific thinking, whose story we have followed through the preceding chapters, now at last come to an end? Is scientific thought finally fulfilling the ancient Epicurean hope of crowding religious ways of thinking off the stage?

Developing the theatrical metaphor an approach to this question might be framed as a drama in three acts, where in each act the events of the same years are seen from a different perspective.

Thus in the last act (*Resolution*), the attempt of science to hog the stage is unmasked as a surprising continuation of the fundamental process that our metaphor of slipstreaming has been trying to describe.

In the second act (*Subversion*) unexpected discoveries cast doubts on whether the pursuit of ultimate questions remains a feasible enterprise.

[1] B.Voelkl et al., 'Matching Times of Leading and Following Suggest Co-operation through Direct Reciprocity during V-Formation Flight in Ibis', *Proceedings of the National Academy of Sciences of the United States of America* 112 (2015), 2115–20.

The first act, however, begins, where we began in the Prologue, with the appearance of straightforward conflict.

Act I: *Conflict*

On the 17th of July 1918 the Russian imperial family—the Tsar and Tsarina together with their five children and attendants—were taken down into the cellar of a house in Yekaterinburg and executed by firing squad. Those that survived the guns were bayoneted to death.

Eighty-three years later, on 11 September 2001, American Airlines Flight 11, carrying 92 passengers and crew and travelling at a speed of 485 miles per hour, was flown into the north tower of the World Trade Centre in New York.

These acts of political violence at the beginnings of the twentieth and of the twenty-first centuries, in very different ways and for rather different reasons, each provided a powerful impetus to a new version of the ancient project, begun by Epicurus, of using science to marginalize religion.

The Bolshevik seizure of power in the Russian Revolution, sealed by the death of the Romanoffs, established the historical materialism of Karl Marx as the official doctrine of an entire state. When the disappearance of religious ideas that this had predicted failed to occur, active methods were established with the aim of ensuring that they did. After the absorption of Eastern Europe into the hegemony of the Soviet Union and the victory of the communists in China, the abolition of religion became, between 1950 and 1990, the official policy of the governments of roughly half the population of the world.

The collapse of the Soviet Union partly mitigated this project in the East, but the September 11th attacks, followed by attacks in London and around the world, brought it back on to the agenda. The fact that these attacks were motivated by so-called 'Islamic jihadism' (though some have argued that it is better described as 'Islamicized Marxism') seemed to many opponents of religion to provide an overwhelming object lesson in the dangers of religious thinking. It gave rise in the decade that followed to a new proliferation of anti-religious writing and campaigning, but this time in the West.

In both instances at the heart of the argument that was made was an appeal to science.

Epicurus Reborn

The development of this new Epicurean project goes back to the middle of the eighteenth century. In his *Enquiry Concerning Human Understanding* the Scottish philosopher David Hume had proposed that 'if we take in our hand any volume of divinity or school metaphysics' we should ask only two questions: '*Does it contain any abstract reasoning concerning quantity or number?* No. *Does it contain any experimental reasoning concerning matters of fact and existence?* No. Commit it then to the flames. For it can contain nothing but sophistry and illusion'.[2]

In the nineteenth century Auguste Comte's *Positive Philosophy* proposed in a similar way that the theological and metaphysical stages of human thought belonged to the past and the only valid form of contemporary thinking was scientific. What Engels referred to as Marx's 'scientific socialism' likewise insisted that any kind of metaphysical thinking that was not based on material causes should be dismissed as 'drunken speculation'.[3]

In the early decades of the twentieth century this mode of thinking was not limited to Marxist Russia. In 1922 a number of philosophers who had previously been meeting in Viennese coffee houses formed themselves into a group that became known as *Der Wiener Kreis*—The Vienna Circle. Their manifesto, published in 1929, was entitled *Wissenschaftliche Weltauffassung. Der Wiener Kreis*—The Scientific Conception of the World: The Vienna Circle—and their goal was expressed by the title of an article by one of the group's leading members, Rudolf Carnap: 'The Elimination of Metaphysics through Logical Analysis of Language'.

'Logical Positivism', as it became known, was introduced to the English-speaking world by the young British philosopher A. J. Ayer in his 1936 book *Language, Truth and Logic*. In it he described the so-called 'verification principle', which held that apart from tautologies, meaningful statements were only those that could be subject to empirical testing, and that consequently 'all metaphysical assertions are non-sensical'.[4]

Ayer's book from its first publication was widely discussed. When it was republished after the war it became a bestseller. Even those who still allowed a place for metaphysics like Bertrand Russel or his disciple (and—according to Russell—plagiarizer) C. E. M. Joad, the great popularizer of philosophy

[2] D. Hume, *An Enquiry Concerning Human Understanding* (1977), 113.
[3] M. Eastman, 'Marxism: Science or Philosophy?', https://www.marxists.org/archive/eastman/1935/science-philosophy.htm, accessed 24 August 2015.
[4] A. J. Ayer, *Language, Truth and Logic* (1946), 41.

at the time, were inclined to present religion as having been supplanted by science.[5]

Although neither materialism nor positivism could themselves claim to be scientific discoveries, from 1860 onwards, Darwin's theory of evolution appeared to provide a justification for this general purge of metaphysics.

The Evolution of Morals

In a letter which he wrote to Engels in December 1860 Marx claimed that *On the Origin of Species* contained 'the basis in natural history for our view'.[6] He sent Darwin an inscribed copy of *Das Kapital* to which he received a polite, though non-committal reply (when Marx's son-in-law Edward Aveling wrote to ask if he might dedicate his *Student's Darwin* to its titular subject, he received a similarly polite but firm refusal to be associated with a direct attack on religion).

The importance of Darwin for Marx was not simply that the English scientist appeared to replace 'the obsolete revelation of Christianity' with 'the rational revelation of science' (as Darwin's French translator had put it). At the heart of Karl Marx's own theory was the idea that while human beliefs might appear to be sustained by an appeal to a transcendental reality, they were in fact the product of social conditions.

Thus 'bourgeois notions of freedom, culture and law etc' were 'but the outgrowth of bourgeois production',[7] while 'law, morality, religion' are to the proletarian, 'so many bourgeois prejudices'.[8] From this perspective human activities could only be understood by understanding the conditions that generated them, and Darwin's hypothesis, Marx, told Ferdinand Lassalle, 'provides me with the basis in natural science for the class struggle in history'.[9]

How though could the struggle for existence account for altruism? Through most of the twentieth century, evolutionary theory had some difficulty in explaining the apparently altruistic behaviour that appears in nature when, for instance, individual ants lay down their lives for the rest of the colony.

In the 1960s an answer to this conundrum began to emerge through the work of the English evolutionary biologist William Hamilton. Hamilton's work on kin selection gave rise to an algebraic inequality (now known as Hamilton's rule) that an agent should rationally perform a costly action if $C < r \times B$

[5] C. E. M. Joad, *God and Evil* (1942); B. Russell, *Religion and Science* (1935).
[6] K. Marx and F. Engels, *Selected Correspondence, 1846–1895* (1942), 125.
[7] K. Marx, *The Essential Left* (1960), 31.
[8] K. Marx, *The Essential Left*, 25.
[9] K. Marx and F. Engels, *Selected Correspondence, 1846–1895*, 125.

(where C is the cost in fitness to the agent, r the genetic relatedness between agent and recipient, and B the fitness benefit to the recipient), which showed why altruistic behaviour would be favoured by natural selection.

When Hamilton's rule was rederived by an American chemist George R. Price (who wrote what became known as Price's equation) and connected by Price and John Maynard Smith to game theory, a larger picture began to emerge which suggested why altruistic behaviours would be a stable feature of whole populations.[10]

In the late 1970s and early 1980s these ideas began to be popularized by Edmund Wilson and Richard Dawkins. They seemed to remove the need for what the American philosopher Richard Rorty described as 'metaphysical skyhooks' to explain human behaviour. 'Soft altruism', according to Wilson, is 'ultimately selfish . . . The "altruist" expects reciprocation'.[11] More broadly morality as a whole, according to the philosopher Michael Ruse, could be seen as 'no more . . . than an adaptation',[12] 'a creation of the genes'.[13]

In fact evolution seemed to some to abolish the need for 'metaphysical skyhooks' altogether. While T. H. Huxley had argued that 'There is a wider teleology that is not touched by the doctrine of evolution',[14] some twentieth-century evolutionary thinkers disagreed.

The Philosophy of Chance

In *Chance and Necessity* published at the beginning of the 1970s the distinguished French biologist Jacques Monod argued that 'Pure chance, absolutely free but blind', was 'at the very root of the stupendous edifice of evolution', and was 'no longer one among other possible or even conceivable hypotheses. It is today the *sole* conceivable hypothesis'.[15] 'Man', in consequence he argued, 'at last knows he is alone in the unfeeling immensity of the universe by which he emerged only by chance'.[16] A similar kind of argument was used by Richard Dawkins in his 1986 book *The Blind Watchmaker* to attack the argument for design put forward by the eighteenth-century anti-slavery campaigner William Paley [Fig. 50.4].

Paley's book *Natural Theology* begins with a parable which has become well known: 'In crossing a heath, suppose I pitched my foot against a stone and were asked how it came to be

[10] J. M. Smith and G. Price, 'The Logic of Animal Conflict', *Nature* 246 (1973), 15–18.
[11] E. O. Wilson, *On Human Nature* (1978), 155.
[12] M. Ruse, *Evolutionary Naturalism* (1995), 241.
[13] M. Ruse, *Evolutionary Naturalism*, 290.
[14] C. R. Darwin, *The Life and Letters of Charles Darwin* (1887), 479.
[15] J. Monod, *Chance and Necessity* (1972), 110.
[16] J. Monod, *Chance and Necessity*, 110.

NATURAL THEOLOGY,

OR EVIDENCES OF THE

EXISTENCE AND ATTRIBUTES OF THE DEITY.

BY THE REV. WILLIAM PALEY, D.D.

WITH ADDITIONS AND NOTES.

WILLIAM AND ROBERT CHAMBERS,
LONDON AND EDINBURGH.

50.4 Title page of William Paley's *Natural Theology*.

there, I might possibly answer that for anything I knew to the contrary it had lain there for ever ... But suppose I had found a *watch* upon the ground, and it should be inquired how the watch happened to be in that place'.[17] The reason the watch seems to require an explanation, Paley argues, is that 'its several parts are framed and put together for a purpose',[18] and with a considerable display of what Dawkins refers to as 'biological scholarship',[19] he goes on to describe the detailed purposiveness of biological structures.

Darwin's account of how random variations could through a process of natural selection produce the complex adaptations of biological organisms appeared to cut the ground from under this argument. If the purposive characteristics of living things can emerge from a kind of biological algorithm why was there any need to invoke a creator? Indeed could not the argument be turned on its head? Did not 'the organised complexity of a deity' assume the existence 'of the very thing we want to explain'[20] and was it not too complex and statistically improbable to be postulated as a given?

[17] W. Paley, *Natural Theology* (1803), 2.
[18] W. Paley, *Natural Theology*, 3.
[19] R. Dawkins, *The Blind Watchmaker* (1986), 5.
[20] R. Dawkins, *The Blind Watchmaker*, 316.

In common with the many other anti-religious books that appeared after 2001, Dawkins' *The God Delusion* attacked religion from a whole variety of different directions. The central argument against theism repeated the theme that the 'God Hypothesis is . . . very close to being ruled out by the laws of probability',[21] that 'any entity capable of intelligently designing something as improbable as a Dutchman's Pipe (or a universe) would have to be even more improbable than a Dutchman's Pipe',[22] and that 'any God capable of designing a universe . . . must be a supremely complex and improbable entity'.[23]

For all the confidence of these assertions, the appeal to science that characterized the new Epicurean movement had been problematic from the start.

Branches and Saws

'Sawing off the branch one is sitting on', as the philosopher Roger Trigg has remarked, 'is not generally regarded as good practice in human life', and when it appears to be the consequence of a philosophical argument it 'must always be seen as a sign that something is going wrong with our reasoning'.[24] Hume's assertion that all writings which did not contain abstract reasoning about quantity or number or experimental reasoning about facts were 'sophistry and illusion' itself contained neither. If it belonged in the same library as the books he advocated burning, what could save it from the flames?

If positivism or materialism could not claim to be scientific discoveries, what were they other than metaphysical theories belonging to a discarded past of 'drunken speculation'? If the verification principle was not itself capable of empirical testing how could it be verified?

Despite the influence of A. J. Ayer's book the contradictions inherent in positivistic thinking soon became apparent, and in 1967 John Passmore reported that 'Logical Positivism is dead, or as dead as a philosophical movement ever becomes'.[25] Ayer himself remarked that the main problem with *Language, Truth and Logic* was that 'nearly all of it was false'.[26] While philosophy exposes incoherence particularly clearly, something of the same kind of problem has bedevilled attempts to reason from particular scientific discoveries to grand anti-metaphysical conclusions.

21 R. Dawkins, *The God Delusion* (2006), 66.
22 R. Dawkins, *The God Delusion*, 120.
23 R. Dawkins, *The God Delusion*, 140.
24 R. Trigg, *Rationality and Science* (1993), 20.
25 J. Passmore, 'Logical Positivism', in *Encyclopedia of Philosophy* (2006), 52–7.
26 'Men of Ideas (BBC): Logical Positivism and Its Legacy', A. J. Ayer, interview by B. Magee (1978).

The Fallacies of Origins

The moral denunciation of capitalist oppression, which is a central theme in the writings of Karl Marx, could, for instance, seem to be undermined by Marx's own insistence that all notions of freedom, culture, law, and morality are 'but the outgrowth of bourgeois production'. If all moral judgements derive from economic conditions, on what basis could such conditions be morally condemned?

The problems of such arguments have sometimes been described as falling within what in 1934 the philosophers Morris Cohen and Ernest Nagel identified as 'the genetic fallacy' or 'the fallacy of origins'.[27] This was the fallacy of supposing that the validity of a conclusion is solely based on its origin (as in 'You only say that because you are a mathematician'), or that the nature of anything is fully captured by an account of its development. In its extreme form it is the fallacy of supposing that an oak tree 'really is' an acorn.

Thus, while Price's equation might demonstrate how unconscious genetic optimization might give rise to 'altruistic' behaviour in ants, it did not follow that conscious ethical altruism 'really is' unconscious genetic optimization.

Both the tools with which we seek to explain things and the things we seek to explain could be presumed to have a genetic origin. Mathematical ability as much as ethical judgement is in this sense 'a creation of the genes'. Genetic origins, however, cannot determine either the validity of mathematics (like Price's equation) or the validity of ethical judgements (like George Price's decision after his conversion to Christianity to give away his property and care for the homeless).

Nor according to this argument could the evolutionary origins of our behaviours necessarily tell us anything about the possibility or otherwise of their metaphysical grounding. The role of chance, which Jacques Monod assumed had answered this question, was in fact more complicated than it appeared.

The Fallacy of Chance

A technical description of chance might be 'an event which does not follow according to law-like precedent from any assignable precursor'.[28] That in itself does not answer the question of whether it was created or uncreated (a biblical text like

27 M. R. Cohen and E. Nagel, *An Introduction to Logic and the Scientific Method* (1934).
28 D. M. MacKay, *Science, Chance and Providence* (1978), 32.

Proverbs 'The lot is cast into the lap, but its every decision is from the LORD' might seem to imply the former[29]). It is only by treating chance as if it were itself a causal agent (like the mythological Tyche) that it could be seen as alternative to theism.

Metaphysical assumptions of this kind are brought to science rather than arising out of it. They do nevertheless continue to affect what scientists look for and what they discover.

In his 1989 book *Wonderful Life* the Harvard evolutionary biologist Stephen Jay Gould made the claim that if the tape of evolution was replayed it would produce entirely different results in which nothing like human beings would appear. Gould's argument was in the nature of a thought experiment, but was not for that reason ungrounded. It was based on the extraordinary proliferation of hitherto unknown phyla that in the early 1970s began to be identified in the remarkably preserved fossils of Burgess Shale (a rock formation in the Canadian Yoho National Park).

The contrary argument made by the Cambridge evolutionary palaeobiologist Simon Conway Morris (whose work on the shale was the basis of Gould's claims) was that evolution navigates towards particular solutions and that something like man may be an inevitable outcome of the process. Conway Morris' argument appeals to the remarkable phenomenon of evolutionary convergence (whereby entirely distinct lines of evolutionary development produce closely similar results because of the way the world works) which he and others have looked for and found in every root and branch of the evolutionary tree.[30]

The two scientists reach their conclusions from contrasting philosophical viewpoints, which at points creep into their writings. Thus, while Gould suggests that the absolute contingence of the evolutionary process points away from any idea of design, Conway Morris concludes that while the pattern of evolution does not prove the existence of God 'all is congruent'.[31]

If, however, attempts to use science as a purge for metaphysics have made little headway, late-eighteenth- and early-nineteenth-century natural theology has proved an easier target.

[29] Proverbs 16:33.
[30] S. Conway Morris, *The Crucible of Creation* (1998); S. Conway Morris, *Life's Solution* (2003).
[31] S. Conway Morris, *Life's Solution*, 330.

Watches and Watchmaking

In the seventeenth century Blaise Pascal had argued that both natural theology and scientific atheism were built on sand: 'we burn with desire to find a firm foundation, an unchanging solid base on which to build a tower rising to infinity', but in the event 'the foundation splits and the earth opens up to its depths'.[32]

In the eighteenth century the great German philosopher Immanuel Kant argued in a similar way that while belief in God was a kind of rational postulate that made sense of the world (and of the moral life in particular) it could not be straightforwardly derived from it.

William Paley's arguments, which ignored such strictures, were in one sense superseded by Darwinian theory. They have nevertheless proved difficult to eliminate altogether, and T. H. Huxley's recognition that 'There is a wider teleology that is not touched by the doctrine of evolution'[33] has proved prescient.

When the Cambridge cosmologist Fred Hoyle first discovered that his theoretical calculation of the resonance that was crucial for the formation of carbon was confirmed by experimental evidence, he expressed the view that the universe had all the appearance of 'a put up job', as though 'a superintellect had monkeyed with physics as well as with chemistry and biology'.[34] In the years that followed Hoyle's discovery many other 'fine tunings' have been identified in the basic constants of reality, all of which were necessary for what another Cambridge cosmologist, Martin Rees, has called a 'biophilic' universe to exist. To ascribe them to chance has seemed increasingly problematic.

One way of explaining these fine tunings have been various kinds of 'multiverse' hypotheses, which suggest that our particular universe is one among an almost infinite ensemble of universes. Each of these, which could be generated by a variety of hypothesized processes, might begin from slightly different initial conditions and come equipped with different laws and different physical constants. Hence what we call 'laws of nature' could, as Rees points out 'in this grander perspective be *local bylaws* consistent with some overarching theory governing the ensemble but not uniquely fixed by that theory'.[35]

This would make the evolution of our particular 'biophilic' universe more probable, but raises the question in turn of where the 'overarching theory' or 'law of laws' itself comes from. There may be innumerable logically possible multiverses which would never produce a fine-tuned universe (however many sets of even numbers you have, none of them will ever

[32] B. Pascal, *Pensées and Other Writings*, 70.

[33] C. R. Darwin, *The Life and Letters of Charles Darwin*, 479.

[34] F. Hoyle, 'The Universe: Past and Present Reflections', *Engineering and Science* 45 (1981), 8–12.

[35] M. J. Rees, 'Numerical Coincidences and "Tuning" in Cosmology', *Astrophysics and Space Science* 285 (2003), 375–88.

produce an odd number), the existence of one that does seems again to require explanation.

In terms of Paley's metaphor if at the far edge of his heath he had come to a range of mountains he might have thought 'they had been there forever'. If though on the far side of that range he happened on a vast complex producing fully automated watch-making factories, the need for explanation would have substantially increased. (Paley himself recognized that if 'the watch before us' had been 'produced by another watch, and that by a former and so on indefinitely . . . contrivance is still unaccounted for. We still want a contriver'.[36])

Reversing the argument with the proposition that the 'God hypothesis is . . . very close to being ruled out by the laws of probability',[37] is, however, equally problematic.

Probability Problems

The difficulty of applying probability of the kind that Pascal developed to this kind of question is the absence of a reference class. If you know that there are nine black balls and one white ball in a jar you can calculate from this reference class that the probability of pulling out a white ball is one in ten. Given that there is no obvious reference class of designers, how is it possible to proceed?

One kind of answer that has been suggested is to invoke the theorem developed by the eighteenth-century Presbyterian clergyman Thomas Bayes. In a paper published by a friend after his death,[38] Bayes derived a system of working backwards to determine how a prior degree of belief should rationally change in the light of evidence. If, for instance, you blindly and at random pick two balls from a jar of 30 mixed black and white balls, Bayes' equation showed how, from a knowledge of the prior probabilities and the observation of the balls you picked out, you can calculate the probability of a given number of black and white balls within the jar. An application of the theorem would be to calculate the probability that you have a particular medical condition given the occurance of the condition in the population and the result of a test which can give a false positive.

Could this be applied to the concept of a designer? If some intrinsic probability is attached to the existence of God (which while no particular number can be specified must presumably

[36] W. Paley, *Natural Theology*, 13.
[37] R. Dawkins, *The God Delusion*, 66.
[38] T. Bayes, 'An Essay towards Solving a Problem in the Doctrine of Chances', *Philosophical Transactions* (1683–1775) 53 (1763), 370–418.

lie somewhere between one [true] and zero [false]), then following this sequence of logic, as the Oxford philosopher Richard Swinburne has argued, the calculus of probability can be continuously updated by further data (philosophical reflection, historical evidence, scriptural revelation, religious experience, etc.) which can be added or subtracted.[39]

The difficulty is that statistical probability of any kind still requires statistics. Hence the limitations of this approach, like that of Professor Dawkins, is that no actual numbers are available. Professor Swinburne himself acknowledges this but argues that what his calculus does is to set out in a rigorous formal way the factors which determine how observational evidence supports some more general theory.

Einstein argued that 'the supreme goal of all theory is to make the irreducible basic elements as simple and as few as possible without having to surrender the adequate representation of a single datum of experience'[40] (sometimes paraphrased as 'make it as simple as possible but not simpler'). The data of experience in this instance would include everything within the universe/multiverse.

Avicenna's argument (described in Chapter Sixteen) was that a necessary being must exist and pointed towards the existence of God. His point was that the explanation of the universe must in some sense go beyond the data within it, since the cause of all being could not itself be contingent (something whose existence was due to something other than itself) but must be necessary (something whose existence was due to itself).

From this would follow the question 'If you postulate the existence of some necessary reality outside the universe/multiverse that brings the laws of nature/law of laws into existence, which is more probable: something like a conscious mind or something like an unconscious material?'

One of the contributions made by the new cognitive science of religion (CSR, which we described in Chapter Seven) has been to show that it seems to be an inbuilt characteristic of the human mind to favour the former alternative. Justin Barrett, the experimental psychologist mentioned in Chapter Seven, argues that a belief in gods 'may be an inevitable consequence of the sort of minds we are born with in the sort of worlds we are born into.'[41]

Estimating probability is not limited to those things which we can attach numbers to and mathematically compute. Maxwell's

[39] R. Swinburne, *The Existence of God* (2004).
[40] A. Einstein, 'On the Method of Theoretical Physics', *Philosophy of Science* 1 (1934), 163–9.
[41] J. L. Barrett, *Why Would Anyone Believe in God?* (2004), 91.

contemporary Hermann von Helmholtz was one of the first to suggest that the brain's ability to organize sensory data could be modelled in terms of probabilistic estimation. It is this approach which has led neuroscientists to the so-called 'Bayesian brain hypothesis'.[42]

The Bayesian Brain

In a seminal paper published in 2004 called 'The Bayesian Brain', David Knill and Alexandre Pouget, two cognitive scientists from the University of Rochester, argued that 'psychophysics' was 'providing a growing body of evidence that human perceptual computations were "Bayes Optimal"' (that is to say, they can be accurately modelled by Bayes' theorem). This led directly to the so-called 'Bayesian coding hypothesis' which suggests that 'the brain represents sensory information probabilistically in the form of probability distributions'.[43]

There were, as Knill and Pouget pointed out, a number of computational schemes which had been proposed as to how this might be achieved in populations of neurons, but the upshot was that one way or another our brains appeared to function as 'Bayesian machines', inferring the nature of the world by continuously updating prior probability assessments. If the outputs of the mental tools identified by CSR researchers were integrated with the same kind of Bayesian coding as sensory inputs, then could theism in one sense be seen as the consequence of a kind of unconscious probabilistic calculation?

Like many of the questions raised by this exciting, diverse, and rapidly developing field, the answers (which will inevitably be modified in the light of accumulating empirical data) could appear to critically narrow the gap between ultimate and penultimate questions. That appearance, however, might be misleading.

The data can be (and are) interpreted by researchers in the CSR field in radically different ways, depending on their own prior religious or anti-religious commitments. Is it evidence of a persistent cognitive error in human thinking or does it reveal an inherent human capacity to respond to the divine?

The factors that determine these prior commitments, however, may not necessarily be available to us.

Barrett argues that the rapid first guesses produced by our unconscious 'non-reflective beliefs' form the basis of conscious

[42] K. Friston, 'The History of the Future of the Bayesian Brain', NeuroImage 62 (2012), 1230–3.

[43] D. C. Knill and A. Pouget, 'The Bayesian Brain: The Role of Uncertainty in Neural Coding and Computation', Trends in Neurosciences 27 (2004), 712–19.

'reflective beliefs'. When a reflective belief is demanded, the conscious mind begins by 'reading off' non-reflective beliefs, and where these conflict, weighing up their different inputs (he suggests we could picture the process in terms of a chairperson trying to determine the consensus of a group about a particular proposition). This does not mean though that 'the process by which we arrive at reflective beliefs is transparent and easily inspected'; it may 'remain largely unavailable to conscious consideration'.[44]

Thus in a religious context, 'while many believers can give numerous explicit reasons why they believe in God ... the process of arriving at this belief probably involved few, if any of these explicit reasons, and even then only in part'.[45] For critics of religion who wish to argue themselves and others out of childhood belief, this semi-conscious accumulation of reasons can seem a fatal drawback.

The philosopher Antony Flew, who from 1950 was for 50 years the most prominent exponent of philosophical atheism, pointed to theists who mistakenly imagine 'that if one weak argument is not good enough then a hundred weak arguments are bound to be better'.[46] The neuroscientist Donald Mackay, on the other hand (who was a colleague of Flew's at the University of Keele), argued that so-called 'gestalt perceptions'—the ability to perceive groupings of features as single objects (a *gestalt* or whole form)—typically do occur in a tentative cumulative way.

Using as an example R. C. James' image of a Dalmatian dog on a leafy background [Fig. 50.5], Mackay argues that while it

50.5 R. C. James' image of a Dalmatian dog.

[44] J. L. Barrett, *Why Would Anyone Believe in God?* (2004), 16.
[45] J. L. Barrett, *Why Would Anyone Believe in God?*, 16.
[46] D. M. MacKay, *Behind the Eye* (1991), 249.

is true that an argument is only as strong as its weakest link if it has a serial form, 'if an argument is a parallel argument—one which involves the integration of much parallel evidence', then while 'all the links individually may be too weak to stand up— they may individually give a probability of less than a half to the conclusion—and yet integration over the Gestalt (as in the case of our Dalmation dog) may provide adequate grounds'.[47] It is possible either to look at the image as a random assemblage of blots—to 'see and see and never perceive'—or to project an image on to the blots in the manner of a Rorschach test (where you puts random ink blots in front of someone and force them to tell you what they remind them of).

Yet when small clues (such as perhaps other people focussing on the same area and all claiming to see something) produce an updating of our prior probability assessments in response to cumulative evidence, the image may suddenly leap out as a coherent perception. And when it has done so, it may subsequently become impossible not to perceive.

This kind of process (small cumulative changes in probability assessments effecting perceptions of reality) may form the basis of childhood theism, but they equally seem to play a part in changes of mind that come as the culmination of a lifetime of philosophical reflection.

Thus, in 1951 C. E. M. Joad announced that (after a lifetime of advocating the opposite) the religious view of the world now seemed to him 'the most rational hypothesis . . . which seems to cover most of the facts and to offer the most plausible explanation of our experience as a whole'.[48] In 1988 A. J. Ayer confided to the doctor who was treating him at London University hospital that in a near-death experience, 'I saw a Divine Being. I'm afraid I'm going to have to revise all my various books and opinions'.[49] In 2004 Flew announced that 'following the argument' had finally lead him to theism, giving as the first reason for his change of mind a reassessment of the probabilities involved.[50]

Can, however, human thinking be trusted to give us a true picture of reality? Doubts about this have been raised both by the discoveries of brain science which can seem to undermine trust in our own cognitive processes and by the strange discoveries of physics which can seem to subvert normal understandings of logic and causality. It is at this point that the curtain rises on Act II.

[47] D. M. MacKay, *Behind the Eye*, 250.
[48] C. E. M. Joad, *The Recovery of Belief* (1952), 13–14.
[49] W. Cash, 'Did Atheist Philosopher See God When He "Died"?', https://variousenthusiasms.wordpress.com/2009/04/28/did-atheist-philosopher-see-god-when-he-died-by-william-cash/, accessed 25 August 2015. Ayer never got around to revising his books and opinions, and revised instead the accounts of his experience.
[50] A. Flew, *There Is a God* (2007), 75.

Act II: *Subversion*

In September 1941 during the Nazi occupation of Denmark, the great atomic physicist Niels Bohr received a visit from his erstwhile student and collaborator Werner Heisenberg. In the 1920s both men had pioneered the application of quantum theory to atomic physics, but now in the middle of the war they were on opposing sides. Heisenberg was in charge of the Nazi's atomic reactor research, while Bohr who was half-Jewish would shortly escape to America and join the Manhattan Project that was building the atomic bomb. The two men subsequently gave contradictory accounts of what they talked about. That disagreement, and the reason for Heisenberg's visit, has been the subject of continuing debate, captured in Michael Frayn's 1998 play *Copenhagen*.

In *Copenhagen* Frayn brilliantly parallels Heisenberg's own uncertainty principle with the uncertainty of what happened at that meeting, and uses it to point towards the impossibility of any kind of certain knowledge. The final line of the play talks of the 'final core of uncertainty at the heart of things'.[51]

In the play the principal source of uncertainty comes from the characters' difficulty in fathoming the intentions of each other or themselves. In his 2006 book of philosophy *The Human Touch* Frayn talks of the 'fatal indeterminability at the source of our actions' and describes how 'the synapse fires, or fails to fire and neither human choice nor the great causal chain determines the outcome'.[52] Whereas quantum uncertainty might seem to raise questions about our grasp of reality, the discoveries of brain science can seem to raise equally hard questions about notions of identity and responsibility.

In 1994, four years before Frayn's *Copenhagen*, Francis Crick published a book called *The Astonishing Hypothesis*, which begins by announcing that 'You, your joys and your sorrows, your memories and your ambitions, your sense of personal identity and free will are in fact no more than the behaviour of a vast assembly of nerve cells'. Crick goes on to proclaim that 'the idea that man has a disembodied soul is as unnecessary as the old idea that there was a Life Force'. 'This', he asserts, 'is in head-on contradiction to the religious beliefs of billions of human beings alive today'.[53]

[51] M. Frayn, *Copenhagen* (2003), 94.
[52] M. Frayn, *The Human Touch* (2006), 11.
[53] F. Crick, *The Astonishing Hypothesis* (1994), 3, 261.

The Astonishing Brain

Understanding of how the brain works has been advancing at an accelerating rate during the past century. A recent book has claimed that more has been learnt about it over the past 15 years (during the course of writing this book) than over the rest of human history.[54] Advances in bottom-up neurophysiology have occurred alongside advances in top-down behavioural psychology. There seems to be no aspect of human mental activity that is not amenable to scientific investigation.[55] A number of writers have followed Crick's lead in regarding these extraordinary advances in neuroscience as opening up a new front against religion not only in undermining the idea of the soul but in subverting such fundamental concepts as free will and moral responsibility.

Biblical scholars, by contrast, have pointed out that the portrait of human beings painted in the Bible (and developed by central Christian thinkers like Thomas Aquinas) is not of 'split-level creations (with, say, a distinct body and soul) but as complex, integrated wholes'.[56] Philosophically minded neuroscientists meanwhile have warned against leaping to large philosophical conclusions from highly specific experimental results.

A notorious example of this was some experiments done in the 1980s by the Californian neurologist Benjamin Libet. These measured an electrical activity in the brain described as the 'readiness potential', which occurred when subjects were asked to choose to perform a particular action. Libet's discovery, confirmed by many subsequent experiments, was that this readiness potential preceded the subjects' conscious awareness of choice by some 300 milliseconds. Was this, as some have claimed, evidence that mind events like 'free will' were merely by-products of brain events and hence illusory epiphenomena?

Among the technical challenges to this work were modified forms of the experiment conducted by Christoph Herrmann and his colleagues at the University of Magdeburg in 2007 and Judy Trevena and Jeff Miller at the University of Otago in 2009. In both these experiments the subjects were instructed to press one of two buttons in response to a particular stimulus. In this case the readiness potential not only appeared, as in the Libet experiment, before the choice to press either button but also significantly *before the stimulus was given*. Hence they concluded

[54] D. F. Swaab, *We Are Our Brains* (2014).
[55] A. R. Damasio, *Descartes' Error* (1994).
[56] M. A. Jeeves, *From Cells to Souls— and Beyond* (2004). N. T. Wright, back cover quotation.

that the readiness potential reflected a 'general expectation'[57] of movement. Not a decision but a state of readiness.

Even were this not the case, philosophers have argued that it is not clear that Libet's experiment would have any direct relevance to the kind of choices between moral alternatives that are normally referred to in the concept of 'free will'. The decisions for which we are held most responsible are those which are carefully considered rather than those taken over a fraction of a second. For a good reason a person faced with a momentous decision may want 'to sleep on it'. Even split-second decisions may reflect established patterns of behaviour, formed out of previous choices and shaped by the various influences and desires that have molded character and developed reflexes.

A more fundamental consideration concerns what Malcolm Jeeves, Emeritus Professor of Psychology at St Andrews, describes as the 'irreducible intrinsic interdependence' between brain events and mind events. While analogies between computer hardware and software are often 'smuggled in as if they were explanations',[58] the reality, Jeeves argues, is more complex.

There is, as Donald Mackay pointed out, a correlation between these two levels of description 'but not a translation'. Mackay gives the example of an equation with two roots embodied in a computer: 'the facts about the equation are not facts about the computer ... Computers don't have roots. And yet if the computer is solving the equation there is a direct physical correlate for the statement "the equation has two roots" and any engineer can tell you what it is'.[59] This would not be grounds for dismissing either the roots or the solution of the equation as 'illusory epiphenomena'. Any complete description of what was going on would need to take into account both the electronics and the mathematics and do full justice to both.

If, however, the vertiginous perspectives of looking into our own brains can seem to subvert our thinking in one way, the strange vistas that appear when, as Maxwell put it, 'we look down through the great millstones of the world' into the depths of physical reality can seem to do so in another.

Quantum Quandaries

On the 7th of October 1900 Max Planck, who was at the time Professor of Theoretical Physics at the University of Berlin, received some new experimental data. In response that evening

[57] C. S. Herrmann et al., 'Analysis of a Choice-Reaction Task Yields a New Interpretation of Libet's Experiments', *International Journal of Psychophysiology* 67 (2008), 151–7.

[58] M. A. Jeeves, *From Cells to Souls—and Beyond*, 241–2.

[59] M. A. Jeeves, *From Cells to Souls—and Beyond*, 242.

he wrote down an equation which was soon recognized as having inaugurated a new era in physical science.[60]

Six years before the turn of the century Planck had been commissioned by some electric companies to discover how to create the maximum output of light from light bulbs for the minimum expenditure of energy. This had led him to the problem of so-called 'blackbody radiation' which had proved bafflingly difficult to solve. By 1900 he had written six papers on the subject, totaling some 162 pages, without solving it. Now he had a solution, but it was a disconcerting one.

Up to this point physicists had cheerfully assumed that the laws governing the behaviour of large-scale systems like the Sun and the planets would still hold when scaled down to atomic dimensions. What Planck had showed was that the physical quantity of 'action' (energy × time) could not be indefinitely scaled down. Designated by the letter h and shortly to become known as the Planck constant, there was it seemed a minimal natural unit of action or 'quantum'. The consequence of this discovery, as Planck's friend Albert Einstein remarked, was to 'set science a fresh task: that of finding a new conceptual basis for the whole of physics'.[61]

One of the first to take on this challenge was the Danish physicist Niels Bohr. While working at Copenhagen University Bohr developed a model of an atom as a kind of miniature solar system with electrons orbiting round the nucleus and showed that the electrons would jump in apparently discontinuous 'quantum leaps' from one energy level to another. The process of observing these leaps would, though, have to be paid for in the same units of energy × time, which set an irreducible limit to the precision with which predictions could be made.

It was this realization that led Werner Heisenberg, who had come to Copenhagen to work as a lecturer and assistant to Bohr in 1926, to conclude that not all the information needed to predict the future from the past was available in principle until after the event. The consequence of this 'uncertainty principle' was that Heisenberg and others (following the path trodden by Maxwell) began to develop a new kind of statistical mechanics (so-called 'quantum mechanics'), which instead of trying to calculate the actual position and speeds of atomic particles used a matrix model to calculate the relative probability of particular kinds of event. Meanwhile Erwin Schrödinger developed a wave model which turned out to be formally indentical.

[60] A. Pais, 'Subtle Is the Lord' (1982), 368.
[61] A. Pais, 'Subtle Is the Lord', 372.

What was the physical reality underlying these 'leaps' and 'probability waves'? For Niels Bohr who had been influenced by the positivist idea that scientific formulations were the only kind of valid statement, this was a meaningless question. For Bohr the question was 'What can one say?' To Albert Einstein, however, positivism was an evasion. For Einstein the question was 'What is the case?' He strongly repudiated any idea that 'God plays dice' and insisted to his friend Max Born in 1924 that 'I find the idea quite intolerable that an electron exposed to radiation should choose of its own free will, not only its moment to jump off but also its direction. In that case I would rather be a cobbler, or even an employee in a gaming house, than a physicist'.[62]

This determination to try to get behind the probabilistic mathematics led Einstein and two colleagues, Boris Podolsky and Nathan Rosen, to formulate the so-called 'EPR paradox'. In 1935 they published a paper entitled 'Can Quantum-Mechanical Description of Physical Reality Be Considered Complete?'[63] The paper put forward a thought experiment which sought to show that either there must be 'hidden variables' at work determining how particles move or two particles must communicate instantaneously. Since propagation of information faster than the speed of light would violate the theory of relativity, the paper presented a paradox. What Einstein called *spukhafte Fernwirkung*—spooky action at a distance—could not take place.

Except that it does.

Since the EPR paradox was published, the radically counterintuitive predictions of quantum mechanics have been consistently verified by increasingly sophisticated experiments.[64] The Briggs group in Oxford have implemented a rigorous mathematical test of macrorealism devised by the Nobel Prize winner Sir Anthony Leggett.[65] With colleagues in Delft they went on to make a practical demonstration of another thought experiment, the so-called 'quantum three-box paradox', using a nitrogen-vacancy defect in diamond excited with a microwave pulse [Fig. 50.6].[66] Meanwhile the 'spooky' phenomenon of quantum entanglement has become central to attempts to develop new quantum technologies.

Many questions remain open.[67] A poll of physicists, philosophers, and mathematicians revealed that the foundations of quantum mechanics remain hotly debated in the scientific community, with a divergence of views on some fundamental questions.[68]

[62] A. Einstein, *The Born–Einstein Letters: Correspondence between Albert Einstein and Max and Hedwig Born from 1916–1955, with Commentaries by Max Born* (1971), 82.

[63] A. Einstein, B. Podolsky, and N. Rosen, 'Can Quantum-Mechanical Description of Physical Reality Be Considered Complete?', *Physical Review* 47 (1935), 777–80.

[64] M. Giustina et al., 'Bell Violation Using Entangled Photons without the Fair-Sampling Assumption', *Nature* 497 (2013), 227–30.

[65] G. C. Knee et al., 'Violation of a Leggett–Garg Inequality with Ideal Non-invasive Measurements', *Nature Communications* 3 (2012).

[66] R. E. George et al., 'Opening up Three Quantum Boxes Causes Classically Undetectable Wavefunction Collapse', *Proceedings of the National Academy of Sciences of the United States of America* 110 (2013), 3777–81.

[67] G. A. D. Briggs, J. N. Butterfield, and A. Zeilinger, 'The Oxford Questions on the Foundations of Quantum Physics', *Proceedings of the Royal Society A—Mathematical Physical and Engineering Sciences* 469 (2013), 20130299.

[68] M. Schlosshauer, J. Kofler, and A. Zeilinger, 'A Snapshot of Foundational Attitudes toward Quantum Mechanics', *Studies in History and Philosophy of Modern Physics* 44 (2013), 222–30.

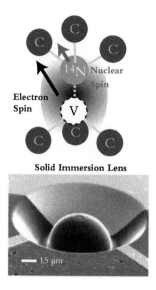

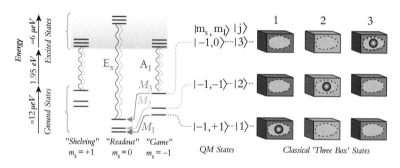

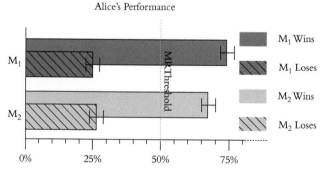

50.6 Quantum three-box experiment carried out by Andrew Briggs' group in collaboration with colleagues in Delft.

The quantum three-box experiment can be described as a guessing game in which if macrorealism holds then Bob will win and if quantum theory holds the younger and more intelligent Alice will win. The system to be measured is a nitrogen atom adjacent to a carbon vacancy in diamond, in which an integrated lens has been milled. Three nuclear spin states constitute the three boxes; the electron spin states provide the means for opening the boxes and looking inside. If macrorealism were applicable, then at best Alice could win half the time. Regardless of whether Bob looked in the first box (M_1) or the second (M_2), Alice does better than that by a comfortable margin.

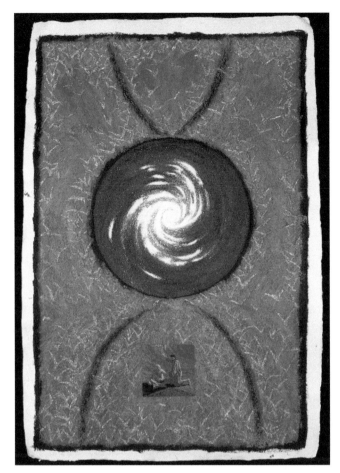

50.7 'Where were you when I laid the foundations of the earth?' From *Out of the Whirlwind* by Roger Wagner.

50.8 'My ears had heard of you but now my eyes have seen you'. From *Out of the Whirlwind* by Roger Wagner.

Scientists have not stopped using quantum mechanics until these questions are resolved (anymore than Maxwell's bell
ringers would give up campanology until they had discovered
the workings of the mysterious belfry). The theory is too robust and useful for that. Neither, however, have they abandoned
the conviction that well-designed experiments can lead us to
a closer understanding of reality. Experiments like those conducted at Oxford show that, like the *via negativa* espoused by
theologians such as Maimonides and Aquinas, experimental
science sometimes shows us what we cannot believe.

Partly as a consequence of the interest stimulated by Frayn's
play *Copenhagen*, a number of documents written by Bohr and
Heisenberg came to light which clarified without completely
resolving some of the details of their meeting.[69] In the Book
of Job, the 179 or so questions posed 'from out of the whirlwind' [Fig. 50.7] are answered by Job first with the under-
standable response 'Surely I spoke of things . . . too wonderful
for me to know', but then with the more unexpected reflection 'My ears had heard of you but now my eyes have seen
you' [Fig. 50.8].[70]

The sidelong approach to truth is one which intractable
questions of all sorts seem to demand. Philoponus talked of
'hypotheses' and Ørsted of 'gedankenexperimente' but metaphorical thought experiments form a kind of common currency all the way from cave paintings to quantum physics. They
can, as both of us in our different paths have discovered, give a
partial access to areas of reality that otherwise remain wholly
out of reach.

Far from undermining or subverting the search for truth, the
necessity (which science shares with art and religion) of living
with the unresolved questions inherent in this approach does
not, however, seem to diminish the quest for answers or the
hope that they can be found.

The philosopher Karl Popper argued that 'that scientific discovery is impossible without faith in ideas of a purely speculative kind, and sometimes even quite hazy; a faith which is
completely unwarranted from the point of view of science, and
which to that extent is *metaphysical*'.[71] Max Planck once remarked that 'Science demands also the believing spirit. Anyone
who has been seriously engaged in scientific work of any kind
realises that over the entrance of the gate of the temple of science are written the words "ye must have faith".'[72]

[69] M. Frayn, *Copenhagen*, 95–149.
[70] Job 42:3, 5.
[71] K. Popper, *Logik der Forschung* (1934), 38.
[72] M. Planck, *Where Is Science Going?* (1933), 241.

The entanglement of this extra-scientific dimension in the day-to-day practice of science brings us back for the last time to the invocations in Oxford and Cambridge and to our final act.

ACT III: *Resolution*

In the Michaelmas term of 1999 an audience began to assemble to hear the inaugural lecture of a new Oxford chair: the Andreas Idreos Professorship of Science and Religion. Gathering in one room at the examination schools, like the suspects at the end of an Agatha Christie novel, were an array of scientists, philosophers, and theologians representing every point on the intellectual compass. At the sides of the room were Peter Atkins and Richard Dawkins, who had both moved outside their research fields of (respectively) physical chemistry and evolutionary biology to become public advocates of atheism. Near the centre were Arthur Peacocke and Alister McGrath who had begun their careers in science (the one as a professor of biochemistry the other as a physical chemist) and ended up as professors of theology.

The new chair was almost unique at Oxford in being interdisciplinary in conception, and the inaugural professor, John Hedley Brooke, joked that the university had perhaps indicated its opinion of the different aspects of his subject by choosing to house him in a redundant vicarage that was shortly to be demolished to make way for a new laboratory. Building this new laboratory would, however, plunge the university into the same storm of controversy that had faced Henry Acland some 120 years earlier in 1883.

Vivisection in Oxford

The 1883 storm had begun when Convocation voted £10,000 for the construction of a laboratory at the museum to accommodate the work of the newly appointed Professor of Physiology, Dr Burden Sanderson. Dr Sanderson held a licence for experiments under the 1876 vivisection act, which immediately sparked fears of 'the establishment at Oxford of a chamber of horrors'.

Acland began to receive letters addressed to THE GREAT PROFESSOR OF SCIENCE, contrasting the compassion of his home with 'the diabolical cruelty' practiced in his laboratory against 'helpless dumb animals'. John Ruskin, the Slade Professor, resigned in protest, and in 1884 a pamphlet entitled *Vivisection in Oxford* was printed to persuade members of Convocation to prevent the sale of land to fund the new laboratory. Acland responded to this by asserting his absolute confidence 'in the humanity of the great men of his acquaintance who practiced vivisection', insisting that he had no doubt that they were morally justified in taking steps that would enable them 'to save mankind from suffering and pain'.[73] On February 4th he wrote a letter to *The Times* appealing (successfully) for members of Convocation to help defeat the wrecking proposal.

The 2004 storm broke when the construction of the new biomedical laboratory on the site of the old St Cross vicarage in South Parks Road was halted after the contractor pulled out in the face of threats from animal rights protestors. A new contractor was eventually appointed but intimidation and vandalism continued (during this period the street where this book was being written was closed several times by police to protect our neighbour Sir Colin Blakemore, then Waynflete Professor of Physiology). In 2006 there was something of a turning of the tide when Laurie Pycroft, a sixteen year old from Swindon, started a counter movement called Pro-Test to support the laboratory. At their first rally (held in Broad Street next to the site of Acland's old house) a succession of medical researchers spoke of how their research might alleviate human suffering [Fig. 50.9].

50.9 Pro-test demonstration in the Oxford Broad.

[73] J. B. Atlay, *Sir Henry Wentworth Acland* (1903), 422.

The Difficult Conclusion

Like the agonizing dilemma that Bohr and Heisenberg had faced in the application of their research to atomic weapons, both these controversies powerfully dramatized the continuing and ineradicable influence of moral considerations in shaping and inspiring the practice of science. Scientific discoveries, far from removing us from the cognitive world of moral choice that *Homo sapiens* had first been thrust into during 'the Garden of Eden moment' in human evolution, seem rather to plunge us ever more deeply into it. In that world the need to discover principles by which to work and stars by which to navigate becomes not less but more urgent.

'Happy is the man who can recognise in the work of Today a connected portion of the work of life, and an embodiment of the work of Eternity'.[74] Maxwell's reflection, written at the age of 23 (and with as much application to an artist as a scientist), expressed the aspiration that his working life should as far as possible be integrated with his fundamental ethical and spiritual identity. The urgent need to act as a whole both as individuals and as societies has, we have suggested, given 'ultimate curiosity'—the need to make sense of the world as a whole—both a central and a permanent position in human life, so that behind all the shifting interests and motivations in modern science it continues to exert its pull.

The vast resources expended on large-scale international projects like the CERN large hadron collider give a contemporary example of this [Fig. 50.10]. The World Wide Web is a spin-off from CERN that is transforming the world. The project itself, however, was set up to answer fundamental questions about the nature of the universe. Science cannot predict its own outcomes, and the discoveries that emerge from the collider (which was being restarted as this book went to press) may change our understanding of the universe. The enthusiasm for such research is shared not merely by the scientists involved but by the national communities that fund it. In neither case is the motivation an economic one.

In 2014 the Director of CERN, Rolf-Dieter Heuer, was appointed guest editor of a special edition of the leading French language newspaper in Switzerland, *Le Temps*, to celebrate the sixtieth anniversary of the project. At his personal request it carried a whole page interview with one of the authors of this book,

[74] B. Mahon, *The Man Who Changed Everything* (2003), 47.

50.10 The CERN Large Hadron Collider.

which was accompanied by Rolf Heuer's own editorial comment that 'Instead of opposing science and religion, you should rather ask if science and religion can respect and accept each other's point of view. In my view it is absolutely possible, and in my experience it is even enriching.'[75]

Despite appearances, penultimate curiosities about the physical world continue to swim in the slipstream of ultimate questions.

That slipstream, it needs to be stressed, is not the special preserve of religions (the biblical claim that 'God has put eternity into the heart of man' recognizes a characteristic shared by all humanity). Consequently while the audience that assembled to hear the new Professor of Science and Religion's inaugural lecture may have held dramatically variegated views on the way to integrate the practice and discoveries of science with the ultimate questions of human existence, they may nevertheless all have been united in their struggle to do so.

Francis Crick's statement, quoted in the Prologue, that his whole scientific career had been shaped by a desire to remove

[75] N. Ulmi, 'Science et foi, comment ne pas choisir', *Le Temps* (27 Sept. 2014), 28.

the supports from religion demonstrates how the focussed investigations of physical science can swim in the slipstream of very different kinds of metaphysical concerns. What he said is in this sense a surprising illustration of our fundamental metaphor. Crick (who installed a model of DNA over the door of his Cambridge house) was well aware that his pugnacious atheism was far from universally accepted. And this was true even in his own former laboratory.

In an article written in 1987 on the history of the Cavendish Laboratory the then Cavendish Professor Sir Brian Pippard first described how 'The great oak doors opening on the site of the original building had carved on them, by Maxwell's wish, the text from Psalm 111 *Magna opera Domini exquisita in omnes voluntates ejus*'.

He then went on to recount how 'Shortly after the move to the new buildings in 1973 a devout research student suggested to me that the same text should be displayed, in English, at the entrance. I undertook to put the proposal to the Policy Committee, confident that they would veto it; to my surprise, however, they heartily agreed both to the idea and to the choice of Coverdale's translation, inscribed here on mahogany by Will Carter' [Fig. 50.11].[76] It reads 'The works of the Lord are great, sought out of all them that have pleasure therein' [Fig. 50.12].

The 'devout research student' was Andrew Briggs. The broad sentiment of that inscription, as surveys have demonstrated,[77] remains more widespread within the scientific community

50.11 A group of senior staff outside the main entrance to the new Cavendish Laboratory (*c.* 1980).

[76] A. B. Pippard, 'The Cavendish Laboratory', *European Journal of Physics* 8 (1987), 231–5.
[77] E. H. Ecklund, *Science vs. Religion* (2010).

50.12 The inscription over the entrance of the new Cavendish Laboratory.

than is generally recognized. While the form in which such thoughts are expressed will change as cultures evolve, the entrenched need of human beings to make sense of the whole depth of their experience of the world remains constant: rooted in the cognitive capacities that at least 90,000 years ago first gave rise to *Homo religiosus*. It follows that the persistent entanglement we have been following throughout this book between penultimate and ultimate curiosity cannot be wished away: it will last as long as humanity lasts.

List of Illustrations

Obtaining permissions to use these pictures was made possible through the support of a grant from Templeton Religion Trust. The opinions expressed in this book are those of the authors and do not necessarily reflect the views of Templeton Religion Trust.

Bibliography

Page numbers where the bibliography entry is cited are given in square brackets.

Acland, H.W. *Memoir on the Cholera at Oxford, in the Year 1854: With Considerations Suggested by the Epidemic.* Oxford: J. H. and J. Parker, 1856. [367]

Acland, H.W. *The Harveian Oration.* London: Macmillan, 1865. [379, 380, 382, 391]

Adamson, P. *Classical Philosophy: A History of Philosophy without Any Gaps. Volume 1.* Oxford: Oxford University Press, 2014. [148]

Adkins, L. *Empires of the Plain: Henry Rawlinson and the Lost Languages of Babylon.* London: Harper Perennial, 2004. [330, 331, 332]

Agathias. *The Histories*, trans. Frendo, J. D. Berlin: W. de Gruyter, 1975. [117, 118]

Albertus, M. S. *Book of Minerals.* Oxford: Clarendon Press, 1967. [157]

Ammonius, H. *On Aristotle Categories.* London: Duckworth, 1991. [104]

Ammonius, H. *Ammonius on Aristotle: On Interpretation 1–8.* London: Duckworth, 1996. [104]

Anaxagoras. *Anaxagoras of Clazomenae: Fragments and Testimonia: A Text and Translation with Notes and Essays*, ed. and trans. Curd, P. London: University of Toronto Press, 2007. [78]

Aquinas, T. *The Summa Theologica: Complete Edition*, trans. Fathers of the English Dominican Province. New York: Catholic Way Publishing, 2014. [157]

Aristophanes. *The Clouds*, trans. Alan H. Sommerstein. Aris and Phillips, 1973. [67]

Aristotle. *De Caelo*, trans. Stocks, J. L., and Wallis, H. B. Oxford: Clarendon Press, 1922.

Aristotle. *Nicomachean Ethics*, trans. Rackham, H. Cambridge, MA: Harvard University Press, 1968. [89, 90]

Aristotle. *The Metaphysics; oeconomica and Magna Moralia*, trans. Tredennick, H., and Armstrong, G. C. Cambridge, MA: Harvard University Press, 1969. [75, 89]

Aristotle. *On the Parts of Animals*, trans. Ogle, W. Oxford: Clarendon Press, 2001. [89]

Armstrong, P. A. *The Piasa: Or, the Devil among the Indians.* Morris: E. B. Fletcher, 1887. [16]

al-Ash'arī, Abū al-Ḥasan ʿAlī ibn Ismāʿīl. *Risālat Istiḥsān al-khawḍ fī ilm al-Kalām.* Ḥaydarābād al-Dakan: Maṭbaʿat Majlis Dāʾirat al-Maʿārif al-Niẓāmīyah al-kāʾinah bi-al-Hind, 1925. [144]

Atlay, J. B. *Sir Henry Wentworth Acland: A Memoir.* London: Smith, Elder, 1903. [xvii, 365, 366, 367, 368, 369, 371, 377, 378, 381, 436]

Atran, S. *In Gods We Trust: The Evolutionary Landscape of Religion.* New York: Oxford University Press, 2002. [54]

Aubert, M., Brumm, A., Ramli, M., Sutikna, T., Saptomo, E. W., Hakim, B., Morwood, M. J., Van Den Bergh, G. D., Kinsley, L., and Dosseto, A. 'Pleistocene Cave Art from Sulawesi, Indonesia', *Nature* 514 (2014), 223. [32, 33]

Augustine. *On Christian Doctrine.* Indianapolis: Bobbs-Merrill, 1958. [111]

Augustine. *Ancient Christian Writers*, vol. I: *The Literal Meaning of Genesis*, trans. Taylor, J. H., S.J. New York: Paulist Press, 1982. [111]

Avicenna. *The Metaphysics of the Healing: A Parallel English-Arabic Text (Al-Ilahīyāt Min Al-Shifaʾ)*, trans. Marmura, M. E. Provo, UT: Brigham Young University Press, 2005.

Ayer, A. J. *Language, Truth and Logic.* London: Gollancz, 1946. [414]

Bacon, F. *The Works of Francis Bacon.* London: Longmans, Green, Reader, and Dyer, 1857.

Bacon, F. *Novum Organum.* New York: Collier, 1902. [234, 236, 240]

Bacon, F. *The Advancement of Learning and New Atlantis.* Oxford: Clarendon Press, 1974. [232, 233, 239]

Bacon, R. *The Opus Majus of Roger Bacon.* Oxford: Clarendon Press, 1897. [169]

Bacon, R. *The Opus Majus of Roger Bacon.* London: Oxford University Press, 1928. [167, 168, 169, 170, 171]

Bahn, P. G. *The Cambridge Illustrated History of Prehistoric Art.* Cambridge: Cambridge University Press, 1998. [6]

Baker, S. W. 'Account of the Discovery of the Second Great Lake of the Nile, Albert Nyanza', *Journal of the Royal Geographical Society of London* 36 (1866), 1–18. [4]

Barnes, J. *Early Greek Philosophy.* Harmondsworth: Penguin, 2002. [78]

Barrett, J. L. 'Metarepresentation, *Homo religiosus*, and *Homo symbolicus*', in *Homo Symbolicus: The Dawn of Language, Imagination and Spirituality*, ed. Henshilwood, C. S., and d'Errico, F., 205–24. Amsterdam: John Benjamins, 2011. [55, 56]

Barrett, J. L. *Why Would Anyone Believe in God?* Oxford: Altamira Press, 2004. [57, 423, 425]

Baur, L. *Die Philosophischen Werke Des Robert Grosseteste, Bischofs Von Lincoln: Zum Erstenmal Vollständig in Kritischer Ausgabe Besorgt Von Ludwig Baur.* Münster: Aschendorff, 1912. [154, 155]

Bayes, T. 'An Essay towards Solving a Problem in the Doctrine of Chances', *Philosophical Transactions (1683–1775)* 53 (1763), 370–418. [422]

Bednarik, R. G. 'First Dating of Pilbara Petroglyphs', *Records of the Western Australian Museum* 20 (2002), 415–29. [21]

Ben-Chaim, M. *Experimental Philosophy and the Birth of Empirical Science: Boyle, Locke and Newton.* Aldershot: Ashgate, 2004. [284]

Berjak, R. and Iqbal, M. 'Ibn Sina–al-Biruni Correspondence (2)', *Islam and Science* 1 (2003), 253–60. [139]

Berjak, R. and Iqbal, M. 'Ibn Sina–al-Biruni Correspondence (5)', *Islam and Science* 1 (2005), 57–63. [139, 140]

Bernard, C. 'Introduction à l'étude de la médecine expérimentale', http://www.cosmovisions.com/textes/Bernard010204.htm#mLcxy51q7sM7yTxF.99http://www.cosmovisions.com/textes/Bernard010204.htm. Accessed 13 August 2015. [238]

Biggar, N. 'What Are Universities For? A Christian View', in *Theology and Human Flourishing: Essays in Honour of Timothy J. Gorringe*, ed. Highton, M., Rowland, C., and Law, J. Eugene: Cascade Books, 2011. [104]

Blackwell, R. J. *Galileo, Bellarmine, and the Bible: Including a Translation of Foscarini's Letter on the Motion of the Earth.* London: University of Notre Dame Press, 1991. [191]

Boesch, C. 'Symbolic Communication in Wild Chimpanzees?', *Human Evolution* 6 (1991), 81–9. [56]

Boyle, R. *The Sceptical Chymist: Or Chymico-Physical Doubts & Paradoxes, Touching the Spagyrist's Principles.* London: Printed by J. Cadwell for J. Crooke, 1661. [257]

Boyle, R. *A Defence of the Doctrine Touching the Spring and Weight of the Air.* London: Printed by F.G. for Thomas Robinson, 1662. [245]

Boyle, R. *The Works of the Honourable Robert Boyle.* London: Printed for J. and F. Rivington, 1772. [245, 246]

Briggs, G. A. D. 'The Search for Evidence-Based Reality', in *The Science and Religion Dialogue: Past and Future*, ed. Welker, M. New York: Peter Lang Edition, 2014, 201–15. [199]

Briggs, G. A. D., Butterfield, J. N., and Zeilinger, A. 'The Oxford Questions on the Foundations of Quantum Physics', *Proceedings of the Royal Society A: Mathematical Physical and Engineering Sciences* 469 (2013), 20130299. [431]

Brodrick, A. H. *The Abbé Breuil, Prehistorian; a Biography.* London: Hutchinson, 1963. [8, 23, 24]

Brooke, J. H. *Science and Religion: Some Historical Perspectives.* Cambridge: Cambridge University Press, 1991. [xv, 286, 376]

Browne, E. J. *Charles Darwin: Voyaging.* London: Jonathan Cape, 1995. [297, 301, 304, 308]

Buridan, J. *Iohannis Buridani Quaestiones super Libris Quattuor de Caelo et Mundo.* Cambridge, MA: Mediaeval Academy of America, 1942. [174]

Burnett, C. 'The Coherence of the Arabic-Latin Translation Program in Toledo in the Twelfth Century', *Science in Context* 14 (2001), 249–88. [150]

Butterworth, B. *The Mathematical Brain.* London: Papermac, 2000. [14]

Campbell, L. *The Life of James Clerk Maxwell: With a Selection from His Correspondence and Occasional Writings and a Sketch of His Contributions to Science*, ed. Garnett, G., and Adams, M. London: Macmillan, 1882. [385, 396, 400, 401, 404, 405]

Cann, R. L., Stoneking, M., and Wilson, A. C. 'Mitochondrial DNA and Human Evolution', *Nature* 325 (1987), 31–6. [29]

Cannon, W. F. 'The Impact of Uniformitarianism: Two Letters from John Herschel to Charles Lyell, 1836–1837', *Proceedings of the American Philosophical Society* 105 (1961), 305. [305, 318]

Cantor, G. N. *Michael Faraday: Sandemanian and Scientist: A Study of Science and Religion in the Nineteenth Century.* Basingstoke: Macmillan, 1991. [396]

Cash, W. 'Did Atheist Philosopher See God When He "Died"?', https://variousenthusiasms.wordpress.com/2009/04/28/did-atheist-philosopher-see-god-when-he-died-by-william-cash/. Accessed 25 August 2015. [426]

Caspar, M. *Kepler.* New York: Dover Publications, 1993. [222, 223, 225, 226, 227, 228, 229]

Chauvet, J.-M., Deschamps, E. B., and Hilaire, C. *Chauvet Cave: The Discovery of the World's Oldest Paintings*, trans. Bahn, P. G. London: Thames & Hudson, 2001. [25]

Chrysostom, J. *Apologist.* Washington, DC: Catholic University of America Press, 2001. [165]

Cicero, *De Natura Deorum*, trans. Rackham, H. Cambridge, MA: Harvard University Press, 1933. [92]

Clagett, M. *The Science of Mechanics in the Middle Ages.* Madison: University of Wisconsin Press, 1959. [174, 208]

Clarke, M. L. *Higher Education in the Ancient World.* London: Routledge & K. Paul, 1971. [110]

Clarke, S. *A Collection of Papers, Which Passed between the Late Learned Mr Leibniz, and Dr Clarke, in the Years 1715 and 1716*, http://www.newtonproject.sussex.ac.uk/catalogue/viewcat.php?id=THEM00224. Accessed 15 August 2015. [274, 275, 281, 282]

Clegg, B. *The First Scientist: A Life of Roger Bacon.* London: Constable, 2003. [160, 161, 162]

Cleveland Coxe, A. *The Ante-Nicene Fathers: Translations of the Writings of the Fathers Down to A.D. 325*, vol. I: *The Apostolic Fathers, Justin Martyr, Irenaeus.* New York: Christian Literature Publishing Co., 1885. [109, 110]

Clement of Alexandria, *The Writings of Clement of Alexandria Volume II*, ed. Roberts, A., and Donaldson, J. Edinburgh: T. and T. Clark, 1869. [107]

Clottes, J. *Return to Chauvet Cave: Excavating the Birthplace of Art: The First Full Report*. London: Thames & Hudson, 2003. [60]

Cohen, M. R. and Nagel, E. *An Introduction to Logic and the Scientific Method*. London: George Routledge & Sons, 1934. [419]

Combier, J. and Jouve, G. 'Nouvelles recherches sur l'identité culturelle et stylistique de la grotte Chauvet et sur sa datation par la méthode du ¹⁴C', *L'Anthropologie* 118 (2014), 115–51. [26]

Conway Morris, S. *The Crucible of Creation: The Burgess Shale and the Rise of Animals*. Oxford: Oxford University Press, 2003. [420]

Conway Morris, S. *Life's Solution: Inevitable Humans in a Lonely Universe*. Cambridge: Cambridge University Press, 2003. [420]

Coxe Stevenson, M. *Ethnobotany of the Zuñi Indians*, Thirtieth Annual Report of the Bureau of American Ethnology to the Secretary of the Smithsonian Institution (1908–1909). [22]

Curley, M. J. *Physiologus*. Austin: University of Texas Press, 1979. [231]

Dalley, S. *Myths from Mesopotamia: Creation, the Flood, Gilgamesh, and Others*. Oxford: Oxford University Press, 1989. [339, 348]

Damasio, A. R. *Descartes' Error: Emotion, Reason, and the Human Brain*. New York: G. P. Putnam, 1994. [428]

Dart, R. A. *Adventures with the Missing Link*. London: H. Hamilton, 1959. [27]

Darwin, C. R. *Notebook D: Transmutation (1838)* Cambridge University Library DAR 123. [316]

Darwin, C. R. *Notebook B: Transmutation (1837–1838)* Cambridge University Library DAR 121. [316]

Darwin, C. R. *The Descent of Man, and Selection in Relation to Sex*. London: John Murray, 1871. [3, 5, 48, 51]

Darwin, C. R. *The Life and Letters of Charles Darwin: Including an Autobiographical Chapter*. London: John Murray, 1887. [375, 416, 421]

Darwin, C. R. *The Foundations of the Origin of Species: Two Essays Written in 1842 and 1844*. Cambridge: Cambridge University Press, 1909. [317]

Darwin, C. R. *The Autobiography of Charles Darwin and Selected Letters*. New York: Dover, 1958. [296, 299, 304, 308, 309, 314, 315, 316, 318, 342]

Darwin, C. R. *The Voyage of the Beagle*. New York: Harper, 1959. [3, 307]

Dawkins, R. *The Blind Watchmaker*. Harlow: Longman Scientific & Technical, 1986. [416, 417]

Dawkins, R. *The God Delusion*. London: Bantam Press, 2006. [418, 422]

d'Errico, F. 'The Invisible Frontier. A Multiple Species Model for the Origin of Behavioral Modernity', *Evolutionary Anthropology* 12 (2003), 188–202. [47]

Desmond, A. and Moore, J. *Darwin*. London: Michael Joseph, 1991. [4, 304, 307]

Desmond, A. and Moore, J. *Darwin's Sacred Cause: Race, Slavery and the Quest for Human Origins*. London: Allen Lane, 2009. [296]

d'Holbach, P. H. T. *The System of Nature*, vol. 1, Chapter 1. 1821. [262]

Diamond, J. M. *Guns, Germs and Steel: A Short History of Everybody for the Last 13,000 Years*. London: Vintage, 1998. [62]

Dinkova-Bruun, G., Gasper, G. E. M., Huxtable, M., McLeish, T. C. B., Panti, C., and Smithson, H. *The Dimensions of Colour: Robert Grosseteste's De Colore*. Durham: Pontifical Institute of Mediaeval and Renaissance Studies, 2013. [154, 155]

Diogenes, L. *Lives of Eminent Philosophers*. London: Heinemann, 1972. [78, 79]

Donne, J. *Essays in Divinity*. Oxford: Clarendon Press, 1952. [341]

Draper, J. W. *History of the Conflict between Religion and Science*. London: Henry S. King, 1875. [xv]

Drozdek, A. *Greek Philosophers as Theologians: The Divine Arche*. Aldershot: Ashgate, 2007. [74]

Dunbar, R. I. M. 'Coevolution of Neocortical Size, Group Size and Language in Humans', *Behaviour and Brain Sciences* 16 (1993), 681–94. [39]

Eastman, M. 'Marxism: Science or Philosophy?', https://www.marxists.org/archive/eastman/1935/science-philosophy.htm. Accessed 24 August 2015. [414]

Easton, S. C. *Roger Bacon and His Search for a Universal Science*. New York: Russell & Russell, 1952. [151]

Ecklund, E. H. *Science vs. Religion: What Scientists Really Think*. New York: Oxford University Press, 2010. [439]

Einstein, A. 'On the Method of Theoretical Physics', *Philosophy of Science* 1 (1934), 163–9. [423]

Einstein, A. *Out of My Later Years*. Westport, CT: Greenwood Press, 1970.

Einstein, A., Podolsky, B., and Rosen, N. 'Can Quantum-Mechanical Description of Physical Reality Be Considered Complete?', *Physical Review* 47 (1935), 777–80. [431]

Eliot, C. W. *The Harvard Classics*. New York: P. F. Collier & Sons, 1910. [217]

Elweskiöld, B. *John Philoponus against Cosmas Indicopleustes: A Christian Controversy of the Structure of the World in Sixth-Century Alexandria*. Lund: Lund University Department of Classics and Semitics, 2005. [121]

Empedocles. *The Poem of Empedocles: A Text and Translation with an Introduction*, ed. and trans. Inwood, B. London: University of Toronto Press, 1992. [74]

Epictetus. *Discourses, Fragments, Handbook*, trans. Hard, R., ed. Gill, C. Oxford: Oxford University Press, 2014. [92]

Epicurus. *The Epicurus Reader: Selected Writings and Testimonia*, trans. and ed. Inwood, B., and Gerson, L. P. Cambridge: Hackett, 1994. [93]

Evans, E. P. *Animal Symbolism in Ecclesiastical Architecture*. London: Heinemann, 1896. [231]

Evelyn, J. *The Diary of John Evelyn*, ed. de la Bédoyère, G. Woodbridge, Suffolk: Boydell Press, 1995. [242]

Faye, H. *Sur l'origine du Monde: Théories Cosmogoniques des Anciens et des Modernes*. Paris: Gauthier-Villars, 1884. [262]

Feynman, R. P., Leighton, R. P., and Sands, M. *The Feynman Lectures on Physics*. Reading: Addison-Wesley, 1965. [93, 384]

Feynman, R. P. *'Surely You're Joking, Mr. Feynman!': Adventures of a Curious Character*. New York: W. W. Norton, 1985. [142]

Finkel, Y. I. *The Ark before Noah: Decoding the Story of the Flood*. London: Hodder & Stoughton, 2014. [345, 346, 350, 352]

Flamsteed, J. *Self Inspections of J.F.* 1667–1671 Royal Greenwich Observatory RGO 1/32/A. [260]

Flandin, E. *Voyage en Perse de MM Eugène Flandin, peintre, et Pascal Coste, architecte*, vol. I. Paris: Gide et Jules Baudry, 1851. [325]

Flew, A. *There Is a God: How the World's Most Notorious Atheist Changed His Mind*. New York: HarperOne, 2007. [426]

Forster, R. and Marston, P. *God's Strategy in Human History*. Eugene, OR: Wipf and Stock, 2001. [165]

Frayn, M. *Copenhagen*. London: Methuen, 2003. [427, 434]

Friston, K. 'The History of the Future of the Bayesian Brain', *NeuroImage* 62 (2012), 1230–3. [424]

Galilei, G. *Dialogues Concerning Two New Sciences*. New York: Macmillan, 1914. [205, 207, 208, 209]

Galilei, G. *Dialogue Concerning the Two Chief World Systems—Ptolemaic & Copernican*. Berkeley: University of California Press, 1953. [202]

Galilei, G. *Discoveries and Opinions of Galileo*, trans. Drake, S. New York: Doubleday, 1957. [183, 191, 199, 200, 209]

Galilei, G. *On Motion, and on Mechanics: Comprising De Motu (ca. 1590)*, trans. Drabkin, I. E., and Drake, S. Madison: University of Wisconsin Press, 1960. [186]

Galilei, G. *The Essential Galileo*. Indianapolis: Hackett, 2008. [184, 187, 188, 191, 193, 194, 195, 205, 207, 209]

George, R. E., Robledo, L., Maroney, O. J. E., Blok, M., Bernien, H., Markham, M. L., Twitchen, D. J., Morton, J. J. L., Briggs, G. A. D., and Hanson, R. 'Opening up Three Quantum Boxes Causes Classically Undetectable Wavefunction Collapse', *Proceedings of the National Academy of Sciences of the United States of America* 110 (2013), 3777–81. [431]

al-Ghazzālī. *Al-Ghazali's Tahafut al-Falasifah (Incoherence of the Philosophers)*, trans. Kamali, S. A. Lahore: Pakistan Philosophical Congress, 1963. [144]

al-Ghazzālī. *The Incoherence of the Philosophers: A Parallel English-Arabic Text* [Tahāfut al-falāsifah], trans. Marmura, M. E. Provo, UT: Brigham Young University Press, 2000. [145]

Gilbert, B. J. 'Puncturing an Oxford Myth'. http://oxoniensia.org/volumes/2009/gilbert.pdf. Accessed 22 August 2015. [371]

Giustina, M., Mech, A., Ramelow, S., Wittmann, B., Kofler, J., Beyer, J., Lita, A., Calkins, B., Gerrits, T., Nam, S., Ursin, R., and Zeilinger, A. 'Bell Violation Using Entangled Photons without the Fair-Sampling Assumption', *Nature* 497 (2013), 227–30. [431]

Goldman, M. *The Demon in the Aether: The Story of James Clerk Maxwell*. Edinburgh: P. Harris, 1983. [392, 401]

Gombrich, E. H. *Art and Illusion: A Study in the Psychology of Pictorial Representation*. London: Phaidon, 1977. [60]

Goodall, J. *Reason for Hope: A Spiritual Journey*. London: Thorsons, 1999. [42]

Goodall, J. *In the Shadow of Man*. London: Phoenix, 1999. [36, 37, 39, 40]

Goodall, J. 'Primate Spirituality', in *The Encyclopedia of Religion and Nature* (2005). [41]

Grant, E. 'Peter Peregrinus', in *Dictionary of Scientific Biography*. New York: Scribners, 1975, 10: 532. [161]

Grosseteste, R. *Commentarius in Posteriorum Analyticorum Libros*, ed. Rossi, P. Firenze: L.S. Olschki, 1981. [153]

Grosseteste, R. *Hexaëmeron*, ed. Dales, R. C., and Gieben, S. Oxford: Oxford University Press for the British Academy, 1982. [153]

Gutas, D. *Greek Thought, Arabic Culture: The Graeco-Arabic Translation Movement in Baghdad and Early 'abbāsid Society (2nd-4th/8th-10th Centuries)*. London: Routledge, 1998. [129, 130]

Guthrie, W. K. C. *A History of Greek Philosophy*. Cambridge: Cambridge University Press, 1962. [72, 73, 74, 75]

Guthrie, W. K. C. *The Sophists*. Cambridge: Cambridge University Press, 1971. [80, 81, 84]

Hamer, D. H. *The God Gene: How Faith Is Hardwired into Our Genes*. New York: Doubleday, 2004. [54]

Hammond, N. *The Cambridge Companion to Pascal*. Cambridge: Cambridge University Press, 2003. [280]

Hannam, J. *God's Philosophers: How the Medieval World Laid the Foundations of Modern Science*. Thriplow: Icon, 2009. [151, 173, 176, 189]

Harrison, P. *The Bible, Protestantism, and the Rise of Natural Science*. Cambridge: Cambridge University Press, 1998. [223, 230, 231, 232, 239, 341, 342]

Harrison, P. *The Territories of Science and Religion*. Chicago: University of Chicago Press, 2015. [xxii, 201]

Heilbron, J. L. *Galileo*. Oxford: Oxford University Press, 2010. [191]

Heraclitus. *The Art and Thought of Heraclitus: An Edition of the Fragments with Translation and Commentary*, ed. Kahn, C. H. Cambridge: Cambridge University Press, 1979. [73, 76]

Herrmann, C. S., Pauen, M., Min, B.-K., Busch, N. A., and Rieger, J. W. 'Analysis of a Choice-Reaction Task Yields a New Interpretation of Libet's Experiments', *International Journal of Psychophysiology* 67 (2008), 151–7. [429]

Herschel, J. F. W. *A Preliminary Discourse on the Study of Natural Philosophy* (1830), 287. Cambridge University Library CCD 118. [297, 298, 306, 308, 309]

Herschel, J. F. W. *Physical Geography: From the Encyclopædia Britannica*. Edinburgh: Adam and Charles Black, 1861. [312, 313]

Herschel, W. *The Scientific Papers of Sir William Herschel*, vol. 1, ed. Dreyer, J. L. E. London: The Royal Society and the Royal Astronomical Society, 1912. [287]

Highfield, R. 'Do Our Genes Reveal the Hand of God?', *The Telegraph*, 20 Mar. 2003. [xvi]

Hillar, M. *From Logos to Trinity: The Evolution of Religious Beliefs from Pythagoras to Tertullian*. New York: Cambridge University Press, 2012. [108]

Holmes, R. *The Age of Wonder: How the Romantic Generation Discovered the Beauty and Terror of Science*. London: Harper Press, 2008. [286, 287, 288, 289]

Hooke, R. *The Diary of Robert Hooke, M.A., M.D., F.R.S., 1672–1680*. London: Taylor & Francis, 1935. [215]

Hooykaas, R. *Robert Boyle: A Study in Science and Christian Belief*. Lanham: University Press of America, 1997. [248]

Hoyle, F. 'The Universe: Past and Present Reflections', *Engineering and Science* 45 (1981), 8–12. [421]

Huffman, C. A. *Archytas of Tarentum: Pythagorean, Philosopher, and Mathematician King*. Cambridge: Cambridge University Press, 2005. [84, 85]

Hume, D. *An Enquiry Concerning Human Understanding; a Letter from a Gentleman to His Friend in Edinburgh*. Indianapolis: Hackett Pub. Co., 1977. [414]

Humphrey, N. *The Inner Eye*. Oxford: Oxford University Press, 2002. [37, 39]

Humphreys, S. C. *The Strangeness of Gods: Historical Perspectives on the Interpretation of Athenian Religion*. Oxford: Oxford University Press, 2004. [81]

Hunter, M. C. W. *Boyle: Between God and Science*. London: Yale University Press, 2009. [242, 243, 249]

Huxley, L. *Life and Letters of Sir Joseph Dalton Hooker*. London: John Murray, 1918. [375]

Ibn al-Nadīm, M. *The Fihrist of al-Nadīm: A Tenth-Century Survey of Muslim Culture*, trans. Dodge, B. New York: Columbia University Press, 1970. [133]

Indicopleustes, C. *Kosma Aigyptiou Monachou Christianikē Topographia* [The Christian Topography of Cosmas, an Egyptian Monk]. London: Hakluyt Society, 1897. [120]

Jaki, S. L. *The Relevance of Physics*. Chicago; London: University of Chicago Press, 1966. [208]

Jaki, S. L. *Science and Creation: From Eternal Cycles to an Oscillating Universe*. Edinburgh: Scottish Academic Press, 1986. [208]

Jansenius, C. *Discours De La Reformation De L'homme Interieur*. Paris: Editions Manucius, 2004. [278]

Jardine, L. *The Curious Life of Robert Hooke: The Man Who Measured London*. London: HarperCollins, 2003. [248]

Jeeves, M. A. *From Cells to Souls—and Beyond: Changing Portraits of Human Nature*. Grand Rapids: Eerdmans, 2004. [428, 429]

Joad, C. E. M. *God and Evil*. London: Faber and Faber, 1942. [415]

Jones, H. B. *The Life and Letters of Faraday*. London: Longmans, 1870. [391]

Jones, R. V. 'James Clerk Maxwell at Aberdeen, 1856–1860', *Notes and Records of the Royal Society of London* 28 (1973), 57–81. [397]

Joordens, J. C. A., d'Errico, F., Wesselingh, F. P., Munro, S., De Vos, J., Wallinga, J., Ankjærgaard, C., Reimann, T., Wijbrans, J. R., Kuiper, K. F., Mücher, H. J., Coqueugniot, H., Prié, V., Joosten, I., Van Os, B., Schulp, A. S., Panuel, M., Van Der Haas, V., Lustenhouwer, W., Reijmer, J. J. G., and Roebroeks, W. 'Homo erectus at Trinil on Java Used Shells for Tool Production and Engraving', *Nature* 518 (2015), 228–31. [32]

Kelemen, D. 'Are Children "Intuitive Theists"? Reasoning about Purpose and Design in Nature', *Psychological Science* 15 (2004), 295–301. [54, 55]

Kepler, J. *Gesammelte Werke*, ed. Caspar, M. München: C. H. Beck, 1937. [221, 223]

Killen, S. S., Marras, S., Mckenzie, D. J., and Steffensen, J. F. 'Aerobic Capacity Influences the Spatial Position of Individuals within Fish Schools', *Proceedings of the Royal Society B: Biological Sciences* 279 (2012), 357–64. [62]

al-Kindī. *The Philosophical Works of al-Kindī*, ed. Adamson, P., and Pormann, P. E. Karachi: Oxford University Press, 2012. [135, 136]

Kirk, G. S. *The Nature of Greek Myths*. Harmondsworth: Penguin, 1974. [73]

Knee, G. C., Simmons, S., Gauger, E. M., Morton, J. J. L., Riemann, H., Abrosimov, N. V., Becker, P., Pohl, H.-J., Itoh, K. M., Thewalt, M. L. W., Briggs, G. A. D., and Benjamin, S. C. 'Violation of a Leggett-Garg Inequality with Ideal Non-Invasive Measurements', *Nature Communications* 3:606 (2012). doi:10.1038/ncomms1614. [431]

Knill, D. C. and Pouget, A. 'The Bayesian Brain: The Role of Uncertainty in Neural Coding and Computation', *Trends in Neurosciences* 27 (2004), 712–19. [424]

Lactantius. *The Works of Lactantius*, trans. and ed. Fletcher, W. Edinburgh: T. & T. Clark, 1871. [164]

Lang, U. M. *John Philoponus and the Controversies over Chalcedon in the Sixth Century: A Study and Translation of the Arbiter*. Leuven: Peeters, 2001. [123]

Lartet, E. and Christy, H. *Reliquiæ Aquitanicæ: Being Contributions to the Archæology and Palæontology of Périgord and the Adjoining Provinces of Southern France*. London: Williams & Norgate, 1875. [11]

Layard, A. H. *Discoveries in the Ruins of Nineveh and Babylon: With Travels in Armenia, Kurdistan and the Desert: Being the Result of a Second Expedition Undertaken for the Trustees of the British Museum*. London: John Murray, 1853. [328, 332]

Leakey, L. S. B. *White African: An Early Autobiography*. London: Hodder & Stoughton, 1937. [35]

Leakey, L. S. B. *Defeating Mau Mau*. London: Routledge, 2004. [35]

Leibniz, G. W. *Sämtliche Schriften und Briefe*, vol. Band 4. Berlin und Leipzig: Deutschen Akademie der Wissenschaften, 1950. [260]

Leibniz, G. W. *Theodicy: Essays on the Goodness of God, the Freedom of Man, and the Origin of Evil*, trans. Farrer, A. M. London: Routledge & Kegan Paul, 1951. [275]

Lewis-Williams, D. *The Mind in the Cave: Consciousness and the Origins of Art*. London: Thames & Hudson, 2004. [16, 19, 20]

Lichtheim, M. *Ancient Egyptian Literature: A Book of Readings*. Berkeley: University of California Press, 1976. [358]

Lindberg, D. C. and Numbers, R. L. (Eds). *God and Nature: Historical Essays on the Encounter between Christianity and Science*. Berkeley: University of California Press, 1986. [108, 228]

Lloyd, G. E. R. *Early Greek Science: Thales to Aristotle*. London: W. W. Norton, 1970. [75, 85]

Lloyd, G. E. R. *Greek Science after Aristotle*. London: Chatto & Windus, 1973. [94, 95]

Lloyd, G. E. R. *Methods and Problems in Greek Science*. Cambridge: Cambridge University Press, 1991. [70]

Locke, J. *A Letter Concerning Toleration*. Indianapolis: Hackett Pub. Co., 1983. [249]

'Logical Positivism and Its Legacy', *Men of Ideas*, episode 6, A. J. Ayer interviewed by B. Magee (televised by the BBC on 23 Feb. 1978). [418]

Lotzer, S. and Schappeler, C. 'The Twelve Articles of the Upper Swabian Peasants (March 1525)', in *The German Reformation and the Peasants' War: A Brief History with Documents*, ed. Baylor, M. G. Boston: Bedford/St. Martin's, 2012. [218]

Luther, M. *Luther's Works*, ed. and trans. Pelikan, J. St. Louis: Concordia Pub. House, 1955.

McEvoy, J. J. *Robert Grosseteste*. Oxford: Oxford University Press, 2000. [155]

MacFarlane, A. *Lectures on Ten British Mathematicians of the Nineteenth Century*. New York: Wiley, 1916. [400]

McGinnis, J. 'Scientific Methodologies in Medieval Islam', *Journal of the History of Philosophy* 41 (2005), 307–27. [138]

Machamer, P. K. *The Cambridge Companion to Galileo*. Cambridge: Cambridge University Press, 1998. [191]

MacKay, D. M. *Science, Chance and Providence: The Riddell Memorial Lectures, Forty-Sixth Series, Delivered at the University of Newcastle upon Tyne on 15, 16, and 17 March 1977*. Oxford: Oxford University Press, 1978. [419]

MacKay, D. M. *Behind the Eye*. Oxford: Basil Blackwell, 1991. [55, 425, 426]

Mahdi, M. 'Alfarabi against Philoponus', *Journal of near Eastern Studies* 26 (1967), 233–60. [137]

Mahon, B. *The Man Who Changed Everything: The Life of James Clerk Maxwell*. Chichester: Wiley, 2003. [384, 387, 396, 398, 399, 400, 403, 404, 437]

Maimonides, M. *The Guide for the Perplexed*. New York: Dover, 1956. [147, 148]

Malinowski, B. *Argonauts of the Western Pacific: An Account of Native Enterprise and Adventure in the Archipelagoes of Melanisian New Guinea*. New York: Dutton, 1961. [13]

Marinus of Samaria. *The Life of Proclus or Concerning Happiness*, trans. Guthrie, K. S. Yonkers, NY: Platonist, 1925. [112]

Maroney, O. 'Information Processing and Thermodynamic Entropy', in *The Stanford Encyclopedia of Philosophy* (Fall 2009 edn), ed. Zalta, E. N., http://plato.stanford.edu/archives/fall2009/entries/information-entropy/. Accessed 27 August 2015. [390]

Marshack, A. *The Roots of Civilization: The Cognitive Beginnings of Man's First Art, Symbol and Notation*. London: Weidenfeld and Nicolson, 1972. [12]

Marx, K. *The Essential Left; Four Classic Texts on the Principles of Socialism*. London: Allen & Unwin, 1960. [414]

Marx, K. and Engels, F. *Selected Correspondence, 1846–1895*. New York: International Publishers, 1942. [415]

Masood, E. *Science & Islam: A History*. London: Icon, 2009. [133]

Matthew, P. *On Naval Timber and Arboriculture*. Edinburgh: Black, 1831. [316]

Maxwell, J. C. *The Scientific Papers of James Clerk Maxwell*, ed. Niven, W. D. Cambridge: Cambridge University Press, 1890. [393, 395, 396, 399]

Maxwell, J. C. *The Scientific Letters and Papers of James Clerk Maxwell*. Cambridge: Cambridge University Press, 1990. [385]

May, S. *Love: A History*. London: Yale University Press, 2011. [362]

Meeks, T. W. and Jeste, D. V. 'Neurobiology of Wisdom: A Literature Overview', *Archives of General Psychiatry* 66 (2009), 355–65. [51]

Midgley, M. *The Ethical Primate: Humans, Freedom, and Morality*. London: Routledge, 1994. [51, 52]

Mills, L. H. and Darmesteter, J. *The Zend-Avesta*. Oxford: Clarendon Press, 1880. [357]

Mithen, S. J. *The Prehistory of the Mind: A Search for the Origins of Art, Religion and Science*. London: Thames & Hudson, 1996. [45, 46, 47]

Monod, J. *Chance and Necessity: An Essay on the Natural Philosophy of Modern Biology*. London: Collins, 1972. [416]

Moorehead, A. *Darwin and the Beagle*. London: Hamish Hamilton, 1969. [299, 306]

Morell, V. *Ancestral Passions: The Leakey Family and the Quest for Humankind's Beginnings*. London: Simon & Schuster, 1995. [35]

Moynihan, M. *The New World Primates: Adaptive Radiation and the Evolution of Social Behavior, Languages, and Intelligence*. Princeton: Princeton University Press, 1976. [43]

Newton, I. *Untitled Treatise on Revelation (Section 1.1)*, National Library of Israel MS 1.1. [268]

Newton, I. *Letter from Newton to Henry Oldenberg*, 18 January 1671–1672. Cambridge University Library, MS Add. 9597.2.18.13. [258]

Newton, I. *Papers Relating to the Dispute Respecting the Inventions of Fluxions 1665–1727*. Cambridge University Library, Department of Manuscripts and University Archives, MS-ADD-03968.37. [260]

Newton, I. *Manna: Transcript of an Anonymous Alchemical Treatise in Another Hand with Additions and Notes by Newton*, 1675. King's College Library, Cambridge, MS 33. [266]

Newton, I. *Original Letter from Isaac Newton to Richard Bentley*, 1692, 10 December. Trinity College Library, 189.R.4.47. [267]

Newton, I. *Draft Letter to Pierres Des Maizeaux*, 1718. Cambridge University Digital Library, MS Add. 3968.41. [269]

Newton, I. *The Mathematical Principles of Natural Philosophy*. London: Printed for Benjamin Motte, 1729. [271, 272]

Newton, I. *Opticks, or a Treatise of the Reflections, Refractions, Inflections and Colours of Light*. London: Printed for William Innys at the West-End of St. Paul's, 1730. [285]

Newton, I. *Sir Isaac Newton's Mathematical Principles of Natural Philosophy; and, His System of the World*, trans. Motte, A., and Cajori, F. Cambridge: Cambridge University Press, 1934. [260]

Newton, I. *The Correspondence of Isaac Newton*. Cambridge: Published for the Royal Society at the Cambridge University Press, 1959. [257, 258, 259, 283]

Newton, I. *The Principia: Mathematical Principles of Natural Philosophy*, trans. Cohen, I. B., and Whitman, A. Berkeley: University of California Press, 1999. [261, 270]

Newton, I. *Draft Chapters of a Treatise on the Origin of Religion and Its Corruption*, ~1690s. National Library of Israel, MS 41. [267]

Newton, I. *Irenicum, or Ecclesiastical Polyty Tending to Peace*, post 1710. King's College Library, MS 3. [265, 266]

Niebuhr, R. *The Nature and Destiny of Man: A Christian Interpretation*. New York: Scribner, 1964. [55]

Numbers, R. L. (Ed.). *Galileo Goes to Jail: And Other Myths about Science and Religion*. London: Harvard University Press, 2009.

Oresme, N. *Le Livre du ciel et du Monde*, ed. Menut, A. D., and Denomy, A. J., trans. Menut, A. D. Madison: University of Wisconsin Press, 1968. [208]

Pais, A. *'Subtle Is the Lord': The Science and the Life of Albert Einstein*. Oxford: Oxford University Press, 1982. [430]

Paley, W. *Natural Theology: Or, Evidence of the Existence and Attributes of the Deity*. London, 1803. [416, 417, 422]

Parfit, J. *The Health of a City: Oxford 1770–1974*. Oxford: Amate Press, 1987. [365]

Pascal, B. *The Provincial Letters*. London: Chatto & Windus, 1875. [278]

Pascal, B. 'Minor Works, of the Geometrical Spirit', Harvard Classics, vol. XLVIII, Part II, 1909–14, 68, http://www.bartleby.com/48/3/9.html. Accessed 15 August 2015. [279]

Pascal, B. *Pascal's Pensées*, trans. Trotter, W. F. London: J. M. Dent & Sons, 1931. [279]

Pascal, B. *Pensées and Other Writings*, ed. and trans. Levi, H., and Levi, A. Oxford: Oxford University Press, 1995. [277, 279, 280, 281, 404, 421]

Pascal, B. *Les Provinciales; Pensées*. Paris: Livre de poche: Classiques Garnier, 2004. [279]

Passmore, J. Logical Positivism in *Encyclopedia of Philosophy* (2006). [418]

Pedersen, O. *The First Universities: Studium Generale and the Origins of University Education in Europe*. Cambridge: Cambridge University Press, 1997. [157]

Pepys, S. *Pepys' Diary*, ed. Kenyon, J. P. London: Batsford, 1963. [215]

Philo. *The Works of Philo: Complete and Unabridged*, trans. Yonge, C. D. Peabody: Hendrickson, 1993. [108, 109]

Philoponus, J. *Joannis Philoponi De Opificio Mundi Libri*. Lips: Lipsiae, 1897. [121, 122, 208]

Philoponus, J. *Against Aristotle: On the Eternity of the World*, trans. Wildberg, C. London: Duckworth, 1987. [117, 118]

Philoponus, J. *Corollaries on Place and Void*, trans. Furley, D. J., and Wildberg, C. London: Duckworth, 1991. [116]

Philoponus, J. *Against Proclus: On the Eternity of the World, 6–8*, trans. Share, M. London: Duckworth, 2005. [113, 114, 115, 116]

Philoponus, J. *Against Proclus: On the Eternity of the World, 12–18*, trans. Wilberding, J. London: Duckworth, 2006. [116]

Philoponus, J. *On Aristotle Physics, 4.1–5*, trans. Algra, K., and van Ophuijsen, J. M. London: Bristol Classical Press, 2012. [116]

Pippard, A. B. 'The Cavendish Laboratory', *European Journal of Physics* 8 (1987), 231–5. [439]

Planck, M. *Where Is Science Going?*, trans. Murphy, J. London: Allen & Unwin, 1933. [434]

Plato. *Cratylus; Parmenides; Greater Hippias; Lesser Hippias*, trans. Fowler, H. N. London: Heinemann, 1926. [78]

Plato. *The Republic*, trans. Lee, H. D. P. Harmondsworth: Penguin, 1955. [85]

Plato. *Protagoras; and, Meno*, trans. Guthrie, W. K. C. Harmondsworth: Penguin, 1956. [80]

Plato. *Gorgias*, trans. Hamilton, W. Harmondsworth: Penguin, 1960. [67, 79, 80]

Plato. *The Laws*, trans. Saunder, T. J. Harmondsworth: Penguin, 1970. [87, 88, 101]

Plato. *The Last Days of Socrates: Euthyphro, Apology, Crito, Phaedo*, trans. Hugh Tredennick. London: Penguin, 1954. [68, 82, 83]

Plato. *Timaeus and Critias*, trans. Waterfield, R. Oxford: Oxford University Press, 2008. [86, 87, 102]

Plott, J. C. *Period of Scholasticism*. New Delhi: Motilal Banarsidass, 2000.

Plutarchus. *Plutarch's Lives: The 'Dryden Plutarch'*. London: J. M. Dent, 1910. [77, 79]

Popper, K. *Logik der Forschung*. Wien: Springer, 1934. [434]

Portugal, S. J., Hubel, T. Y., Fritz, J., Heese, S., Trobe, D., Hailes, S., Wilson, A. M., Usherwood, J. R., and Voelkl, B. 'Upwash Exploitation and Downwash Avoidance by Flap Phasing in Ibis Formation Flight', *Nature* 505 (2014), 399–402. [62]

Pritchard, J. B. *The Ancient Near East: An Anthology of Texts and Pictures*. Princeton: Princeton University Press, 1958. [358]

Purver, M. *The Royal Society: Concept and Creation*. London: Routledge and Kegan Paul, 1967. [234, 242, 247]

Rawlinson, G. *A Memoir of Major-General Sir Henry Creswicke Rawlinson*. London: Longmans, Green and Co., 1898. [328]

Rawlinson, H. 'X—Notes on Some Paper Casts of Cuneiform Inscriptions Upon the Sculptured Rock at Behistun Exhibited to the Society of Antiquaries', *Archaeologia* 34 (1851), 73–6. [327]

Rees, M. J. 'Numerical Coincidences and "Tuning" in Cosmology', *Astrophysics and Space Science* 285 (2003), 375–88. [421]

Repcheck, J. *Copernicus' Secret: How the Scientific Revolution Began*. London: Simon & Schuster, 2009. [219, 221]

Richard, N. 'De l'art ludique à l'art magique; Interprétations de l'art pariétal Au XIXe siècle', *Bulletin de la Société préhistorique française* 90 (1993), 60–8. [10]

Rodríguez-Vidal, J., Cáceres, L. M., d'Errico, F., Queffelec, A., Pacheco, F. G., Blasco, R., Finlayson, G., Fa, D. A., Finlayson, S., Finlayson, C., Rosell, J., Jennings, R. P., Bernal, M. A., López, J. M. G., Carrión, J. S., Jiménez, S. F., and Negro, J. J., 'A Rock Engraving Made by Neanderthals in Gibraltar', *Proceedings of the National Academy of Sciences of the United States of America* 111 (2014), 13301–6. [47]

Ruse, M. *Evolutionary Naturalism: Selected Essays*. London: Routledge, 1995. [416]

Russell, B. *Religion and Science*. London: Thornton Butterworth, 1935. [xv]

Sabra, A. I. 'Ibn al-Haytham', *Harvard Magazine* (2003), 54–5. [141, 142]

Sala, N. et al., 'Lethal Interpersonal Violence in the Middle Pleistocene', *PLoS ONE* 10 (2015), e0126589. doi:10.1371/journal.pone.0126589. [34]

Salusbury, T. *Mathematical Collections and Translations*. London: London, 1661. [204]

Sandars, N. K. *Poems of Heaven and Hell from Ancient Mesopotamia*. Harmondsworth: Penguin, 1971. [340, 347, 348]

Sanders, R. '160,000-Year-Old Fossilized Skulls Uncovered in Ethiopia Are Oldest Anatomically Modern Humans', http://www.berkeley.edu/news/media/releases/2003/06/11_idaltu.shtml. Accessed 7 August 2015. [34]

Savage-Rumbaugh, E. S. and Fields, W. M. 'The Evolution and the Rise of Human Language: Carry the Baby', in *Homo Symbolicus: The Dawn of Language, Imagination and Spirituality*, ed. Henshilwood, C. S., and d'Errico, F., 13–48. Amsterdam: John Benjamins, 2011. [55]

Scally, A. and Durbin, R. 'Revising the Human Mutation Rate: Implications for Understanding Human Evolution', *Nature Reviews Genetics* 13 (2012), 745–53. [30]

Schlosshauer, M., Kofler, J., and Zeilinger, A. 'A Snapshot of Foundational Attitudes toward Quantum Mechanics', *Studies in History and Philosophy of Modern Physics* 44 (2013), 222–30. [431]

Shadwell, T. *The Virtuoso*. London: Edward Arnold, 1966. [214, 215]

Sharratt, M. *Galileo: Decisive Innovator*. Oxford: Blackwell, 1994. [192, 199]

Smith, G. *Assyrian Discoveries: An Account of Explorations and Discoveries on the Site of Nineveh, during 1873 and 1874*. London: Sampson Low, Marston, Low, and Searle, 1875. [335, 337]

Smith, J. M. and Price, G. 'The Logic of Animal Conflict', *Nature* 246 (1973), 15–18. [416]

Smith, L. P. *The Life and Letters of Sir Henry Wotton*. Oxford: Clarendon Press, 1907. [188]

Snobelen, S. D. 'Isaac Newton, Heretic: The Strategies of a Nicodemite', *British Journal for the History of Science* 32 (1999), 381–419. [265, 266]

Sommers, T. 'Interview with Frans de Waal', http://www.believermag.com/issues/200709/?read=interview_dewaal. Accessed 7 August 2015. [50]

Sorabji, R. *Philoponus and the Rejection of Aristotelian Science*. London: Duckworth, 1987. [117, 121, 185]

Sorabji, R. *Aristotle Transformed: The Ancient Commentators and Their Influence*. London: Duckworth, 1990. [112]

Spalding, F. *John Piper, Myfanwy Piper: Lives in Art*. Oxford: Oxford University Press, 2009. [xvi]

Spence, J. *Anecdotes, Observations, and Characters, of Books and Men: Collected from the Conversation of Mr. Pope, and Other Eminent Persons of His Time*. London: John Murray, 1820. [255]

Spencer, N. *Darwin and God*. London: SPCK, 2009. [318]

Sprat, T. *History of the Royal Society*. London: Routledge & Kegan Paul, 1959. [248, 249, 250, 251, 257, 286]

Stukeley, W. *Revised Memoir of Newton*, 1752. Royal Society Library, MS 142. [268, 269]

Swaab, D. F. *We Are Our Brains: From the Womb to Alzheimer's*. London: Penguin, 2014. [428]

Swammerdam, J. *The Letters of Jan Swammerdam to Melchisedec Thévenot*. Amsterdam: Swets & Zeitlinger, 1975. [239]

Swinburne, R. *The Existence of God*. Oxford: Oxford University Press, 2004. [423]

Synesius. *The Letters of Synesius of Cyrene*, trans. Fitzgerald, A. Oxford: Oxford University Press, 1926. [111]

Talbot, W. H. F., Hincks, E., Oppert, H. C., and Rawlinson, H. C. 'Comparative Translations', *Journal of the Royal Asiatic Society of Great Britain and Ireland* 18 (1861), 150–219. [333, 334]

Táran, L. *Parmenides: A Text with Translation, Commentary, and Critical Essays*. Princeton: Princeton University Press, 1965. [74]

Tennyson, A. *In Memoriam*. London: Penguin, 1991.

Tertullian, F. Q. S. *The Writings of Quintus Sept. Flor. Tertullianus*, ed. Kaye, J., Thelwall, S., Holmes, P., and Wallis, R. E. Edinburgh: T. & T. Clark, 1882. [164]

Tertullian, F. Q. S. *Quinti Septimi Florentis Tertulliani Opera*, ed. Reifferscheid, A., and Wissowa, G. Vindobonae: Hoelder-Pichler-Tempsky, 1890. [110]

The Bible: New International Version.

The Koran: A New Translation, trans. Dawood, N. J. Harmondsworth: Penguin Books, 1956. [128, 130]

Thomson, J. J. *James Clerk Maxwell: A Commemoration Volume, 1831–1931*. Cambridge: Cambridge University Press, 1931. [403]

Thucydides. *History of the Peloponnesian War*, trans. Warner, R. Harmondsworth: Penguin, 1972. [81]

Trezise, P. J. 'Aboriginal Cave Paintings; Sorcery versus Snider Rifles', *Journal of the Royal Historical Society of Queensland* 8 (1968), 546–51. [21]

Trigg, R. *Rationality and Science: Can Science Explain Everything?* Oxford: Blackwell, 1993. [418]

Ulmi, N. 'Science et foi, comment ne pas choisir', *Le Temps* (27 Sept. 2014), 28. [438]

Valladas, H., Tisne'rat-Laborde, N., Cachier, H., Arnold, M., Oberlin, C., and Evin, J. 'Bilan des datations carbone 14 effectuées sur des charbons de bois de la grotte Chauvet', *Bulletin de la Société préhistorique française* 102 (2005), 109–13. [26]

Voelkl, B., Portugal, S. J., Unsöld, M., Usherwood, J. R., Wilson, A. M., and Fritz, J. 'Matching Times of Leading and Following Suggest Cooperation through Direct Reciprocity during V-Formation Flight in Ibis', *Proceedings of the National Academy of Sciences of the United States of America* 112 (2015), 2115–20. [412]

Wadley, L. 'Complex Cognition Required for Compound Adhesive Manufacture in the Middle Stone Age Implies Symbolic Capacity', in *Homo Symbolicus: The Dawn of Language, Imagination and Spirituality*, ed. Henshilwood, C. S., and d'Errico, F. Amsterdam: John Benjamins, 2011. [11, 12]

Wagner, R. *In a Strange Land*. Oxford: Besalel Press, 1988. [343]

Wagner, R. *The Book of Praises: A Translation of the Psalms*. Oxford: Besalel Press, 1994. [360, 361]

Wallace, W. A. 'Dialectics, Experiments and Mathematics in Galileo', in *Scientific Controversies*, ed. Machamer, P., Pera, M., and Baltas, A. New York: Oxford University Press, 2000. [187]

Walter, A. *Evolutionary Psychology and the Propositional-Attitudes*. Dordrecht: Springer, 2012. [47]

Watson, J. D. *The Double Helix: A Personal Account of the Discovery of the Structure of DNA*. London: Weidenfeld and Nicolson, 1968. [xiv]

Weale, M. E. 'Patrick Matthew's Law of Natural Selection'. *Biological Journal of the Linnean Society* (2015). doi: 10.1111/bij.12524. [317]

White, M. *Isaac Newton: The Last Sorcerer*. London: Fourth Estate, 1997. [261]

Wickham, C. 'Notes on Depuch Island', *Journal of the Royal Geographical Society of London* 12 (1842), 79–83. [20]

Wigner, E. P. 'The Unreasonable Effectiveness of Mathematics in the Natural Sciences', *Communications on Pure and Applied Mathematics* 13 (1960), 1–14. [399]

Wilberforce, S. 'Review of Darwin's *On the Origin of Species*', *The Quarterly Review* (Jul. 1860), 225–64. [375]

Wilkins, J. *A Discourse Concerning a New World and Another Planet*, vol. I. London: Printed for I. Maynard, 1640. [241]

Wilson, E. O. *On Human Nature*. London: Harvard University Press, 1978. [416]

Winchester, S. *Bomb, Book and Compass: Joseph Needham and the Great Secrets of China*. London: Viking, 2008. [xx]

Wiseman, D. J. *Chronicles of Chaldaean Kings (626–556 B.C.) in the British Museum*. London: British Museum, 1956. [344]

Woolhouse, R. S. *Locke: A Biography*. Cambridge: Cambridge University Press, 2007. [249]

Xenophon. *Memorabilia; Oeconomicus*, trans. Marchant, E. C., and Todd, O. J. Cambridge, MA: Harvard University Press, 2013. [83]

Index

Page numbers with the suffix *il* indicates an illustration; *n* indicates a footnote, followed by the footnote number. Punctuation marks as orthography for the /Xam language are sorted to the top of the index. Indexing for works cited in footnotes is provided in the Bibliography on pages 447–56.